Henry George Madan

An elementary treatise on heat

Henry George Madan

An elementary treatise on heat

ISBN/EAN: 9783337275679

Printed in Europe, USA, Canada, Australia, Japan

Cover: Foto ©berggeist007 / pixelio.de

More available books at **www.hansebooks.com**

AN ELEMENTARY TREATISE

ON

·HEAT·

BY

H. G. MADAN, M.A., F.C.S.

FELLOW OF QUEEN'S COLLEGE, OXFORD; LATE ASSISTANT MASTER
AT ETON COLLEGE

WITH ILLUSTRATIONS

RIVINGTONS
WATERLOO PLACE, LONDON
MDCCCLXXXIX

INTELLIGATUR HOC QUOD DIXIMUS DE MOTU (NEMPE UT SIT INSTAR GENERIS AD CALOREM[1]), NON QUOD CALOR GENERET MOTUM AUT QUOD MOTUS GENERET CALOREM (LICET ET HAEC IN ALIQUIBUS VERA SINT); SED QUOD IPSISSIMUS CALOR, SIVE QUID IPSUM CALORIS, SIT MOTUS ET NIHIL ALIUD. — BACON, *Novum Organum*, Book II. Aphorism XX.

[1] That is, Motion is the genus of which Heat is a species; or, in plain words, Heat is a particular kind of Motion.

PREFACE

THE branch of Physics which includes the study of the laws and phenomena of Heat seems to have recommended itself to most, if not all, teachers of Natural Science as a very suitable and fruitful subject for school instruction. The phenomena of Heat have so close and obvious a connection with our everyday life, that an interest in them hardly requires to be awakened; it already exists in nearly every mind. The laws of Heat are capable of being stated very simply and precisely, and of being illustrated by very convincing and easy experiments.

The present short Treatise represents, with some few additions, a course of instruction which has been found suitable for boys who have already made some acquaintance with Physiography and the Elementary Dynamics of solids and fluids, before they proceed to the more difficult[1] subject of Chemistry.

The treatment of the subject throughout is intended to be of an elementary and practical character. Especial stress has been laid upon the results and applications of the general laws of Heat to the arts and manufactures, and to the explanation of phenomena of common occurrence.

---

[1] If, that is, the subject is to be treated in a really scientific spirit, and not merely as an excuse for producing brilliant effects of light and colour, startling explosions, and unsavoury smells.

These laws seem worthy of better treatment than to be considered as little more than subjects upon which arithmetical questions may be set for the purposes of a competitive examination (after the manner of numerous text-books of the present day). Various numerical problems, however, chiefly of such kinds as practically occur in scientific work, are stated and explained in Appendices to several chapters, and a pretty full set of Questions and Exercises will be found at the end of the book. Mathematical formulæ are occasionally employed, but only after a full statement in ordinary language of the facts they express.

A rather large number of experiments has been introduced, with such details as may suffice to ensure their successful execution. No experiment is given which has not been repeatedly performed in exact accordance with the description of it: hence it is hoped that with moderate care and skill failures will seldom occur. A 'juvenile auditory' is in some respects an exacting one. No allowance is made by untrained minds for delays which unavoidably arise in the execution of an experiment, or for deviations from the theoretical results which every reasonable man accepts as inevitable consequences of the imperfection of human methods, and at which adult audiences have usually too much good-breeding and self-restraint to express dissatisfaction and impatience. Mischances, no doubt, *will* happen, in spite of every precaution; and the work of science lecturers often seems to be little more than a perpetual struggle against the malice of Nature. Hence not a few interesting experiments have to be omitted even by skilled lecturers, as too tedious or too uncertain for the conditions of a lecture-room.[1]

Most of the illustrations in this book have been photographed direct on the wood from negatives, taken by the author, of the apparatus actually used: and many of the fine old well-worn engravings which have adorned so many text-books will, it is feared, be looked for in vain.

The stereotyped forms of apparatus sold by dealers have in various instances been departed from, as in the case of Leslie's Differential Thermometer, Ferguson's Pyrometer, apparatus for demonstrating expansion of gases, conductivity, convection, etc.; with a view of rendering them more convenient in use, and of illustrating general laws in a plainer and more striking manner. It is too often the case that those who make the apparatus know nothing about the use of it, and those who use it care nothing about the making of it.

The Metric System of measures and weights, the Centigrade thermometer-scale, and the C.-G.-S. system of units are now in such general, almost universal, use, that no apology is necessary for employing them consistently and exclusively throughout the book. Tables are given in the Appendix for reduction to English measures and Fahrenheit degrees, but these latter are only in the rarest instances mentioned in the text.

It has been thought worth while in many cases to give references to the original memoirs on such subjects as

---

[1] M. Regnault was once asked by an eminent English chemist how he succeeded in the manipulation of a certain experiment which was very graphically described in his *Cours élémentaire de Chimie*. He replied, '*Monsieur, c'est une expérience qu'on ne fait pas.*'

specific heat, radiation, etc. Good scientific libraries containing such standard works as the *Philosophical Transactions of the Royal Society*, the *Philosophical Magazine*, the *Annales de Chimie et de Physique*, etc., are within the reach of many students at the present day. To read the narratives of the original investigators of a scientific subject, and note the current of thought which led to and guided their researches, the patience and ingenuity with which they overcame experimental difficulties, and the caution with which they received and interpreted results, is not only very interesting, but also affords an excellent training for younger minds in the methods and aims of scientific research.

H. G. MADAN.

OXFORD, *Easter* 1889.

CONTENTS

CHAPTER I.

	PAGE
INTRODUCTORY,	1

CHAPTER II.

THE NATURE OF HEAT,	6

CHAPTER III.

THE SOURCES OF HEAT—

SECTION I.—Natural Sources of Heat,	12
SECTION II.—Artificial Sources of Heat,	16

CHAPTER IV.

THE EFFECTS OF HEAT UPON SUBSTANCES—

SECTION I.—Characteristics of the Three States of Matter,	39
SECTION II.—Expansion of Solids by Heat,	42
SECTION III.—Expansion of Liquids by Heat,	64
SECTION IV.—Expansion of Gases by Heat,	79

CHAPTER V.

	PAGE
THE MEASUREMENT OF TEMPERATURE—THERMOMETRY,	90
APPENDIX.—Problems on Expansion by Heat,	114

CHAPTER VI.

THE MEASUREMENT OF QUANTITY OF HEAT—CALORIMETRY—

SECTION I.—The Relative Capacities of Substances for Heat—Specific Heat,	120
SECTION II.—Methods of determining Specific Heats,	127
SECTION III.—Explanation of the Differences in Specific Heats,	135
SECTION IV.—Results and Applications of Specific Heat,	140
APPENDIX.—Problems relating to Specific Heat,	142

CHAPTER VII.

PHENOMENA CONNECTED WITH CHANGE OF STATE—

SECTION I.—Fusion and Solidification,	144
SECTION II.—Vaporisation and Condensation,	160
SECTION III.—Evaporation at different Temperatures,	173
SECTION IV.—Liquefaction of Vapours and Gases,	198
SECTION V.—The Densities of Vapours and Gases,	204
SECTION VI.—Distillation,	208
SECTION VII.—The Spheroidal Condition of Liquids,	215
SECTION VIII.—Hygrometry,	220
APPENDIX.—Problems relating to Hygrometry,	236

CHAPTER VIII.

THE CONVEYANCE OR TRANSMISSION OF HEAT—

SECTION I.—Conduction of Heat,	242
SECTION II.—Convection of Heat,	264
SECTION III.—Radiation of Heat,	290

CONTENTS.

CHAPTER IX.

THE RELATIONS BETWEEN HEAT AND MECHANICAL MOTION—
THERMO-DYNAMICS—

	PAGE
SECTION I.—Conversion of Mechanical Motion into Heat,	349
SECTION II.—Conversion of Heat into Mechanical Motion,	358
APPENDIX.—Problems in Thermo-Dynamics,	362

CHAPTER X.

HEAT-ENGINES—

SECTION I.—General Principles,	367
SECTION II.—Steam-Engines,	371
SECTION III.—Hot-air Engines,	407
SECTION IV.—Gas-Engines,	409
SECTION V.—The Power and Duty of Heat-Engines,	412
CONCLUSION.—The Depreciation or 'Degradation' of Energy,	418

GENERAL APPENDICES.

APPENDIX A.—The Metric System of Measures and Weights,	423
APPENDIX B.—Thermometer Scales,	432
APPENDIX C.—List of Apparatus and Materials,	433
QUESTIONS AND EXERCISES,	441
INDEX,	455

The system of measures and weights used exclusively in this book is the Metric System.

The system of units is that known as the C. G. S. (centimetre-gramme-second) system.

An account of these systems will be found at the end of the book.

Temperatures are always expressed on the Centigrade scale. An explanation of this will be found in Chap. v.]

CHAPTER I.

INTRODUCTORY.

The subject of Heat is one in which all living beings are almost compelled to take an interest. Our bodies are so constituted that we can only exist comfortably within a rather narrow range of temperature. When the temperature of the air falls a little below the average, we are benumbed with cold; when it rises a little above the average, we are almost equally oppressed by heat. Neither in the Arctic regions nor within the Tropics can a really healthy, vigorous life be maintained for any length of time without some more or less conscious effort, and the application of some artificial means to modify the unsuitable temperature of the climate.

Let us consider, to begin with, what we precisely mean in calling a body 'hot' or 'cold.' We judge primarily by our own feelings. We stand in front of a fire and call it hot because it communicates a particular sensation to our bodies. We take up a handful of snow and call it cold because it gives us another very different sensation. But a few very simple experiments will show that our feelings cannot be always depended upon to give us correct ideas respecting the amount of heat in a body.

Experiment 1.—Let three basins, fig. 1 (next page), be taken, A containing cold water, B containing hot water, and C containing tepid or 'lukewarm' water.

(*a*) If the hand is placed for a minute in A and then transferred to C, the water in C will feel *warm* to that hand.

(*b*) If the same hand is next placed for a minute in B (the hot water) and then put into C, the latter feels *cold* to it.

Thus the *same* water appears, as judged by our sensations, both hot and cold.

Meaning of 'Hot' and 'Cold.'

Again, suppose that a room is kept at a moderate, equable temperature all the year round. Coming into this room on a sunny day in summer, we should call it cool. Coming into the

| Cold Water. | Hot Water. | Tepid Water. |

Fig. 1.

same room on a frosty winter day, we should call it warm. A man working in a well comes out into the sun to warm himself; a man working in the stoke-hole of a steam-ship comes out into the same sun to cool himself.

How can we reconcile these contradictory judgments? When we examine accurately what was taking place in the experiment with the three basins of water, we find that when the water in C felt warm to the hand (which had been in A), it was communicating heat to that hand, which rapidly became warmer itself. When it felt cold to the hand which had been in B, it was taking away heat from the hand, which rapidly became colder. The same is true in the other examples given, and we may say generally that,

(*a*) Whenever a body 'feels hot,' it is *giving up heat* to our bodies.

(*b*) Whenever a body 'feels cold,' it is *taking away heat* from our bodies.

This tendency of a substance to impart heat to, or take heat from, other bodies depends on that state of it which is called its 'temperature.' The meaning of the term may be illustrated in the following way: Take two vessels, A and B, fig. 2, connected by a flexible tube so that either can be raised or lowered; and let both of them be partly filled with water. Then the tendency of water to flow from one to the other vessel depends entirely upon the relative level of the water in each. If A is higher than B, water will tend to flow from A to B; if A is put lower than B, water will tend to flow from B to A. If both vessels are placed at the same level, there will be no tendency of the water to flow in either direction through the connecting tube.

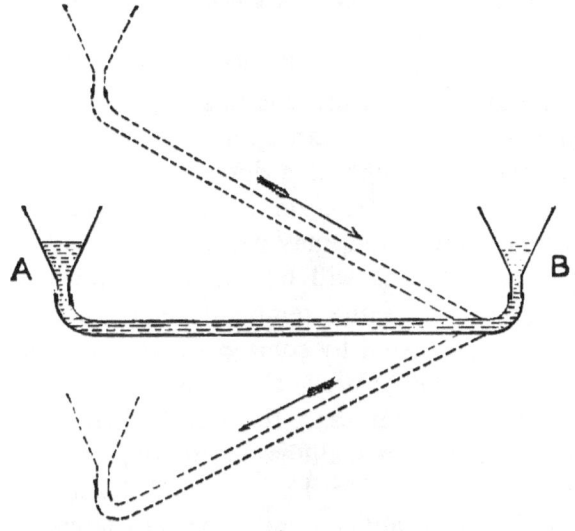

Fig. 2.

So with regard to heat. When heat passes, or tends to pass from one body to another, as from the water to the hand, and *vice versâ*, the two bodies are said to be at different heat-levels or 'temperatures.' The body *from* which it passes is said to be 'at a higher temperature,' or 'hotter,' than the body *to* which it passes; the latter body is said to be 'at a lower temperature,' or 'colder,' than the former. Thus, with regard to the water in the three basins, A would be said to be at a relatively low temperature, B at a high temperature, and C at an intermediate temperature.

Our feelings, then, enable us to judge roughly whether anything is hotter or colder than we ourselves are at the time. They tell us, as a rule, nothing whatever about the absolute amount or intensity of the heat which is present in a substance.

Consider, for example, the case of a large basin of warm water and a small cupful of water taken from it. Both will feel equally warm to the hand; yet the water in the cup must have far less heat in it than that in the basin. This may be proved by leaving them both in the same cold room; the water in the cup will get cold much the sooner of the two, showing that it had less heat in it to lose.

Another fact which shows the unreliability of our sense of feeling as a judge of temperature is this. Place the hand successively upon a piece of metal and upon a piece of wood near it, such as the handle or hinge of a door and the door itself, or a metal inkstand and the table on which it is placed. The two substances, metal and wood, may be fairly assumed to be at the same temperature (for, as will be hereafter shown, things soon acquire and retain the same temperature as other things near them): they may be tested by some accurate, reliable apparatus, such as a thermometer, to prove this; and yet the metal will feel to the hand much colder than the wood. The explanation of this depends upon the great difference in the conducting power of the metal and the wood for heat, and will be more fully given hereafter; the point to notice here is that our feelings would lead us to consider two things to be at very different temperatures, which may be demonstrated to be at the same temperature.

Since, then, our feelings are such imperfect and misleading judges of the amount and intensity of heat, recourse must be had to some other methods for detecting the presence of heat and ascertaining its amount. Now, when heat is transferred from one body, A, to another, B, it is found that (with hardly an exception) A gets smaller and B larger in size. Hence by observing any change in size which takes place, we can tell with certainty whether any heat is passing between two bodies, *i.e.* whether there is any difference in temperature between them. The exact laws of the expansion of bodies by heat will be considered in a future chapter: meanwhile, it will be sufficient to say that all ordinary thermometers are constructed on the above principle, and on the following general plan. Mercury is put into a glass bulb connected with a narrow tube, in sufficient quantity to fill the bulb and part of the tube at ordinary temperatures. Then if, on putting the bulb in contact with any body, the column of mercury in the tube is observed to rise, it shows that the metal has expanded, and we may be sure that heat has been transferred to it, and therefore that the body which touched it was hotter than the mercury. If, on the contrary, the level of the mercury in the tube sinks, the metal must have contracted, owing to heat having been transferred from it to the body which touched it; therefore the latter must be colder than the mercury. If no change takes place in the level of the mercury, then both it and the body which touched it are at the same temperature.

In a similar way we can compare the temperatures of several bodies by applying the thermometer to each in succession; that which causes the top of the mercury column to stand highest in the tube will certainly be the hottest, and the temperatures of the rest will be in the order of their effects in expanding the mercury.

It is further found that the amount of expansion of the mercury is exactly proportional to the amount of heat which has passed into it; so that if the water in one basin causes the mercury column to rise 1 cm., and that in another basin causes it to rise 2 cm., we may be sure that the difference in temperature between the two portions of water and the mercury was twice as great in the latter case as in the former. The 'scales,' or rows of divisions, applied to thermometers are intended to express these differences of temperature in terms of some arbitrarily chosen 'unit,' or standard of reference, and will be more fully explained in Chapter v.

Another method applied to the measurement of temperature is based on the fact that when rods of different metals are joined together, and the point where they touch is heated more than the rest of the rods, a current of electricity is developed, tending to flow in a definite direction; and as there are means of detecting and measuring extremely small electrical currents, such a pair (or series of pairs) of rods will form a most delicate thermometer. Instruments constructed on this principle are called 'Thermopiles.'

CHAPTER II.

THE NATURE OF HEAT.

HEAT has been spoken of above as something which can pass or be transferred from one substance to another. We must next consider what this something is.[1] Is Heat a particular kind of matter, a fluid like water, with the flow of which we have compared it? Up to about one hundred years ago it was universally considered to be so. Heat was defined as an extremely thin, subtle, invisible fluid, to which the name 'Caloric' was given, and which made its way into the spaces or 'pores' between the particles of which bodies are composed, just as water soaks into a sponge, or air passes through the foliage of a forest. It was admitted that we could not detect this fluid by any of the ordinary instruments used in sharpening, as it were, our perceptions, such as the microscope and the balance: but this was put down to the imperfections of the instruments then in use. As time went on, however, and experiments were made more carefully and the results observed more accurately, it became clear that heat, whatever it might be, possessed none of the characteristics of ordinary matter.

In the first place, it was found impossible, even with the most delicate and sensitive balances, to detect the slightest difference in absolute weight between a body when cold and the same body when intensely heated. But one of the most invariable and characteristic properties of every kind of matter known to us is,

---

[1] It will not be necessary to consider 'cold' as anything more than an expression for the *absence of heat;* just as darkness merely means the absence of light, dryness is the absence of moisture, etc.

that it is affected by the force of gravitation; a consequence of which is that there is a mutual attraction between the earth and all forms of matter on its surface which causes them to exert a pressure towards the earth's centre; in other words, to show 'weight.' If, then, heat was a fluid form of matter, we should expect that a body when hot would show more weight than when cold, corresponding to the amount of heat-fluid which had entered it. This increase in weight might be so slight as to escape detection by the rough balances formerly constructed, but since even the most refined modern instruments fail to show it, there certainly arises a presumption that heat does not possess this property of weight, and therefore cannot be any form of ordinary matter.

But there is another and more undoubted proof that heat is not any material substance such as a fluid must be. Matter cannot be created or destroyed by us. All scientific research tends to show that not a single particle of any kind of matter has ever been actually annihilated by natural or human agencies; nor, on the other hand, has a particle of new matter been called into existence. All that can be done is to change existing matter from one shape to another. Thus, when a candle burns, the matter (chiefly consisting of carbon and hydrogen), combines with the oxygen of the air to form compounds called carbon dioxide ('carbonic acid') and water, which happen to be invisible and so might escape notice, but from which we can by proper means obtain every particle of the original constituents of the candle. The candle can no more be destroyed by us than it can be made, in the strict sense of the words. When we talk of the 'making' of a candle we merely mean the putting into a new form materials which already exist, *viz.* the tallow or wax. Nor was the wax 'made' by the bee: it was elaborated from the food obtained from flowers, and the carbon and hydrogen present in the flowers were (in part, at any rate), derived from the carbon dioxide and water produced by the burning of candles. Thus we have a complete round or 'cycle' of changes, but no creation of new matter.

With regard to heat the case is different. We can take a cold substance and make it hot without transferring to it any heat from other bodies; we can, in fact, call heat into existence without having any previous supply of it to draw upon. When a cold lucifer match is gently rubbed on a cold piece of sand-paper, so much heat is produced that the match catches fire. When a cold bullet strikes on a cold iron target, enough heat is developed to melt the lead. When a horse's shoe strikes upon the hard stones of a road, sparks are seen,—a sufficient proof of the creation of heat,—even though both the iron and the stone are cold and wet.

On the other hand, we can absolutely destroy heat, and not merely transfer it to other things. When water is boiled over a fire, a very large amount of heat disappears entirely, as heat. The steam which is formed is no hotter than the water from which it is formed, though a great deal of heat is constantly being supplied by the fire. When spirits of wine or eau de Cologne is dropped on the skin, the latter becomes colder; but neither the liquid nor the vapour of the spirit become any hotter than the skin: heat, in fact, has been destroyed.[1] When a quantity of air which has been compressed into a small bulk in a vessel, such as the reservoir of an air-gun, is allowed to expand, the whole mass of it becomes colder; in other words, heat disappears.

Since, then, heat shows neither weight nor indestructibility—does not, in fact, possess any of the characteristic properties of matter—we are forced to the conclusion that it is not a material substance at all; and we must turn our thoughts in some totally different direction to seek for an explanation of its nature.

Now, since it is only recognisable in association with matter (we can, in fact, only deal with heat in connection with a hot substance), it seems clear that it must be some condition assumed by matter, just as waves are a condition assumed by the sea, and not anything distinct from the water itself. A reference to the

---

[1] Of course energy in some form is present in the steam and the alcohol-vapour, as will be more fully explained in a future chapter; but the point to notice here is, that nothing which possesses the distinctive properties of heat is present in them.

examples of the production of heat already given may help to suggest what this condition may possibly be. In them and in many others we notice that whenever heat makes its appearance there is some stoppage of mechanical motion. It is when the motion of the bullet is stopped by the target, and not till then, that heat is developed. Sparks fly from the horse's hoofs, not when he is pawing the air, but when the motion of the hoof is stopped by contact with a stone. Moreover, whenever heat disappears in such cases as the expansion of compressed air, there is a development of mechanical motion in the particles of air, as shown by their outward rush.

Clearly, then, in the instances above given, heat has something to do with mechanical motion; and it is easy to take a step forward in conjecture and ask, Why should not Heat be a kind of motion itself,—not of the extensive, palpable kind which constitutes the motion of a bullet; not, in fact, the motion of masses of matter collectively (for that is what we mean by 'mechanical motion'),—but a movement of each individual particle of a mass on its own account, quick when the intensity of heat is great, slow when it is slight? This introduces us to the modern view of the nature of heat, called 'The Kinetic Theory'[1] (Gk. κίνησις, 'motion'), which completely and easily explains all the phenomena of heat hitherto observed—a very convincing proof of its truth.

Put shortly, the theory is this. Masses of matter are made up of a great many extremely small particles called 'molecules' (Lat. *molecula*, 'a small mass'), held together more or less strongly by the force of cohesion, but capable, even in the hardest solid, of a certain amount of free, independent movement. In a hot body these molecules are each and all in extremely rapid motion, probably revolving in very small paths or orbits, like those of the moon and planets on an immensely greater scale; or, to take a more homely comparison, like those of dancers in a ballroom. In a heated gas, the molecules are moving in straight lines,

---

[1] Sometimes called the 'Dynamical Theory.'

darting about in all directions in the vessel which contains them, knocking against each other and against the surface of the vessel, and thus having the direction and rate of their motion changed many times in a second. The quickness of this motion determines the intensity of the effects of heat, *i.e.* the temperature of the body, while the quantity of heat depends upon the number of molecules which are in motion. When heat is transferred to a body the molecules of the latter are simply set in more rapid motion; the molecules of the hotter body losing a corresponding amount of their motion, and consequently becoming colder.

It may be worth while to consider for a moment what the rapidity of this movement of the molecules must be; at any rate in the case of a gas, in which state of matter the force of cohesion is overcome entirely, and the molecules can move without being impeded by their mutual attraction. A gas exercises a certain amount of pressure against the surface of the vessel which contains it; and according to the kinetic theory this pressure is due to the constant striking of the rushing molecules against the surface, as bullets strike on a target and press it backwards. The amount of this pressure can be accurately measured, and the 'mass' of the molecules of most substances is approximately known; so that we have the necessary data for determining the velocity which they must have in order to produce the observed effects of pressure. In this way it is estimated that in a portion of a gas at the temperature of freezing water the molecules must be moving at the following rates:[1] —

Hydrogen, . . .	1859 metres per second.
Oxygen, . . .	465 ,, ,,
Carbon dioxide, . .	396 ,, ,,
('carbonic acid gas').	

The objections to the older view do not apply to the Kinetic Theory now under consideration.

1. In setting a body in motion we do not alter the action of

---

[1] See the articles of Professor Clerk Maxwell in *Nature*, vol. viii. pp. 298, 437.

other forces upon it: we do not, for instance, alter its weight. A bullet weighs no more and no less when it is flying through the air than when it is motionless in the rifle. A top weighs no more or less when spinning than when still. A violin may, while it is being played upon, be carried from one room to another without any alteration in its weight, although several kinds of motion are going on simultaneously in the strings; for instance, (1) their transference with the whole instrument from one place to another; (2) the swinging of each from side to side so as to produce a wave of sound.

2. Though we cannot make or destroy matter, we can change its condition from one of rest to one of movement: we can produce and destroy motion. By doing work of some kind we can send a previously motionless cricket-ball flying through the air; and we can (also by doing work) stop its motion with our hands or a bat. It is fair to infer that, by applying energy in an appropriate way, we may be able to increase or lessen (within limits which have not yet been reached) the motion of the individual molecules of a body in such a manner as to render them more or less capable of producing heat-effects.

Thus the Kinetic Theory accounts in the easiest and completest manner for that appearance and disappearance of heat which is so frequently observed. Instances of the former phenomenon, *viz.* the production of heat, will form the next subject for consideration.

CHAPTER III.

THE SOURCES OF HEAT.

THESE may be divided into two great classes :—

A. Those which already exist in the universe, and over which we have no control, though we can utilise them more or less for our own purposes. These may be called natural or cosmical sources (κόσμος, *the universe*).

B. Those which depend on methods devised by man himself for producing heat when the natural supplies of it are not available or sufficient. These may be called artificial sources of heat.

SECTION I.—Natural Sources of Heat.

1. *The Sun.*—The great supplier of heat to the solar system is the sun; indeed, practically, the whole of the heat on the earth's surface, except what we produce ourselves, comes from thence. The great globe is constantly giving out on all sides an amount of heat so enormous that we have some difficulty in forming an adequate conception of it. In the first place it must be remembered that the earth only gets a very small fraction of the total heat emitted by the glowing surface of the sun. Suppose that a large hollow shell of a diameter equal to that of the earth's orbit (300,000,000 kilometres) surrounded the sun: this would receive all the heat emitted, but the earth only gets an amount corresponding to the space which its area takes up out of the whole surface of the shell. This space is no more than $\frac{1}{2,100,000,000}$ of the whole surface, about equal to that of a fly sticking to the inner surface of a globe equal in diameter to the dome of St. Paul's Cathedral.

Yet even this insignificant fraction of the sun's heat is sufficient to produce great effects. Ice requires, as we shall see, a very large quantity of heat merely to turn it into a liquid; but if the earth was covered completely with a layer of ice 31 metres thick, the whole of this mass would (if all the heat which reached the earth was retained in it) be melted in a year, and in another fifteen months the ocean of water thus produced would be raised to the boiling-point.

When the heat which falls upon 15 square metres of the earth's surface in New York is concentrated upon an iron boiler, 3 metres long and 15 centimetres in diameter, enough steam is generated to work an engine of 2 horse-power.

This may give some idea of the quantity of heat which reaches us. The intensity of this heat, *i.e.* the temperature of the sun's surface, is estimated as being at least 20,000°; in fact, at least twelve times higher than the temperature of melting iron. We have good evidence, indeed, that iron, copper, gold, and some other metals exist, not as solids or liquids, but as vapours, close to the surface of the sun.

We obtain proofs of this intensity from the effects produced by concentrating the sun's rays by a convex lens or 'burning glass.' This does not, of course, increase the intensity of the heat; it only gathers together the rays falling on a comparatively large surface and throws them all upon a small spot or 'focus.' Every one knows that paper and wood can be set fire to by placing them in the focus of even a small pocket magnifier; but when a lens one metre or so in diameter is employed, all metals may be readily melted, and even rocks, such as slate, are fused into a glassy mass.

Such experiments give us some idea of what the actual temperature of the sun must be. Yet we suffer no damage, and are scarcely even inconvenienced by it, because the heat is spread over a comparatively large surface; just as the swiftly falling water in a rain-storm does little or no damage compared with what it would do if it fell with the same velocity in one mass on one spot.

2. *The Interior of the Earth.*—Not only does heat reach us from above, but there are also immense stores of it under our feet. The earth we tread upon was, within a measurable period of time, a ball of liquid as hot as the sun; and it is now cool enough to be a comfortable dwelling-place, only because its mass is much less than that of the sun, so that it had much less heat to give out and has therefore lost it sooner.[1] But its heat is not yet all lost: the interior of the earth is still so hot that all the rocks which we are conversant with as difficultly fusible solids are at no great depth below our feet in the state of glowing liquids.[2] Let us see what evidence can be given that this is the case.

(1) Volcanoes are formed over cracks in the earth's crust, from which we get samples, as it were, of the materials which fill up the interior in the shape of torrents of liquid lava.

(2) Geysers and other less violent boiling springs derive their heat from the rocks through which they pass, and thus give an indication of the high temperature at which these rocks must be.

(3) The temperature of the rocks low down in deep mines is invariably found to be very high, and to be higher in exact proportion to their depth below the surface. Thus, by carefully observing the rate of this increase of temperature with depth, we can gain a fair idea of the heat which must exist at distances below the earth's surface which we cannot hope to reach by boring.

We have, in fact, been able to penetrate only to an insignificant depth into the earth. The deepest mine-shaft in Europe, and probably in the world, is at Charleroi in Belgium; it is 1039 metres in depth (about two-thirds of a mile). Compared with the whole diameter of the earth (nearly 13,000,000 metres), this is a mere pin-scratch on a football. The temperature of the rocks in the lowest workings of this mine is about 40°, the average temperature of the rocks at the surface being 11°. Borings for water have been made to rather greater depths. At Sperenberg

---

[1] It is pretty certain that the planet Jupiter, the mass of which is nearly 1000 times that of the earth, is still red-hot at its surface.

[2] Unless, as seems probable, the enormous pressure of the rocks above them prevents the materials from assuming the liquid state (see note, p. 15).

near Berlin, for instance, a boring has been carried to a depth of 1300 metres, and the water which comes up from the bottom is at a temperature of 47°, about the temperature of water in a warm bath.

Careful observations have been made of underground temperatures at various depths by taking the readings of thermometers inserted in the soil and rocks, with the following general results:—

At the actual surface the temperature of the soil varies, of course, with day and night, summer and winter; but this variation becomes imperceptible at a depth of about 20 metres, where the rocks are found to have a constant temperature of 11·6°. At greater depths than this the temperature of the earth is found to increase fairly regularly at the rate of 1° for every 27 metres. Now, if this rate of increase is maintained at still greater depths (and there is no reason for supposing the contrary), at a depth of 3 kilometres the rocks must be as hot as boiling water, at 40 kilometres below the surface they must be as hot as melting iron, and at all depths much greater than this the materials of the earth must be so intensely hot that (if no cause existed to prevent it [1]) they must be in a state of liquid, or even vapour.

We see, then, on what a thin shell of cold materials we must be living; and it may naturally be asked why this intense heat of the earth's interior is so little felt by us that we ignore its existence, except when the eruption of a volcano forces it upon our notice.

The reasons are—

(1) That the rocks are very bad conductors of heat, so that the heat of the interior only reaches the surface with extreme slowness.

(2) That the earth's surface, like that of the sun, very readily gives off heat by radiation (as will be more fully explained in a

---

[1] As will be hereafter shown, most substances, when under heavy pressure, do not melt until a higher temperature is reached than that which is sufficient to liquefy them under ordinary conditions; and the weight of the rocks above them may keep the deep-sunk materials of the earth in a solid state, in spite of the intense heat.

later chapter); and hence no sooner does any heat reach the exterior than it passes off into the intensely cold regions of space, and we have nothing more to do with it.

The final result is that the earth's surface is only raised in temperature by about $\frac{1}{70}$ of a degree in consequence of the heat which reaches it from below; a quite imperceptible effect.

Section II.—Artificial Sources of Heat.

These all depend upon the conversion of some form of existing Energy into that particular form of it called Heat.

Energy may be shortly explained as being a particular condition of matter, in virtue of which it is capable of doing *Work:* work, in the most general sense, consisting in the setting in motion, or displacing, a body, or altering its rate of motion. Thus, a moving rifle-bullet is endowed with energy, in virtue of which it can make a hole in a board or a man, displacing the particles of the object hit, in spite of the resistance opposed by their cohesion and inertia. There is energy in the muscles of a horse, in virtue of which it can move itself and drag heavy loads from place to place. Now, the whole course of scientific research leads to the conclusion that this work-producing condition cannot be absolutely created or destroyed by human means, any more than the matter in which it resides; it can only be transferred from one portion of matter to another; taking, according to the conditions of its transfer, the form of Mechanical Motion, or Heat, or Electricity, or Chemical Action. All the phenomena classed under these heads are now believed to be due to different kinds of motion of the molecules of matter; and it is usually pretty easy (sometimes almost too easy) to convert other kinds of motion into Heat-motion.

1. *Conversion of Mechanical Motion into Heat.*—Whenever the motion of a bullet is stopped and energy transferred from it by opposing some resistance to its progress, more or less of this energy always passes into the form of heat. When, for instance, it pierces a hole in a board, some energy, no doubt, continues in the form of mechanical motion in the splinters of wood which

are driven aside and in front of it, but a very large proportion of the total energy lost by the bullet appears as heat. When an iron armour-plate is pierced by one of the enormous projectiles used in modern artillery, a flash of light is seen at the moment of impact, and the edges of the hole show signs of fusion. If the ball strikes on a strong target which it cannot pierce, practically the whole of the energy which it possessed as mechanical motion during its flight, takes the form of heat both in the ball and in the target.[1]

When a horse dashes his hoof with all the energy of his muscles against a hard stone, the energy which caused the mechanical motion of the leg partly does the work of tearing off portions of the iron shoe (as being softer than the stone), partly throws the molecules of the detached particles into such rapid heat-motion that they are raised to their melting-point, and glow, forming the sparks.

The same explanation, of course, applies to the old 'flint and steel,' used not only for firing guns, but also extensively for lighting fires and candles before lucifer matches were invented.

Expt. 2. Obtain a piece of flint (one with a freshly fractured jagged surface is best) or a bit of gritty paving-stone, and strike the back of a knife forcibly against it in a slanting direction, as if endeavouring to scrape off bits of the stone. Sparks will be seen at each blow, if the latter is properly directed.

[An old flint gun-lock, which may still be purchased without difficulty, is very suitable for showing this effect.]

Expt. 3. Take a narrow strip of thick sheet lead, about 20 cm. long, wrap a bit of flannel round one end of it (to prevent heat passing between it and the hand), and flatten it on an anvil with eight or ten quick, heavy blows of a hammer: then hold the flattened part to the hand, or to the bulb of a thermometer (preferably an air-thermometer, if the effect is to be visible at a distance). It will be found decidedly warm, or even hot.

---

[1] As will be hereafter shown, when a leaden bullet, moving at the rate of not less than 360 metres per second, has its motion entirely stopped by an obstacle, enough heat is generated to melt the lead if it is all concentrated in the bullet.

MECHANICAL SOURCES OF HEAT.

A blacksmith will, by skilfully applied blows of a hammer, make the end of a small iron bar actually red-hot: sufficiently so to light his forge fire with it, although few would take the trouble to do so at the present day.

Fig. 3.

The same result occurs if an elastic gas, such as air, is taken to receive the blow: as is well shown by the so-called 'fire-syringe,' fig. 3, which is nothing more than an accurately bored tube, closed at one end, and fitted with a solid piston having a small cavity at its inner end, in which a piece of tinder is placed. Thus by forcing down the piston the air contained in the tube is compressed and hammered flat, as it were, and the energy expended in doing so appears as heat in proportion as the mechanical motion of the piston disappears.

Expt. 4. Place in the cup at the end of the piston of a fire-syringe a small bit of the yellow cord used for cigar-lighters,[1] pressing it down slightly so that it may not be shaken out; introduce the piston into the barrel of the syringe, and force it down with a sudden heavy pressure. Withdraw it *quickly*, and the tinder will be found to be on fire: if it is now blown upon, the glow will spread, and a match or a pinch of gunpowder may be lighted with it.

A syringe with glass barrel is much preferable if it can be obtained: a flash of light is then seen at the moment of greatest compression of the air. In fact, the air contained in the syringe has so small a mass that the heat produced is sufficient to make it red-hot.

It is not necessary, of course, that the *whole* of the mechanical motion of a body should be destroyed in order that heat should be generated, as in the above instances. Whenever *any* of it is

---

[1] This answers better than the German tinder usually sold for the purpose: especially if it is slightly and superficially charred by lighting it for a moment and then putting it out by pressing the finger or a bit of wood upon the cup in which it lies.

destroyed,—in other words, whenever a moving body is made to move slower,—an amount of heat can be produced, which is (as will be shown hereafter) the exact equivalent of the lost motion. This explains the sparks which fly from brake-blocks when pressed against the wheels of a railway carriage in stopping the train at a station, and the similar stream of sparks from a steel knife or chisel when held against a dry, quickly-rotating grindstone for the purpose of sharpening it. In these cases the carriage-wheels or grindstone turn less rapidly than the force applied to them would make them do if the wood or the steel were not pressing against them; and all the energy which does not appear as mechanical motion, appears as heat.

Expt. 5. Take a wooden or ivory button (or the handle of a paper-knife will do), and rub it for half a minute backwards and forwards, quickly and with strong pressure upon a piece of cloth, such as the sleeve of the coat, or a strip of board. The substance will soon become quite hot, as may be proved by holding it to the back of the hand, or to the bulb of a thermometer.

A stream of sparks may be shown by attaching a small grindstone or emery-wheel, about 15 cm. in diameter, to the spindle of a 'whirling table' (an apparatus in common use for illustrating some mechanical laws) and pressing an old knife or file against it while quickly rotating (no water should be put on the stone).

Many similar instances of the production of heat will occur to the thoughts. Thus, we make our hands warm in winter by rubbing them together. Savages get enough heat to kindle a fire by cutting a notch in a piece of dry wood, placing in it the blunt-pointed end of a similar but smaller stick, and quickly twirling the latter round in the notch, as if the object was to make a hole in the wood, fig. 4. Smoke soon appears, and little bits of hot charred wood make their way out from under the borer; these are allowed to fall on a tuft of dry grass, and blown upon, and the whole is shortly in a blaze. Any one who has turned wood or metal in a lathe, or simply bored holes with a gimlet or bradawl, will have noticed how hot the chisel or boring tool is liable to become, so that in heavy work water is made continually to drop on the cutting edge to keep it cool. It was by observing the large amount of heat produced in this way during the boring of cannon at Munich that Count Rumford was led to conclude that heat could not be any form of matter, and to determine roughly the exact relation of heat to mechanical motion, which will be more fully considered in a later chapter.

Fig. 4.

The axles of railway engines and carriages are, if insufficiently greased, liable to become red-hot and set fire to the wood-work, or even get melted into their bearings and twisted off.[1]

On many clear nights, brilliant, swiftly moving specks of light appear for a second or more in the sky, and sometimes glowing masses of rock or iron fall from the regions of space upon the earth, usually bursting into fragments as they approach the ground. These 'shooting stars' or 'meteorites' are masses of matter flying through space at the rate of twenty or thirty kilometres per second, and originally as cold as the parts of space they move in. When in the course of their flight they come near the earth, they are attracted by it and swerve from their course; and when they pass within the limits of the earth's atmosphere, their motion is checked by it and a large amount of heat is generated; so much so that the materials of which they are composed are melted and even vaporised superficially, as shown by the trail of light they often leave behind them in the air, and by the rounded, drop-like shape assumed by their fragments.

It is more than probable that the temperature of the sun is in a great degree maintained, in spite of the immense amount of heat

---

[1] For the details of an experiment in which ether or water may be boiled in a tube attached to the spindle of a whirling-table, by clasping the tube, while quickly rotating, between two pieces of wood hinged together, and for many other illustrations of a similar kind, see Tyndall's Lectures on *Heat as a Mode of Motion*.

constantly given out by it, by the incessant collision of swarms of meteorites with its surface. It has, indeed, been calculated that a mass of matter equal to that of the earth would, if it fell from the distance of the earth (150,000,000 kilometres) upon the sun, generate, in consequence of the stoppage of its motion when it reached the surface, as much heat as would make good the amount lost by the sun in ninety-five years.

2. *Conversion of Chemical Energy into Heat.*—This is by far the most important source of heat to us: in fact, nearly all our supplies of artificial heat come from hence.

It is found that, with hardly an exception, whenever two things combine to form a chemical compound, heat makes its appearance. As a simple illustration of this law we may take the combination of phosphorus and iodine, two elements which have a strong chemical affinity for each other.

Expt. 6. Cut off, under water, a small thin chip of phosphorus from a stick of the substance, dry it by pressing it gently between folds of blotting-paper (avoiding all rubbing), and put it on a plate. Place upon it one or two crystals of iodine; in a few seconds so much heat will be produced that the whole will melt and catch fire. A jug of water should be at hand, to put out the flame when the above result has been well observed.

Here the energy of chemical affinity between the atoms[1] of phosphorus and iodine has, when they have united, appeared in another form, *viz.* heat. The red compound formed, phosphorus tri-iodide, does not show the same chemical affinity for other substances as was shown by the elements before combination, but the lost chemical power is fully accounted for in the heat which is generated.[2]

---

[1] An atom is the smallest particle of a substance which can take part in a chemical action.

[2] Compare this with what happens when a piece of iron is placed near a horse-shoe magnet. The magnetic energy shown by their mutual attraction passes into the form of mechanical motion as they approach each other: and when they are in contact they show little or no attractive power for other magnets or pieces of iron. When we separate them, expending (or rather transferring) some form of energy in so doing, magnetic energy again appears.

The ignition of a lucifer-match is caused in a very similar way. The end of the match is coated with a mixture of phosphorus with potassium chlorate, a salt containing and readily yielding oxygen, between which latter element and phosphorus there is almost as strong an affinity as between iodine and phosphorus. When this mixture is rubbed or struck, chemical action is set up; the phosphorus takes oxygen from the potassium chlorate, and the energy shown in the attraction of the atoms for each other is, when they have united, no longer available in that particular form, but manifests itself as heat, sufficient in quantity to set the match on fire.

This 'catching fire,' or 'burning,' of the wood itself is due to a similar cause. The carbon and hydrogen which wood contains have a strong affinity for oxygen, and when the temperature is sufficiently raised by contact with the heated phosphorus, chemical combination is set up between them and the oxygen of the air. A great deal of heat is thus produced, so that the action continues until all the carbon and hydrogen of the wood have united with oxygen to form the two compounds, carbon dioxide (formerly known as 'carbonic acid gas') and water.

The chemical action which is illustrated on a very small scale in the burning of a match is identical in character with that which produces the heat (and light) of all ordinary fires and candles, of the largest smelting furnace no less than of the smallest lamp. In all of them carbon in some form, whether nearly pure as in the case of charcoal and coke, or combined with hydrogen and other elements as in the case of coal, oil, wax, spirit of wine, etc., is (as well as the hydrogen, if present) combining with oxygen plentifully supplied by the air. The energy which is manifested in the powerful affinity between these elements passes into the form of heat, which is utilised for innumerable purposes by us.

The heating power of some of the principal kinds of fuel is given in the following table.

Combustion.

Heating Power of Different Fuels.

[The figures express the weight of water which can be raised from the freezing-point to the boiling-point by the combustion of 1 kilogramme of the substance.]

Fuel.	Weight of Water.	Fuel.	Weight of Water.
	Kilogrammes.		Kilogrammes.
Coal	82	Wood	36
Coal-gas . .	116	Charcoal . . .	81

The following experiments will serve to illustrate what goes on during the burning of a candle (and also of fires). When a burning match is held to the wick, the first effect is to heat some of the wax with which the wick is saturated until it turns into vapour; and this vapour is further heated until chemical combination begins between the carbon and hydrogen in it and the oxygen of the surrounding air. Thus a flame (which is simply the glowing vapours and gases) is produced, and the heat generated vaporises more wax, so that the action travels down the wick to the candle; a pool of liquid wax is formed round the base of the wick and rises up it by capillary action, and thus the flame is maintained. We have now to examine what becomes of the materials of the wax, which were formerly thought to be utterly destroyed.

Expt. 7. Light a candle and observe the points above mentioned, the spread of the chemical action down the wick and the formation of the pool of liquid material at its base. When it is burning steadily, hold over it for a few seconds a wide-mouthed bottle (holding about three-fourths of a litre), so that the flame is just within the mouth of the bottle. Observe that a dew of fine liquid particles is deposited on the glass. This liquid when collected is found to have all the properties of water, and to contain all the hydrogen of the candle united with oxygen.

Remove the bottle, replacing the stopper or cork at once, and proceed to try whether anything besides water is formed during the combustion. Pour about 20 or 30 c.c. of 'lime-water'[1] into a clean bottle similar in size to the one already used, and shake it up with the air in the bottle. No perceptible alteration will take place in the liquid, showing that this amount of ordinary air contains no appreciable quantity of anything which causes a visible change in the solution. Now pour the liquid into the bottle within which the candle has been allowed to burn, and shake it up. It will become milky, a cloud of solid particles being formed in it, which are found to be chemically identical with chalk (calcium carbonate). There is only one commonly occurring gas which produces this effect on lime-water, *viz.* carbon dioxide; and so the result proves that this gas, as well as water, is formed when a candle burns in air. It is found, on analysing it, that the carbon dioxide contains all the carbon originally present in the wax, united with oxygen.

Carbon dioxide, then, and water, are the products formed during the burning of a candle, and the smoke of fires and furnaces is found to consist essentially of the same substances.

The principle involved in blowing a fire consists in bringing an additional supply of oxygen into contact with the carbon and hydrogen of the fuel, and thus causing more chemical combination to occur in a given time and space, and therefore more heat to be generated. It is possible, of course, to blow too much air on a fire or candle, in which case the result is to convey away so much heat from the burning materials that they are cooled down below the temperature at which chemical combination is possible. Then the fire burns less intensely, and the candle-flame goes out.

Not only does chemical combination thus supply us with sufficient heat to make us more or less independent of the sun, but the temperature of our bodies is in a great degree maintained by internal actions of a similar character. Our bodies are, so to speak, furnaces in which heat is produced by the combination of

---

[1] This is the popular name of a solution of calcium hydrate, the substance which is formed by the combination of quicklime with water, as in the slaking of lime for making mortar. Directions for making it are given in the Appendix.

the carbon and hydrogen of the food we eat with the oxygen of the air we breathe. This oxygen, entering the lungs, dissolves in the blood passing through them, and is carried by it to all parts of the body. Whenever work is being done—whether internal (such as the action of the heart and lungs) or external (such as walking, running, carrying loads, playing cricket, etc.)— the necessary energy is obtained by the combination of the materials of the food with the oxygen contained in the blood, carbon dioxide and water being produced just as in the burning of coal. These products are carried in the blood as it returns to the lungs, where they pass into the air we breathe out; and it will be easy to prove their presence in the breath in the following way :—

Expt. 8. Take a clean dry bottle, about half a litre in capacity, and, after filling the lungs with air, breathe this air through a glass tube into the bottle. Observe that a dew is deposited on the sides of the tube and the bottle, showing that the air exhaled contains more water-vapour than the ordinary air. Next, pour some lime-water into the bottle and shake it up. It will become milky in appearance, proving, as already mentioned, the presence of carbon dioxide.

Now, whenever energy is thus being supplied to our muscles, only a portion of it is available for doing mechanical work, which is the conscious, ostensible purpose for which we eat food. At least one-half of it takes necessarily the form of heat; and in this way the temperature of the blood and of the whole body is maintained pretty steadily at $37°$, or raised considerably higher when much physical exertion is being made. It is thus easily seen why we get so hot when we run or wrestle or lift heavy loads. Energy must be supplied for such purposes in larger amounts than usual; the chemical actions above indicated go on more rapidly and abundantly, and a proportionately greater amount of heat is generated in the system.

The following table shows (though only approximately) the comparative value of different foods for the above purposes.

Food.	Heat produced by its Oxidation.	Mechanical Work.
1 Gramme of	Calories.[1]	Centimetre-grammes.[2]
Butter ⎫		
Bacon ⎬ . . .	7000	300 millions.
Cocoa ⎭		
Sugar . . .	3300	150 ,,
Bread . .	3000	130 ,,
Beef . . .	2000	90 ,,
Apples . . .	600	30 ,,

In doing work the materials of the muscles themselves and other tissues wear away, of course; and the chemical actions involved in their destruction contribute in some measure to the heat of the body, but the amount has been proved by experiments to be quite insignificant in comparison to the result of the main action above explained.[3]

3. *Conversion of Electrical Energy into Heat.*—This has not come into much practical use at present (we are more concerned with the light given out by an electric lamp than with its heat), but it is worthy of notice as affording the very highest temperature as yet obtained by artificial means.

Whenever electrical energy is transferred along a body some

---

[1] For an explanation of the meaning of a 'calorie,' see p. 126.

[2] A centimetre-gramme is the work implied in lifting a mass of one gramme one centimetre high against the force of gravitation. See p. 351.

[3] It is worth noting what a fallacy it is to suppose that bread and beef are absolutely the most nourishing articles of food. It is true that they contain much nitrogen, an element which must be supplied in order to repair the worn-out muscular tissues, etc.; but this wear and tear of muscle (like the wear and tear of the parts of a machine) is of quite a subordinate character, and the main source of our strength is that which has been above explained, *viz.* the chemical combination of the carbon and hydrogen of the protoplasm in the tissues with oxygen. Foods, then, which consist mainly of carbon and hydrogen, such as butter, bacon, sugar, cocoa, etc., are the best sustainers of human life and energy. See Dr. Frankland's paper in the *Journal of the Chemical Society*, vol. xxi. p. 33.

resistance is always offered to its passage; and this implies, not only that the inertia of the molecules has to be overcome in order to set them in electrical motion, but also that some of the energy passes into the form of heat. Some bodies are of such a nature that, in order that any electrical energy may pass as such, they cause the conversion of a very large amount of it into heat: then we say that they are 'bad conductors,' or that they 'offer great resistance' to the current. Thus, while a thick copper wire transmits the energy comparatively unaltered in amount as electricity, a thin wire of iron or platinum, or a thin filament of certain kinds of carbon, causes the conversion of a large amount of the energy into heat, and glows brightly, or is even raised to the melting-point. This is the principle of all the incandescent lamps now in use, which, however (as already mentioned), are of practical value from the light they give out, and not from the intense heat of which that brilliancy is a proof.

When a very intense current is passed through thick carbon rods, the ends of which are in contact, it is found that these ends may be separated for a short distance and the current will still continue to pass through the intervening space, being chiefly carried across by particles of carbon torn off from the ends of the rods; and that these latter as well as the detached particles become intensely hot. The temperature of the space lying directly between the ends of the rods is estimated at no less than 4700°, and the majority of metals when brought into it are turned into vapour. Quite recently this method has been utilised in the manufacture of aluminium.

Apparatus practically used for applying the Heat produced by Chemical Combination.

Nearly all our artificial supplies of heat are obtained by the burning of either coal and wood, or the gases and liquids (*e.g.* coal-gas and paraffin-oil) produced by heating coal in close vessels. But to simply set fire to a heap of coal is neither an economical nor an effective way of utilising the heat generated by its combination with the oxygen of the air. Heat shows so great a tendency to diffuse itself that a great deal is lost (for practical purposes) unless the burning fuel is enclosed in a fireplace or furnace of brick, or some such badly conducting material. This not only ensures the concentration of the heat upon whatever we want to make hot, but also enables the supply of air to be regulated and increased if necessary. This latter purpose is effected in most cases by the chimney, which does not merely serve to carry off the smoke and gases formed in the burning of the fuel, but also by the upward current of the gases in it [1] causes much more air to enter the fire than would otherwise be the case; since the pressure in the furnace is lessened by the withdrawal of gases up the chimney, and air, from its elasticity, always moves in the direction in which there is least pressure.

In order to obtain the full effect of the air in producing heat, the whole of it should be caused to pass through the fuel by closing the furnace doors, or in the case of an ordinary fire by putting a sheet of iron so as to block up the space between the grate and the top of the fireplace. Every one will have noticed the effect of this expedient in quickening a dull-burning fire; a result due to two causes—(1) the prevention of the entry of a mass of cold air above the fuel into the chimney, which would cool down the gases in it and so lessen the draught; (2) the forcing all the increased supply of air that enters to pass through the whole mass of fuel, so that a much larger amount of chemical combination takes place.

In many furnaces used in smelting and working metals, the substance to be heated is kept quite apart from the fuel, and only exposed to the flame. One of these so-called 'reverberatory furnaces' is shown in fig. 5. It will be seen that the actual fireplace A is at one end, and between it and the chimney there is a long low chamber of brick-work, B, on the floor of which (called the 'hearth') is placed the ore or metal to be heated. The peculiar form of the low arched roof causes the up-rushing flame to be

---

[1] The reason of this up-rush of the hot gases will be explained in a later chapter.

beaten back or *reverberated* upon the hearth, and thus to heat intensely whatever is placed there, while all injury from actual contact with the fuel is avoided.

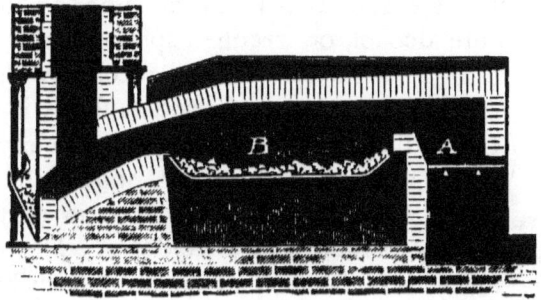

Fig. 5.

In cases where a still higher temperature is required, as in iron smelting, and where it is not convenient to use a high chimney, the air is driven through a jet into the fire by bellows or some other form of blowing machine. This expedient has the advantage of compressing a large mass of air into a small space, and thus a great deal of oxygen is made to act on the fuel in a given time, and its action is concentrated upon a small area of the combustible substance. Such furnaces are called 'blast-furnaces.'

Heating Apparatus in which Liquid Fuel is employed.—The two chief kinds of liquid fuel which are practically available are alcohol, or 'spirit of wine,' and paraffin-oil, the latter including several of the more volatile products of the distillation of coal-tar. These have hitherto been used only on a small scale, and fig. 6 represents the usual form of 'spirit lamp' adapted for burning alcohol, which was extensively used in laboratories before coal-gas was available, and is still of some value as a portable and cleanly heating apparatus. It is so simple in construction as hardly to require description, consisting of a reservoir for holding the alcohol, at the top of which is a metal screw cap carrying a short tube through which a wick made of a bundle of loose cotton threads passes down into the spirit. The latter rises between the cotton fibres by capillary action, and is kindled at the top of the wick, burning with a pale bluish flame which has the great advantage of depositing no soot on things held in it.[1] A glass cap *a*, is fitted on the neck of the lamp, and serves to prevent loss of the spirit by evaporation when the lamp is not in use.

Instead of pure alcohol, or 'rectified spirits,' methylated spirit, a mixture of real alcohol with one-tenth its volume of wood-naphtha, is now generally used for burning, as being much less expensive: but it should be examined before use, to see that no resin has been added, as is often done.[2] The temperature of the flame of alcohol is fairly high, though not equal to that of most gas-burners, and spirit lamps may be used instead of the latter in many of the experiments described in this book, if gas is not available.

Fig. 6.

Paraffin-oil lamps are, in general, more adapted for lighting than for heating purposes. They must necessarily have chimneys to ensure the complete combustion of the oil, and a great deal of soot is deposited upon any object held actually in the flame, owing to the large amount of carbon the substance contains. Nevertheless, several good forms of stove, fitted with paraffin-burners, are now obtainable; and it is highly probable that furnaces in which the same material is utilised instead of coal will shortly be employed in steam-ships and locomotive engines.

Heating Apparatus in which Gas is employed.—Coal-gas is now almost universally used for heating as well as for lighting purposes in laboratories, and to some extent in private houses. Indeed,

---

[1] The reason why no soot or smoke is formed is that ordinary alcohol contains comparatively little carbon, so that the whole of this (as well as the hydrogen present) obtains sufficient oxygen from the surrounding air to combine with, and none is liberated in the form of soot.

[2] A teaspoonful evaporated to dryness (cautiously, at some distance from a flame, on account of its inflammability) ought not to leave any decided amount of gummy residue. If it does so, it must be rejected, as likely to clog the wick.

wherever coal-gas can be obtained it is, though somewhat expensive, a far more effective and cleanly fuel than any other. Much, however, depends on the mode in which the combination of the gas with the oxygen of the air is caused to take place, and a short account of the best forms of gas-burners will be next given; the aim in all of them being to burn as much gas as possible in a given time and space, and to burn it as completely as possible.

(1) *The Argand Burner.*—The simplest form of burner, a plain jet at the end of a tube, is ineffective and wasteful, since only a small quantity of gas can be burned in a reasonable time; and if the jet is made large, the combustion is imperfect, since air cannot reach the interior of the stream of gas. The great improvement introduced for oil-lamps by M. Argand (of Geneva, A.D. 1780) was to arrange a number of small separate jets in a circular ring, as shown in fig. 7, admitting a free supply of air to the inside as well as the outside of the ring, and increasing the stream of air by a chimney placed round it. Thus a continuous circular hollow flame is obtained, giving great intensity both of light and heat. In the usual form of Argand burner for heating purposes shown in fig. 7, only a short metal chimney is employed, to protect the flame from irregular draughts of air.

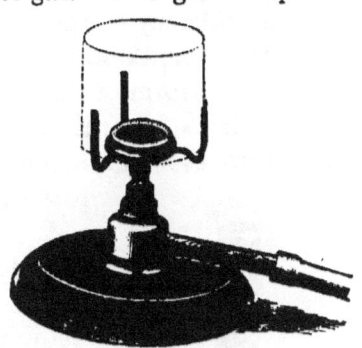

Fig. 7.

As a source of heat this burner is chiefly applicable where a steady, only moderately high temperature is required, as for keeping water boiling in flasks, etc., since its flame is perfectly under control, and can be regulated to the lowest possible point by diminishing the supply of gas. In using it the flame should never be allowed to actually touch the vessel which is being heated; otherwise a coating of soot will be deposited which obstructs the passage of heat.

The reason of this deposit of soot should be clearly understood, since the explanation of it will show not only the cause of the brightness of ordinary gas-flames (and candle-flames also, for candles are simply miniature gas-works), but also why we cannot get the full heating effect of coal-gas when burnt under ordinary conditions. Coal-gas consists mainly of hydrogen and compounds of hydrogen with carbon. Both these elements unite with the oxygen of the air, producing heat in doing so; but oxygen has more affinity for hydrogen than it has for carbon, and hence when (as is the case where a jet of coal-gas issues into the atmosphere) insufficient oxygen is supplied to combine with both the elements, all the hydrogen is burnt, but only *some* of the carbon. The rest of the carbon is separated in the flame, and it is these particles of carbon which, heated to whiteness, give the brilliancy to the flame, and which form the deposit of soot when cooled down by contact with a cold surface, or as they escape in the form of smoke at the top of the flame.

It is plain then that only a limited amount of gas can be burnt with even approximate completeness at an ordinary jet when simply surrounded by air; and hence an Argand burner, or indeed any form of illuminating burner, is by no means the best when a great quantity and intensity of heat is required.

Fig. 8.

(2) *The Bunsen Burner.*—The principle of this most useful burner is, to mix air with the gas before it arrives at the jet, and thus to ensure its complete combustion. It consists, in its original and simplest form, fig. 8, of a brass tube about 1 cm. in diameter, and 10 or 12 cm. in length, having one or more openings close to its lower end, which can be closed, when requisite, by a revolving cap. Gas is brought in through a small central jet, the orifice of which is just above the holes. Thus the gas is allowed to mix freely with the air which enters through the holes; the supply

of air being increased by the upward rush of gas in the same way as the draught in the chimney of a locomotive engine is increased by the blast of escaping steam. The mixture of gas and air is kindled at the top of the main tube, and here it meets with an additional quantity of oxygen from the surrounding air, so that altogether there is enough oxygen supplied to burn *all* the carbon as well as the hydrogen of the coal-gas.[1] The flame thus produced is not, as usual, white and luminous, but of a faint bluish colour, none of the carbon being separated in a solid form; as may be proved by holding a white plate for a moment in the flame. No soot at all will be deposited, but only drops of condensed water vapour, the carbon dioxide escaping as an invisible gas.

The temperature of the Bunsen burner flame is extremely high, being estimated as at least 2000° (nearly the melting-point of platinum); but a great deal of heat is lost by radiation unless the flame is enclosed in a furnace. A piece of copper wire, about 1 or 1·5 mm. in diameter, readily melts when held in the flame a little below the top.

This burner is so extensively used in various forms that it may be well to mention one or two points to be observed in order to get the best results from it.

1. The supply of air allowed to enter through the holes must be properly proportioned to the amount of gas issuing from the jet. It should be about *one-half* of the total volume of air required to burn the gas completely; the remainder coming from the air surrounding the flame at the top of the tube. If much

---

[1] It must be observed that the above is not a complete account of the theory of the Bunsen burner. Other causes besides the one mentioned in the text, which is the chief one, contribute to give the flame its peculiarities; such, for instance, as the dilution of the gas by the large quantity of inactive nitrogen in the air admitted. Its molecules are thus spread over a wider space, and hence (like a scattered army) attacked more effectively by the oxygen.

For a fuller account of the burner, Professor Thorpe's lecture to the Chemical Society (*Chem. Soc. Journal* for 1877, vol. xxxi. p. 627) may be consulted.

more than this proportion is admitted, the mixture in the tube becomes too explosive, and the flame is liable to pass suddenly down the tube, the gas continuing to burn at the little central jet with a small smoky flame and an unpleasant smell. In such a case, the gas must be entirely turned off and the holes partially closed by turning round the regulating cap; the gas may then be turned on again and rekindled. In fact, it is always best to close the holes altogether before lighting the burner, and then by slowly turning the cap to admit just sufficient air to destroy the luminosity of the flame. In this way even a small flame, 2 or 3 cm. high, can be safely obtained.

2. The gas must be thoroughly mixed with the air before reaching the top of the tube. This is effected, in a properly made burner, by the peculiar shape of the gas-jet. If the flame is luminous, even though the full supply of air is admitted through the holes, there is reason to suspect that the jet is out of order, possibly clogged by dust or some fused substance which has dropped down the tube. The main tube should be unscrewed, and the jet and air-holes examined and cleaned. If this does not remedy the defect, the jet itself is of faulty construction, and must be replaced by a better one. It is remarkable how slight a variation in the form of the jet makes all the difference between a good burner and a bad one.

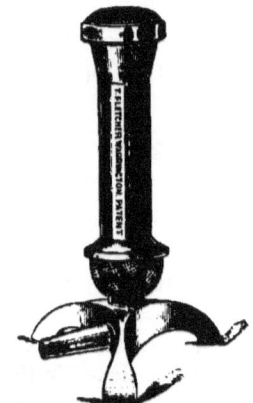

Fig. 9.

Since the introduction of this burner by Professor Bunsen of Heidelberg, in 1855, it has been modified and improved in various ways, especially by Mr. Fletcher of Warrington, one or two of whose burners are shown in figs. 9 and 10. Fig. 9 shows a burner, the top of which is slightly enlarged and covered with a piece of fine wire-gauze. This not only increases the breadth of the flame, but also effectually prevents the flame passing down the main tube (for a reason which will be explained in the chapter on Conduction of Heat), even when the

Gas Furnaces.

supply of gas is so far checked as to give the least possible flame.

Fig. 10 represents a ring-burner made of cast-iron, the gas and air having ample time to mix in the large horizontal tube and

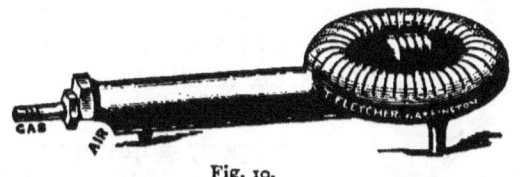

Fig. 10.

hollow ring. In the circumference of the latter a number of fine slits are cut, the small flames from which unite (as in the Argand burner) to form a continuous ring of extremely hot, smokeless flame.

In fig. 11 we have a good example of a gas furnace. The body is made of porous fire-clay, an extremely bad conductor of heat. Below it is seen a large horizontal Bunsen burner, the flame of which, issuing from the up-turned jets, fills the whole of the furnace, enveloping the crucible completely, and passes up the chimney at the side, which is about 2 metres or more in height.

Fig. 11.

In this furnace a kilogramme of brass, copper, or silver can be readily melted.

(3.) *The Gas Blowpipe.*—The principle of this apparatus is to increase still further the amount of chemical combination taking place (and therefore the amount of heat obtained) in a given time, by forcing condensed air (or pure oxygen) into a jet of burning gas by means of bellows.

A common form of blowpipe, invented by Mr. Herapath, is represented in fig. 12. The gas enters at the side of a tube about 1.5 cm. in diameter, at the end of which it is kindled, while a blast of air is driven into the centre of it through the internal jet shown in the figure. The on-rushing particles of air carry with them a large quantity of gas, while a great deal of oxygen is present in even a small volume of the highly compressed air; and an extremely hot flame is the result of their union. The size and shape of the flame may, by properly adjusting the supplies of gas and air, be varied from a small pointed blue cone to a large, brush-like, roaring flame: the temperature of either being so high that a piece of rather thin iron wire (about 0·5 mm. in diameter, No. 26 wire-gauge) is readily melted when held in it, near the top. It will be found, however, impossible to melt a similar piece of platinum wire, although it is raised to a glowing white heat.

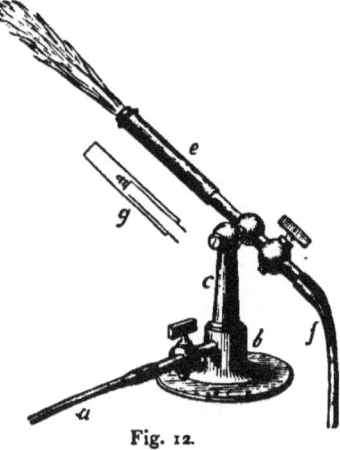

Fig. 12.

When the flame is enclosed in a fire-clay chamber, as in fig. 13, which represents one of Fletcher's injector furnaces, a kilogramme of cast-iron can be readily melted in less than a quarter of an hour, and even crucibles of the most refractory fire-clay can be softened and formed into a kind of enamel.

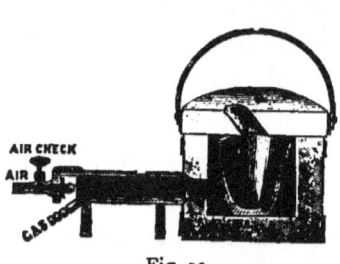

Fig. 13.

In order to obtain the highest possible temperature by the combustion of coal-gas, pure oxygen must be used in the blow-pipe instead of air. Ordinary air contains only about one-fifth its volume of oxygen, the remaining four-fifths being nitrogen, a substance which shows hardly any tendency to combine directly with carbon or hydrogen, and which is therefore useless as a heat producer; or rather, it is worse than useless, since (*a*) it dilutes the oxygen and weakens its effect, (*b*) it carries off some of the heat of combustion.

When pure oxygen is blown into the gas-flame through the jet (which should be rather smaller than that used for air), the flame contracts in size but increases in temperature. A piece of platinum wire introduced into it melts like wax; clay, such as the end of a tobacco-pipe, fuses into a white enamel; and a piece of steel, such as a bit of watch-spring, or the end of a small file, melts and burns with showers of brilliant sparks. When the flame is directed upon a cylinder of quicklime (a substance which has never been melted by any artificial source of heat), the parts of the latter within the area of the flame glow with a most intense light. The 'lime-light' used in optical lanterns and for theatrical effects is produced in this way.

In the manufacture of platinum large quantities of the metal are fused by means of one or more blowpipes such as have been described above. The metal to be melted is placed in a cavity hollowed out in a block of quicklime (the only easily procurable material which will stand the heat without melting); and this is covered with a similar block in which one or more holes are cut for the purpose of introducing blowpipes, the flames of which are directed downwards upon the metal. An opening is cut in the side of the lower block to allow the products of combustion to escape, and through this opening also the liquid metal is poured into moulds of quicklime. In such an apparatus a mass of 150 kilogrammes (about 3 cwt.) of platinum can be melted in less than an hour.

(4) *The Oxy-hydrogen Blowpipe.*—Just as a higher temperature is obtained with the ordinary blowpipe by using oxygen in place of air, so when pure hydrogen is substituted for coal-gas (a rather indefinite mixture of gases), we get the highest temperature which can be obtained by any process of chemical combination, *viz.* about 2800°.[1]

---

[1] This is Professor Bunsen's estimate of the actual temperature of the oxy-hydrogen flame. Theoretically the union of the two pure gases would produce a temperature of 8060°.

The Oxy-Hydrogen Blowpipe.

When pure oxygen and hydrogen are used, the form of the blowpipe is slightly modified, as shown in fig. 14. The gases are brought from separate reservoirs by the tubes O and H respectively, and pass into a short brass cylinder, A, filled with discs of wire-gauze or a bundle of fine brass wires : the object being to mix them thoroughly before they arrive at the platinum jet, B. A vertical adjustable holder, C, is sometimes attached, on which a lime-cylinder can be supported.

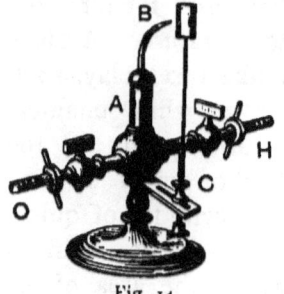

Fig. 14.

CHAPTER IV.

THE EFFECTS OF HEAT UPON SUBSTANCES.

Section I.—Characteristics of the Three States of Matter.

When Heat is transferred to, or taken from, any substance, the resulting changes may be broadly classified under the two following heads :—
1. *Change of Size.*
2. *Change of State.*

The latter kind of change is the one which we most generally and easily take notice of. Every one has observed, for instance, that ice when heated turns into water, and that water when further heated turns into steam; the three substances, ice, water, and steam, though in physical properties so unlike each other, being absolutely identical in chemical composition. It may be well, before further examining the precise mode in which heat produces these changes, to notice briefly the characteristics of the three states, solid, liquid, and gaseous, in which matter can exist.

1. A solid, such as ice, iron, iodine (a substance with which experiments will shortly be made), has a definite shape and size, which it does not change unless some force is applied. It is bounded, in fact, by surfaces which are clearly and sharply marked out, as is well shown in the case of crystals, such as those of iodine and alum.

A liquid, such as water, alcohol, mercury, shows hardly any tendency to preserve a definite shape (under ordinary conditions), but readily adapts itself to the shape of any solid vessel which contains it. It may, however, have a distinct boundary or surface above, such as water shows when it only partially fills a basin or bottle; so that it is always easy to see exactly where the liquid is, and where it is not, in the vessel.

A gas, such as steam, air, vapour of iodine, has no definite shape or size at all: it does not even show a level upper surface like a liquid. If only a little of a gas is put into a bottle it quickly spreads over the whole space within the bottle; and if the mouth is open, most of it soon escapes from the bottle and travels in all directions to great distances. In fact, no limit can be assigned to this power of spreading (or 'diffusion,' as it is termed) of a gas; and if its escape is prevented by closing the bottle, it always exercises a pressure in every direction against the solid walls which confine it.

The characteristics of the three states of matter, and the relation of heat to them may be well shown in the case of iodine as follows:—

Expt. 9. Put a few crystals of iodine into a large globular flask about 18 or 20 cm. in diameter, noticing their rhombic, diamond-like shape,[1] their opacity and almost metallic lustre, and their comparative hardness, as shown by their rattling against the sides of the flask when it is shaken. Heat the bottom of the flask gently over an Argand burner (or spirit lamp), waving it about over the flame so as to distribute the heat. Very shortly the rattling of the crystals in the flask will cease as they lose their sharp edges and flat surfaces, and they will sink down into drops of an almost black liquid, spreading over, and taking the shape of, the bottom of the flask. Almost immediately, as the temperature rises higher, a deep-coloured vapour will be observed rising from the liquid and spreading through the flask, which will soon be entirely filled with the splendid purple vapour of iodine.[2]

Thus we have traced the passage of a substance from the solid to the liquid, and then to the gaseous state. But since the upper

---

[1] The diamond, however, does not crystallise in precisely the same form as iodine. The crystals of iodine are, strictly speaking, oblique prisms with a rhombic base.

[2] Iodine obtained its name (Gk. ἴον, a violet; ἰοειδής, violet-like), from this colour of its vapour.

parts of the flask are comparatively cool, the vapour will here soon give up so much heat as to become liquid and finally solid again, condensing about the neck of the flask in minute crystals of the same form as those originally taken.

Now, it is easy to see how heat, if it is a peculiar movement of the molecules of a body, may act in overcoming their natural cohesion.

In solids, this motion of the molecules probably takes place through very small spaces and with comparative slowness, each molecule moving but a very little way from its normal position, so that it never escapes from the range of attraction of other molecules. Thus their mutual cohesion asserts its full force, and the whole mass of them preserves its definite shape, like a battalion of well-disciplined soldiers. As more and more heat-motion is communicated to them, their velocity increases and the range of their movement gets larger; and so the whole mass of the heated body increases in size, and the molecules (though still keeping within a certain tether, as it were) approach more nearly the limits of distance within which alone cohesion can exert its influence in keeping them together. At a certain point of temperature these limits are so nearly reached that a very slight external force is all that is needed to tear the mass asunder in any part, or alter its shape. The solid, in fact, becomes a liquid.

In liquids the heat-motion is so rapid that, as just mentioned, cohesion is almost entirely overcome, partly on account of the increased distance of each molecule from its neighbours, but mainly because of their increased velocity. The molecules of matter in the liquid state make wider excursions from their original positions; moving, when unimpeded, in straight lines, though still greatly hampered, not only by the attraction of, but also by constant collisions with other molecules. They may be roughly compared to a close swarm of gnats, the individuals of which keep, as a rule, within the bounds of the swarm, but move with a certain amount of freedom to all parts within these bounds, and sometimes fly off altogether into less crowded spaces beyond, and enter other swarms.[1] These increased movements of the molecules demand an increased average space for movement, and the total volume of the liquid gets larger as the temperature rises. Finally a point is reached at which cohesion is altogether overcome, and the liquid loses its property of showing a well-defined upper surface, and becomes a gas.

In gases the communication of heat has given the molecules such a velocity that they are no longer kept within any bounds by their mutual cohesion: they move in straight lines only, incessantly coming into collision with each other and with the sides of the vessel which contains them. They may be roughly compared to a swarm of active bees imprisoned in a hive against their wills. It is, in fact, these incessant collisions with any solid or liquid substance that comes in their way, which occasions the outward pressure observed when a gas is confined in a vessel. As the temperature rises, the velocity of the flying molecules increases, and so also does the force with which they strike upon the barriers which confine them. If the resistance of the latter is limited (in other words, if the pressure upon the gas is simply maintained constant and not proportionately increased) the impact of the molecules will drive them back, and the gas will expand.

Having thus gained a general idea of the way in which heat acts in producing changes in the matter with which it is associated, we may proceed to examine the exact nature and extent of these changes, taking first the changes of size.

Section II.—Expansion of Solids by Heat.

Solids expand so little, even when strongly heated, that we hardly notice their change of size at all. The fact, however, that they do expand may be readily shown in the following ways:—

Expt. 10. Take a brass bar, A, fig. 15, which has been cut

---

[1] That this is the state of things in liquids is shown by the tendency they have to mix with each other, called their power of diffusion. Wine, for instance, mingles with water; and a drop of ink let fall into water soon spreads through the whole mass and colours it uniformly.

EXPANSION OF SOLIDS BY HEAT. 43

to such a length that it just fits easily into the long notch cut out of the brass plate, B C, when both are at the same temperature. Heat the bar for a few seconds in the flame of a Bunsen burner or spirit lamp, and try to fit it again in the notch. It will now be found too long to go into its place, proving that the communication of heat has expanded it in length. To make sure that heat and nothing else has wrought this change, dip it into cold water, which will simply take heat from it, and try it again in the gauge; it will be found to fit in as easily as at first. Now, since the simple addition of heat has made it longer, and the simple withdrawal of heat has made it shorter, it follows pretty certainly that heat alone has caused the change in length.

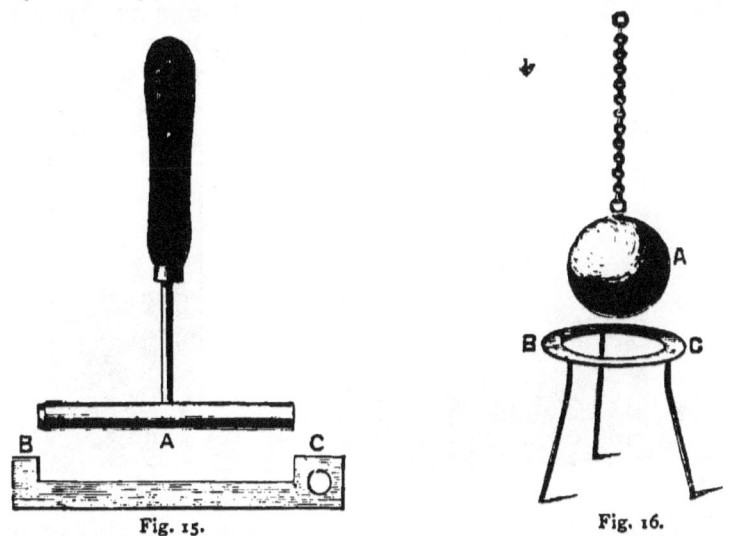

Fig. 15.   Fig. 16.

That heat causes a body to expand, not merely in one direction, e.g. in length, but in all directions, may be proved as follows :—

Expt. 11. Take a brass ball, A, fig. 16, which has been accurately turned so as just to pass through the opening in the flat brass ring, B C, when both are at the same temperature. Dip the ball into boiling water for half a minute,[1] and place it in the

---

[1] This is the safest way of heating the ball, if it is made as they usually are, of two hollow hemispheres soldered together with soft solder. If it is made of a single hollow casting, which is much the best way, it may be heated in a Bunsen burner flame.

ring. It will now be found to have become too large to pass through the opening, in whatever direction it may be tried,[1] proving that heat has expanded it in all directions. If it is cooled to its former temperature by being dipped into cold water, it will shrink and easily pass through the ring, as at first.

We may next take other substances, such as iron and glass, and see whether they also change in size when heated and cooled, using a method by which any slight change may be magnified enormously so as to be visible to others at a distance.

Expt. 12. Take a flat rod of iron about 30 cm. long, 1 cm. broad, and 2 or 3 mm. thick (a piece of ordinary 'hoop-iron' will do), and support it at each end upon wooden blocks about 10 or 12 cm. high, as shown in fig. 17. Place a weight

Fig. 17.

upon one end to prevent the bar moving in that direction, and under the other end (*i.e.* between the metal and the wooden block) place a small needle, to the eye-end of which a light pointer of straw or thick paper has been cemented at right angles by sealing-wax. Then, whenever the bar changes in length, it must roll upon the needle and turn it round; and since

---

[1] The brass balls usually sold for this experiment are very often not turned accurately into spheres. The ball should be tested previously, to see that it just drops through the ring, not merely when the suspension-ring is uppermost, but when it is inclined to the perpendicular in various directions.

EXPANSION OF SOLIDS BY HEAT. 45

the circumference of the needle is small, a very slight variation in the length of the bar placed upon it will turn it some way round, such motion being made easily visible by the variation in position of the straw index, like the hand of a clock.

Heat the bar by a Bunsen burner or spirit lamp, passing the flame along it so as to heat nearly its whole length pretty uniformly. The expansion of the metal will be readily evident from the movement of the pointer.

Next, replace the iron bar by a strip of glass of about the same size, and repeat the experiment.[1] The expansion of this material also will be easily demonstrated. Bars of other substances such as copper, zinc, etc., may, if at hand, be tested in the same way.

By such experiments the general law may be established that all solids expand when heat is communicated to them, and contract when it is taken away from them: and the next question is,—Do they all expand equally when equally increased in temperature? If we take, for instance, a rod of brass 1 decimetre long, and a rod of iron of the same length, and heat them equally, will their increase in length be equal?

The following apparatus may be used to answer this question by experiment in the simplest way.

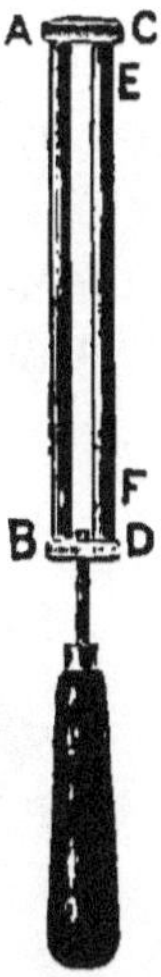

Expt. 13. A B, fig. 18, represents a bar of brass about 15 cm. in length, and about 7 or 8 mm. in thickness, having short brass arms, A C and B D, projecting at right angles from each end. Between these arms is placed the iron bar E F, similar in size and of just such a length as to be gripped pretty firmly by them when the whole is at the same temperature. Now let the two bars be heated uniformly by passing them through the flame of a Bunsen burner. If the brass and iron rods lengthen equally (or if the iron expands more than the brass), the iron bar will be held firmly by the arms, however

Fig. 18.

---

[1] Care should be taken in heating the glass, on account of its liability to crack with sudden changes of temperature. The flame of the burner should not be allowed to rest on one point for any length of time; and no more heat should be applied than is necessary to observe the effect of expansion clearly and unmistakably.

much the temperature is increased. But it will be found that the iron bar falls out as soon as the whole gets hot; proving that the brass expands so much more than the iron that the arms fail to pinch the latter between them.

The difference in expansion of various substances may be

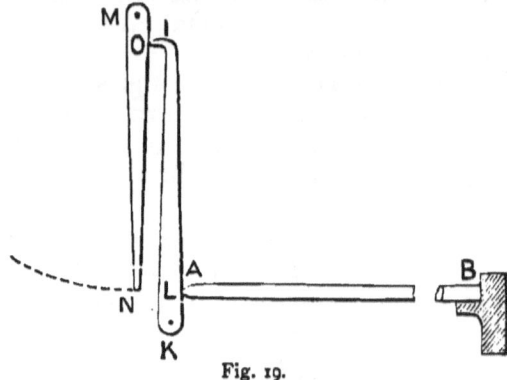

Fig. 19.

shown more strikingly, though less simply, by an instrument generally known as 'Ferguson's Pyrometer,' in which levers are used to magnify the very small actual expansion of solids. Fig. 19 is a diagram illustrating its principle, and the actual apparatus as

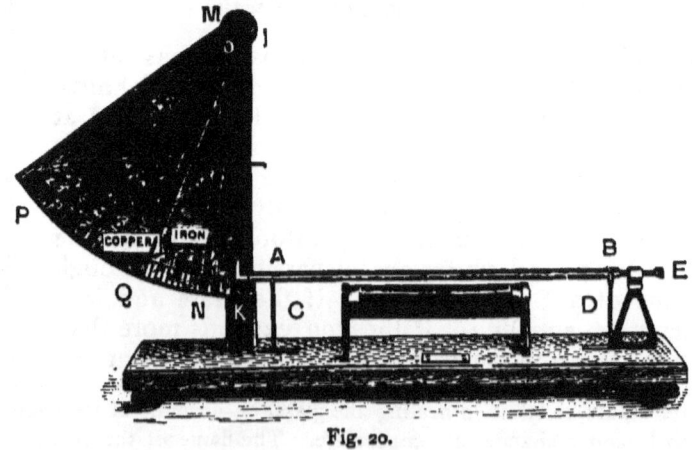

Fig. 20.

arranged for comparing the expansion in length of two different solids is shown in fig. 20. The same letters refer to both the figures.

Expansion of Solids by Heat.

A B is a bar of the substance to be examined. It rests in notches cut in the upper ends of two supports, C and D, fig. 20, and the end B bears against a screw E, passing through a strong, firmly fixed pillar F, so that it cannot move in that direction. The other end A touches a lever I K (see fig. 19) at a point L, very near its fulcrum. Now, from the well-known mechanical principles of the lever, if a very slight motion is imparted to the point L, the outer end I moves through a much greater distance: in fact, if the distance I K is ten times the distance L K, the point I will move through ten times the space that the point L does. To magnify the movement still further, the end I of the lever is made to touch a similar lever M N at a point O near its fulcrum M: so the outer end N of this lever moves through a much greater space than I does, and forms a pointer traversing the divided scale P Q.

The principle being now understood, the actual instrument, fig. 20, in which there are two rods, one of each of the substances to be tested, and two sets of levers arranged side by side and moving over the same scale, may be used.

Expt. 14. Place two bars, one of copper or brass, the other of iron, about 7 or 8 mm. in diameter, side by side in the notches, so that each may lie between one of the levers and its corresponding screw. Turn the screws at the right hand end until the pointers just begin to move up the scale, showing that the ends of the bars are in contact with the levers. Pour a little common methylated spirit into the trough R, and set fire to it. The flame will soon heat both bars fairly uniformly, and the amount of their expansion will be shown by the movement of the pointers on the divided scale. It will be noticed that the pointer connected with the copper moves much farther than that connected with the iron, proving conclusively that copper expands by heat much more than iron.

The expansion of iron may be compared in a similar way with that of zinc, brass, and glass.[1]

---

[1] A tube of difficultly fusible glass should be taken, sealed at one end so that the screw may bear upon it, and smoothed at the other end where it is to touch the lever. Neither it nor the zinc should be heated very strongly, or they may soften and bend.

48 EXPANSION OF SOLIDS BY HEAT.

One other method of showing the unequal expansion by heat of different solids may be noticed, since it illustrates an important application of the fact to the construction of thermometers and the compensation of watch balance-wheels (p. 63). It depends upon the alteration in shape, when heated, of a bar composed of two flexible metals firmly joined together throughout their length.

Let us consider for a moment what happens when a straight rod of any material is bent into a curve: obviously the molecules on the outside of the curve are forced farther apart. Now, conversely, if by any force the molecules on one side of a straight rod are forced farther apart, the rod must become curved, that side of it where the distance between the molecules has been increased forming the outside of the bend.

Expt. 15. Take a compound bar (fig. 21) made up of two strips,[1] of iron and copper respectively, about 40 cm. long, 3 or 4 cm. broad, and 2 mm. thick, riveted or brazed firmly together throughout their length; test its straightness by applying a ruler or by laying it upon the table; then, holding

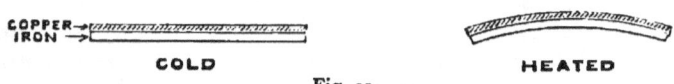

COPPER →
IRON →
COLD   HEATED
Fig. 21.

it with a pair of pincers, heat it uniformly in a large lamp-flame, passing it through the flame from end to end. If its form is now again tested by the ruler, it will be found to have become curved, and the copper bar will be observed to form the outside of the curve; proving, as above explained, that it must have expanded in length more than the iron.

Before passing on, we may notice the enormous mechanical force with which the expansion of solids takes place. It is equal, in fact, to the force which would be required to compress

---

[1] Such a strip may be readily made by soldering together with soft solder a strip of thick tinned iron (sold as 'tin plate'), about 2 cm. broad, and a similar strip of copper; but in this case only a slight heat must be applied for fear of melting the solder.

Expansion of Solids by Heat.   49

the solid into its original dimensions. For instance, a bar of iron 1 centimetre square and 1 metre long at the freezing-point of water, expands in length 1·2 mm. when heated to the boiling-point of water. In order to shorten such a bar by 1·2 mm. a mechanical force of about 3000 kilogrammes is required, and this same force of 3000 kilogrammes represents the force with which the iron, when heated, presses against any obstacle which resists its expansion. Similarly, when a heated bar is cooled, it contracts with the same intense force; and the fact may be shown in the following way:—

Expt. 16. A strong cast-iron frame must be procured of the shape shown in fig. 22. A, B are two massive supports,

Fig. 22.

having deep notches cut in them. C, C' are two <-shaped projections, about 6 cm. apart, on the outer side of A. D E represents a bar of wrought-iron not less than 1 cm. square, and rather longer than the space between the uprights of the frame, which may be about 25 cm. At one end, E, a screw is cut and fitted with a strong iron nut, having 'wings' on opposite sides so that it can be turned round by the fingers. The other end, D, of the bar is formed into a loop or 'eye,' with an aperture of about 1 cm. Through the loop a short bar of cast-iron is passed, the ends of which, when the long bar is laid in the frame in the position shown in the figure, rest against the projections C C'. Thus when the nut E is screwed up against the support B, the wrought-iron bar is stretched, as it were, between the supports, and any contraction in its length must either break away the massive supports, or strip off the threads of the screw, or break the cast-iron bar placed in the loop,—whichever is the weakest part of the apparatus.

The wrought-iron bar should be heated to dull redness in every part except the ends by laying it across some glowing coals in an ordinary fire (or, better, by placing it in a small gas furnace, such as is used for heating tubes), and then laid in the notches of the frame. The cast-iron bar should then be slipped quickly into its place as shown in the figure, and the nut at the other end screwed up pretty tightly against the support. As the bar cools (the cooling may be hastened by pouring cold water over it) the contraction will take place with such force that the cast-iron bar will break in two at the loop.

By such methods as have been above indicated, the following general laws may be established regarding the alteration in volume which solids undergo when their temperature is changed.

I. All solids increase in size when heated, and diminish in size when cooled.

II. This change of size is different in the case of different materials: no two substances expanding to exactly the same proportional extent for the same increase of temperature.

III. This change of size is strictly proportional to the change in the temperature, and is a definite fraction of the size of the solid.

Thus, if a bar the length of which at 0° is 1 metre expands 1 mm. in length when heated from 0° to 100°, it will expand 2 mm. in length when heated from 0° to 200°, 3 mm. when heated from 0° to 300°, and so on. Again, a bar of the same substance 2 metres long at 0° will become 2 mm. longer when heated from 0° to 100°; a bar 3 metres long at 0° will become 3 mm. longer when similarly heated, and so on.

This enables us to express the change of size caused by heat in a body in a very concise and exact way, *viz.* by simply stating once for all *the fraction of its size which it expands when heated, say, 1° of temperature.* Thus, in the case of the above bar, 1 metre long at 0°, since it expands 1 mm. when heated

COEFFICIENT OF EXPANSION.

100° it will expand $\frac{1}{100}$ mm. when heated 1°. But since 1 mm. is $\frac{1}{1000}$ of a metre, the amount of expansion of the bar for 1° is ($\frac{1}{100}$ of $\frac{1}{1000}$ =) $\frac{1}{100,000}$ of its length at 0°, and this fraction will be the same whatever the actual length of the bar is.

COEFFICIENT OF EXPANSION.

This is the general name given to *the fraction of its size which a body expands when heated from* 0° *to* 1° *C.*

Thus, in the example above given, the coefficient of expansion in length of the bar would be said to be $\frac{1}{100,000}$, or 0·00001.

The exact determination of these coefficients of expansion is a matter of great importance, especially to engineers: and very careful investigations of the subject have been made by several distinct methods, one or two of which may be briefly described here.

1. Roy and Ramsden's method.

This is interesting from its having been originally devised and used for finding the coefficient of expansion of the iron measuring-rods used in the determination of the 'base-line' of the first accurate trigonometrical survey of Great Britain.[1]

The principle of it is, to measure the displacement of the image of an object in the field of view of a telescope, caused by the object-glass of the telescope being shifted in position by the expansion of a bar of the substance to be tested. If a telescope is pointed at any object so that the latter is seen exactly in the middle of the field of view, and if the outer end of the telescope (where the object-glass is placed) is pushed to the right hand, for instance, the image shifts its position in the field by an amount proportional to the change in position of the object-glass.

In Roy and Ramsden's apparatus three long troughs were

---

[1] A full account is given in the *Philosophical Transactions of the Royal Society* for 1785, p. 461. Original accounts of scientific researches should always be referred to when practicable.

placed parallel to each other, as shown in fig. 23. In the two outer ones were placed iron bars, A B and C D: in the middle one was put the bar to be tested, X Y. The ends A, X, and C, rested against projections in the troughs, so that they could not move in that direction; but the other ends, B, Y, and D, were free to move. Upon each end of each bar was fixed a short upright pillar, and on the top of the pillars at A and B were fixed the eye-lenses of telescopes; while on X and Y were fitted the corresponding object-glasses of the telescopes. On C and D

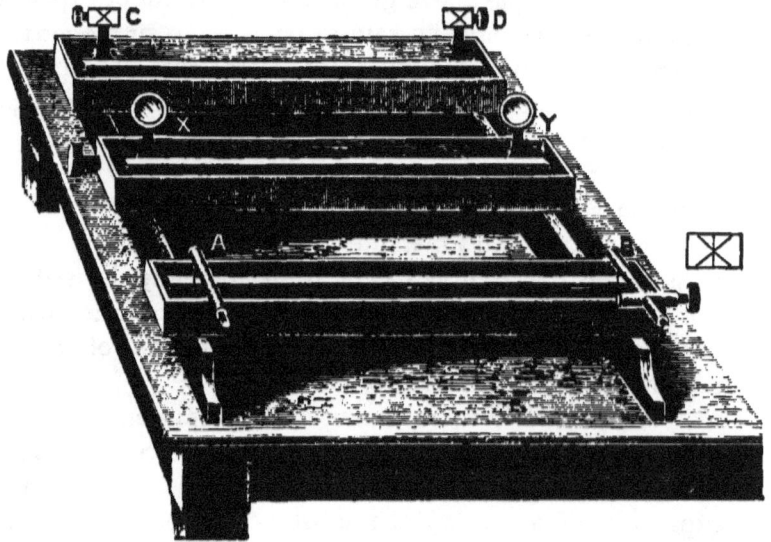

Fig. 23.

were fixed brass frames, across which fine wires were stretched, intersecting like the letter X. The position of the bars was so adjusted that when they were all at the same temperature (which was secured by filling the troughs with melting ice) the centres of the eye-lenses, the object-glasses, and the brass frames, were in the same straight line, so that the image of the cross wires appeared in the exact centre of the field of view to an observer looking through the eye-lens of either telescope. A single wire was placed vertically in each eye-piece to mark the exact centre of the field, and was made to pass through the point of intersection of the images of the two wires, as shown in the small diagram at the side of the figure. The vertical wire in the eye-piece at B could be moved to the right or left by a most accurately made screw, and by counting the number of turns of this screw the distance through which the wire was moved could be measured very exactly.

The ice in the middle trough was now replaced by boiling water; and the bar in that trough expanded in the only direction in which it was free to move, *i.e.* towards the right hand. Thus the object-glass at Y was shifted to the right, and the image of the cross wires no longer appeared in the middle of the field, but towards the right-hand side of it. While the bar was kept steadily at the temperature of boiling water, the vertical wire was moved by its screw until it passed through the point of intersection of the images of the cross wires in their new position, and the distance through which it had to be moved was ascertained by counting the number of turns and parts of a turn which the screw had made. From this it was easy to calculate the distance through which the object-glass at Y had moved; and this distance is, of course, the space through which the bar had expanded in length when heated from the temperature of freezing water to that of boiling water. The coefficient of expansion for $1°$ will be $\frac{1}{100}$ of this.

Coefficient of Expansion in Volume of Solids.—This can be ascertained in two ways—

1. *By Calculation*, when the coefficient of expansion in length is known.

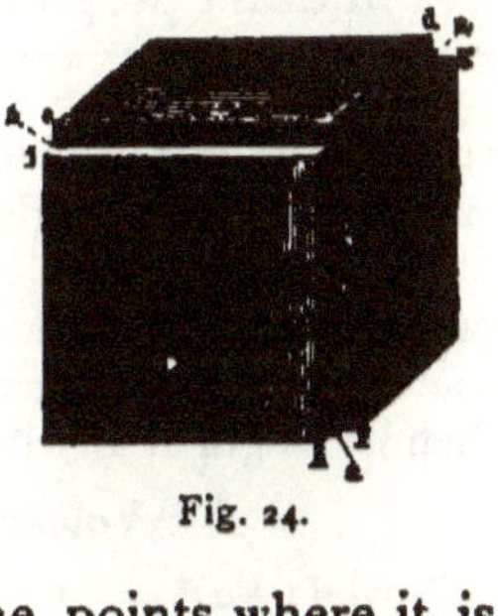

Suppose a cube of the solid *a b c*, fig. 24, is laid in the corner of a box so that it can only expand in the directions opposite to the points where it is in contact with the box. Then, if it is heated, it will expand in height to the extent shown by the lightly shaded part *d e*: in length to the extent shown by *f g*; in breadth to the extent

Fig. 24.

shown by $h\ i$.[1] Thus it is easy to see that the total increase in volume in the three directions will be very nearly three times its expansion in one direction: *i.e.* three times its linear expansion. Thus we may take it as a rule that the coefficient of expansion in volume (or 'cubical expansion') of a solid is almost exactly three times its coefficient of linear expansion.[2]

2. *By Experiment.*

Two distinct methods have been employed for this purpose.

(*a*) Dulong and Petit's method.

The solid is put into a glass vessel, resembling a 'weight thermometer,' fig. 51, which is then filled up completely with mercury at the freezing-point of water. Then the whole is put into boiling water; some of the mercury will of course be forced out by the expansion of the solid, and the amount which overflows is carefully measured. Allowance being made for the expansion of the vessel and of the mercury itself, the remainder will express the expansion in volume of the substance immersed in the mercury.

(*b*) Matthiessen's method.[3]

This consists simply in finding the alteration in density of a solid produced by a definite alteration of temperature.

Density means the quantity of matter in a given volume of a substance; and if the size of a body is increased, its density gets less in exact proportion to the increase in size.

---

[1] It will expand also in the direction of the corners, but the increase in volume due to this will be extremely small compared with the others, and may be practically neglected.

[2] It is easily provable that the volumes of bodies are to one another in the proportion of the cubes of one of their dimensions. Hence if a denotes the coefficient of linear expansion, and the length of a body at 0° be taken as $= 1$, then its length at 1° will be $1 + a$. Thus we have

$$\text{Volume at } 0° : \text{Volume at } 1° : : 1^3 : (1+a)^3.$$

Now $(1+a)^3 = 1 + 3a + 3a^2 + a^3$, and since a is in the case of solids a very small fraction, its square and cube will be so extremely small that they may be neglected. Thus we may practically represent the increase in volume or 1° by the two first terms only, *viz.* $1 + 3a$.

[3] *Philosophical Transactions*, 1866, Part I.

COEFFICIENT OF EXPANSION.

Thus, suppose that a cubic centimetre of iron, A B C D, fig. 25, has its volume doubled so that it measures 2 c.c., as represented by E B C F. Then, obviously, the quantity of iron in 1 c.c. will be only half what it originally was; *i.e.* the density of the block will be half its original amount.

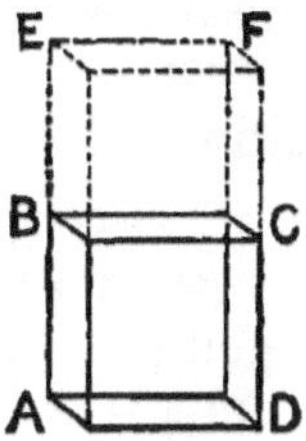

Now, conversely, if we find that the density of the block at a certain high temperature is one-half what it is at a certain lower temperature, we know that the size of the block must have been doubled by the increase in temperature. In fact, it may be stated generally that density varies *inversely* with volume, so that from the proportion,—

Fig. 25.

Density at 1° : Density at 0° : : Volume at 0° : Volume at 1°, the increase in size of the substance when heated from 0° to 1° can be ascertained.

COEFFICIENTS OF EXPANSION OF SOLIDS.

[The figures denote the fraction of its size which the substance expands when heated from 0° to 1° centigrade.[1]]

Substance.	Coefficient of Expansion.	
	In Length.	In Volume.
Zinc,	·00003 $(= \frac{1}{33,000})$	·00009 $(= \frac{1}{11,000})$
Brass,	·000018 $(= \frac{1}{55,000})$	·000054 $(= \frac{1}{17,000})$
Copper,	·000017 $(= \frac{1}{58,000})$	·000051 $(= \frac{1}{19,000})$
Iron,	·000012 $(= \frac{1}{83,000})$	·000036 $(= \frac{1}{28,000})$
Platinum,	·0000088 $(= \frac{1}{113,000})$	·0000264 $(= \frac{1}{38,000})$
Glass,	·0000086 $(= \frac{1}{116,000})$	·0000258 $(= \frac{1}{38,000})$

Practical Results and Applications of the Alteration in Volume of Solids by Change of Temperature.—This change of size is a fact which we cannot (in most cases) prevent or avoid the conse-

---

[1] For the sake of simplicity, the approximate values only are given. The exact linear coefficient of zinc, for instance, would be (according to Matthiessen) ·00002976.

56   APPLICATIONS OF EXPANSION OF SOLIDS.

quences of. It causes sometimes great inconvenience and even danger, while in other cases it is applied to very useful practical purposes. We will first of all consider a few of the latter applications.

1. The 'tyres' or iron rims of wheels are fixed firmly in their places, so as to bind the whole frame-work of the wheel securely together, in the following way:—

The tyre is made of an iron ring very slightly too small to fit round the wheel at ordinary temperatures. Then it is heated to redness, or nearly so, in a furnace, and thus expanded sufficiently to drop easily over the wheel; the latter being laid horizontally on a flat surface. As soon as it is in its place, water is dashed over it to cool it and prevent its burning the wheel if the latter is of wood. As it cools it contracts, and grips the wheel with a force and firmness which could hardly be secured in any other way.

2. The different parts of which the enormous guns now constructed are built up, are held together by similar means. Fig. 26 shows the construction of one of the guns now made at

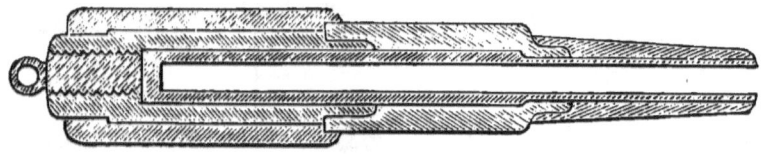

Fig. 26.

Woolwich, and often weighing as much as 100,000 kilogs. (equal to more than a hundred tons). It will be seen that the centre or 'core' of the gun is a comparatively thin tube of steel, round which are placed thick hollow cylinders made up of wrought-iron coils welded together. These cylinders are, like the wheel-tyres above mentioned, turned slightly too small to fit over the tube at ordinary temperatures. Each one is then heated until it has expanded sufficiently to allow it to pass easily over the tube: and when it is in its place, it is cooled by jets of water. The consequent contraction binds the whole most firmly to-

Applications of Expansion of Solids. 57

gether, and enables the gun to withstand the enormous outward pressure of the exploding gunpowder.

3. When two plates of metal are to be riveted together, as shown in fig. 27, the iron rivets are usually heated to redness before they are passed through the holes in the plate, not merely to soften them so as to make it easier to hammer the projecting part into a flat head, but because their contraction, when they

Fig. 27.

cool, binds the plates together with much greater firmness than if cold rivets were used.

4. It often happens that a glass stopper sticks so firmly into the neck of the bottle that no ordinary force can loosen it. Now, if the outside of the neck is cautiously heated (by holding it over a lamp-flame for a few seconds and constantly turning it round, or by pouring hot water over it), the neck will expand before the heat reaches the stopper, and the latter will generally be so far loosened that a slight blow on each side of it will bring it out.

Perhaps, however, the most generally noticed effects of the expansion and contraction caused by heat and cold are the cases in which this alteration in size, and the enormous force with which it takes place, are a disadvantage and an evil which it requires all our ingenuity to counteract.

1. For instance, the iron beams, or 'girders,' which are often used in buildings to support floors and brace together the framework in roofs, are constantly altering in temperature with the season, and by their expansion and contraction (which may amount to at least 2 or 3 mm. for every 10 metres in length) are liable to become loose, or even displace the walls into which they are built. Similarly, the bars of grates and furnaces, the solid iron fronts of fireplaces, and the long lengths of iron pipe used in heating buildings by hot water or steam, from their

exposure to very great differences of temperature will cause serious damage unless space is allowed for their free movement. In hot-water pipes this is often done by making them slide into one another like the tubes of a telescope : the joint being made water-tight by a stuffing-box,—a contrivance which will be explained in the chapter on the Steam-Engine.

2. In larger metal structures, such as bridges, which are often of very great length and mass, special arrangements are made to prevent displacement of the piers or distortion of the framework. The Britannia Bridge, for example, which consists of two enormous tubes of wrought-iron, 457 metres (about 1500 feet) long, spanning the Menai Strait between England and Anglesea, alters in length to the extent of at least 2 decimetres with the heat of a summer sun and the cold of a winter frost. Hence its ends are supported on rollers, so that they can move backwards and forwards without injury to the piers.[1]

3. A similar allowance for expansion has to be made in laying the iron rails of a railway. These, like the bridges, are exposed to great alterations of temperature, and a rail 6 metres long alters in length about 3 mm. for a range of 40° of temperature. This amount of expansion, though small in itself, becomes a serious matter when hundreds of kilometres of rails have to be laid in one line. For example, the distance from London to Edinburgh is 644 kilometres ; and if the rails between the two places were laid with their ends in contact on a cold frosty day, and were free to expand under a hot sun, the line of iron would become nearly 400 metres longer. If this movement was prevented, the enormous force with which the expansion tends to take place would throw the rails out of the true straight line and parallelism with each other into a series of curves, and tear them away from their fastenings. Hence (as can be noticed on any railway) the rails are always laid with a small interval

---

[1] It is observed that the bridge curves perceptibly upwards when the sun's direct rays fall on the top of it. The reason of this will be plain from what has been said (p. 48) respecting the curvature of the compound strip of iron and copper.

Results of Expansion of Solids.

between their ends, so that each can expand some little distance before it comes into contact with the next one. Several accidents have happened (and more have only just been prevented by timely observation) through neglect of this precaution by careless workmen.

4. Again, signals and 'points' on a railway are worked by long iron rods extending to levers in a 'signal-box,' which may be more than 400 metres distant. These rods, although they may be of the proper length for working the signal at one particular temperature, may alter in length with heat and cold so far as to give a totally wrong direction to the signal or points without any movement of the lever by the signalman. The arrangement by which the two ends of the rod are kept at the same invariable distance from each other, however the temperature may vary, is of the following kind:—Midway between the signal and the working lever,[1] the rods are jointed together by a short cross-arm, as shown in fig. 28. Then, if any rise of temperature

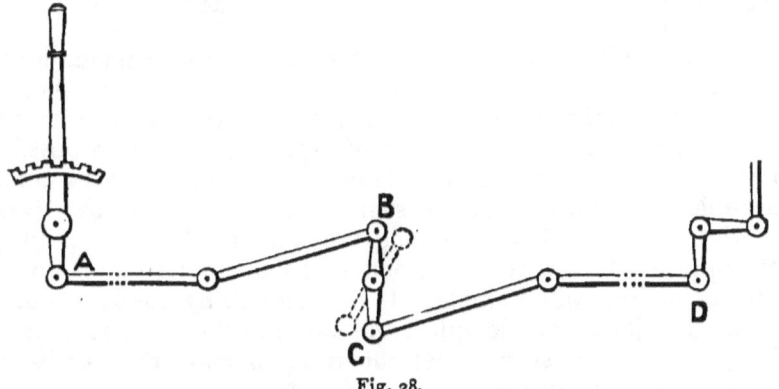

Fig. 28.

occurs which causes the rod A B to expand towards the right hand, B, the other rod C D can also expand towards the left hand, C; and thus the only effect of the alteration in length of the whole system of rods is to throw the cross-arm into such a position as is shown by the dotted lines, and the points and signal can still be worked correctly when the lever is moved.

---

[1] Or, if the distance is great, at several intermediate points.

5. The last inconvenient result of the alteration in size caused by heat in solids which will be noticed here, is one which many will have noticed for themselves, without, perhaps, being able to explain the reason of it. It is a fact that ordinary clocks and watches alter in their rate with the temperature to which they are exposed. A clock will go slower in summer and faster in winter than its average rate; and a watch will lose time when it is carried in the pocket, though it may have been regulated correctly when hanging up in the maker's shop. Since this variation in rate is found strictly to follow variations in temperature, it is fair to infer that the latter must in some way be the cause of the former. Now, a clock is really only an apparatus for registering the number of swings made by a pendulum, the latter being the true definer of intervals of time. Thus, a pendulum, which is almost exactly 1 metre long,[1] makes one swing from side to side in a second of time. But it does this only so long as its length remains the same; if it is longer than a metre it swings slower, and if shorter it swings quicker.

Expt. 17. Hang a heavy weight (such as the iron ball mentioned in the list of apparatus) by a string from a fixed support, such as a hook projecting from the wall, making the length from the centre of the weight to the point of support as nearly as possible 1 metre, and set it swinging. Then it will represent a 'seconds pendulum,' and will make one swing in each second of time (very nearly). Now, without stopping the motion of the pendulum, draw up the string over the hook so as to shorten the length. The pendulum will immediately begin to swing quicker; but if restored to its former length it will swing at the same rate as at first; and if more string is let out so as to make the pendulum longer, it will swing slower still.

We thus see how a clock (though of the best possible workmanship) must necessarily, if an ordinary pendulum is used, vary in rate as alterations in temperature make the pendulum longer or shorter; in fact, if the pendulum rod is made of brass, a varia-

---

[1] Strictly speaking, the length of such a pendulum is, in the latitude of London (51° 29′ 8″), 0·9941 metre—*i.e.* about 6 mm. less than a metre.

COMPENSATED PENDULUM.

tion in temperature of only 20° will cause it to alter in length nearly 0.4 mm., and cause the clock to vary in rate about sixteen seconds a day, an amount of very serious importance, especially in astronomical work. Since this expansion and contraction of solids cannot (as we have already seen) be practically prevented from taking place, the earlier clockmakers, astronomers, and navigators, must have almost despaired of their instruments.

At the beginning of the last century, however, the offer of a large reward by the Admiralty for the discovery of practical methods of determining longitudes, which involve the use of extremely accurate timepieces, stimulated inventors; and means were found, if not of preventing, at any rate of completely counteracting the effect of the alteration in length of the pendulum-rod, by taking advantage of the *difference* in the coefficients of expansion of different substances. One of the first of these 'compensated pendulums,' as they are called, was constructed in 1730 by Harrison, an English chronometer-maker.

Fig. 29 will serve to explain the principle of this pendulum. Suppose that an iron rod, A B, is fixed at the point A, but free to move at B. At this latter point let a short arm, B C, be attached at right angles, to the end, C, of which a zinc rod, C D, is fixed parallel to A B, and free to move at the end D. Then, if the whole apparatus is exposed to heat, the iron rod will expand downwards towards B, so as to lower the cross-arm B C. At the same time the zinc rod will expand upwards in the direction of D. Now, the coefficient of expansion in length of zinc is very nearly $2\frac{1}{2}$ times that of iron (see the table on p. 55), so that a zinc rod will expand as much as an iron rod $2\frac{1}{2}$ times as long. Hence if the iron rod A B is made $2\frac{1}{2}$ times as long as the zinc rod, the latter will expand upwards just as far as the iron rod will expand downwards, and the distance between A and D, shown by the dotted

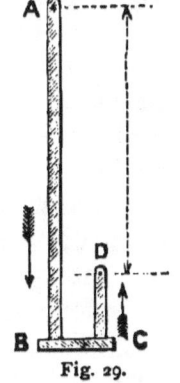

Fig. 29.

line, will remain the same, whatever variations of temperature may occur. If, then, a heavy pendulum-bob is attached at D, of very great mass compared with that of the rods (so that the effect of the mass of the latter may be neglected), and if A is taken as the point of suspension of the pendulum, we shall have what we want—a pendulum of which the acting length does not vary with its temperature.

In practice it would be awkward to give a pendulum the exact form of fig. 29: and fig. 30 represents the actual construction of the pendulum used in some of the best observatory clocks. The main rod is made of steel (or sometimes wood) and has a circular plate at its lower end, supported by a screw below. Upon this plate rests a hollow cylinder of zinc (or sometimes lead), sliding freely on the rod; and the relative lengths of the rod and cylinder are so adjusted that as the latter expands upwards the centre of its mass is raised exactly as much as the plate on which it rests is lowered by the downward expansion of the steel rod.

Another form of compensated pendulum, Fig. 30 used in commoner clocks, and called the 'gridiron pendulum,' is shown in fig. 31. Here steel and brass (instead of zinc or lead) are the metals used; and since the apparatus would be unwieldy if only one long rod of each metal was employed, the rods are cut up into short pieces and connected together as shown in the figure (in which the dark parts represent the steel rods and the lighter parts the brass rods), attention being paid to the following points:—

Fig. 31.

1. That the total length of the steel rods shall be just as much greater than that of the brass ones as the coefficient of expansion of brass is greater than that of steel.

2. That the steel rods shall be so connected as to be free to

COMPENSATED BALANCE. 63

expand downwards only, while the brass rods are free to expand upwards only.

[Another form of compensated pendulum will be found described in the next Section.]

In watches there is, instead of a pendulum, a wheel with a heavy rim called the balance-wheel, which swings to and fro under the control of a coiled elastic spring. This wheel of course alters in diameter with temperature, and the larger it is the slower it swings; moreover, the elasticity of the controlling spring becomes weakened when it is heated, and from both these causes the rate of the watch is altered. In order to compensate this the rim of the wheel is made of two rings, one of steel, and the other of brass fitting outside the steel ring and firmly soldered to it. This rim is attached to the centre of the wheel by two spokes only, as shown in fig. 32, and is cut through in two places, A and B, near which heavy screws are passed through it.

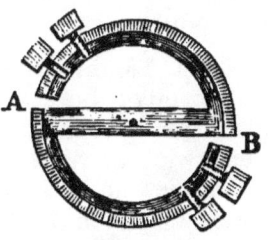

Fig. 32.

The action of this compensation balance is as follows:—Brass has a higher coefficient of expansion than steel, and hence (as shown already with the compound bar, p. 48) an increase of temperature causes the free ends, A and B, of the rim to curve more sharply inwards. Thus the weight of those portions of the rim and of the screws is brought nearer to the centre round which the wheel swings, and the path of the 'centre of inertia'[1] of the whole mass, which determines the rate of oscillation, is kept at the same distance from the centre of rotation, whatever temperature (within a moderate range) the watch may be exposed to. The same mode of compensation is applied to the best chronometer-balances.

---

[1] For an explanation of this term a book on Dynamics must be consulted.

Section III.—Expansion of Liquids by Heat.

The changes in volume which liquids undergo in consequence of changes of temperature, cannot be quite so simply examined as in the case of solids, since a liquid does not keep its shape but has to be enclosed in a solid vessel, and the change in size of this latter has to be allowed for in making experiments. If, for instance, the material of the containing vessel were to expand just as much as the liquid on the application of heat, then the expansion of the latter would not show itself at all—a completely filled vessel remaining exactly full, without either sinking of the surface of the liquid or overflow.

The following experiment, however, will show more than one important fact respecting the effect of heat in altering the volume of liquids.

Expt. 18. Fit a good cork to a flask about half a litre or more in capacity; bore a hole in the cork,[1] and fit into it a piece of glass tubing of small bore (about 1 mm.),[2] 30 or 35 cm. in length, bent twice at right angles as shown in fig. 33, and drawn out to a fine jet at the end *a*. Fill the flask completely with cold water, and then press down the cork into its place, so that some of the liquid may be forced into the tube and fill it entirely. The end *a* of the tube should, while this is being done, be dipped under the surface of some water in a jug, and not withdrawn until the cork is tightly fitted, so as to ensure the whole being filled with water. Support the flask on a piece of wire-gauze over a lamp, place an empty beaker or cup under the jet, and light the lamp. The first visible effect will be that the liquid retreats in the tube, as though it had contracted instead of expanding with rise of temperature. Very shortly, however, the end of the column of liquid will move forward again in the tube, and some water will overflow from the jet as the contents of the flask get warm.

This result shows two things :—

---

[1] This is best done by the regular cork-borers sold for the purpose, but a small round file may be used, the point being urged with a screwing motion through the cork.

[2] The tubing used for making alcohol thermometers is most suitable.

(1) That water expands on being heated.
(2) That it expands more than glass does under similar conditions.

The slight apparent contraction noticed at first will be easily seen to be due to the fact that the heat of the lamp expanded the glass vessel before it reached the liquid within, and thus made more room for the latter, which therefore retreated in the tube.

The next question is,—Do all liquids expand to an equal extent for an equal rise of temperature, or has each (like a solid)

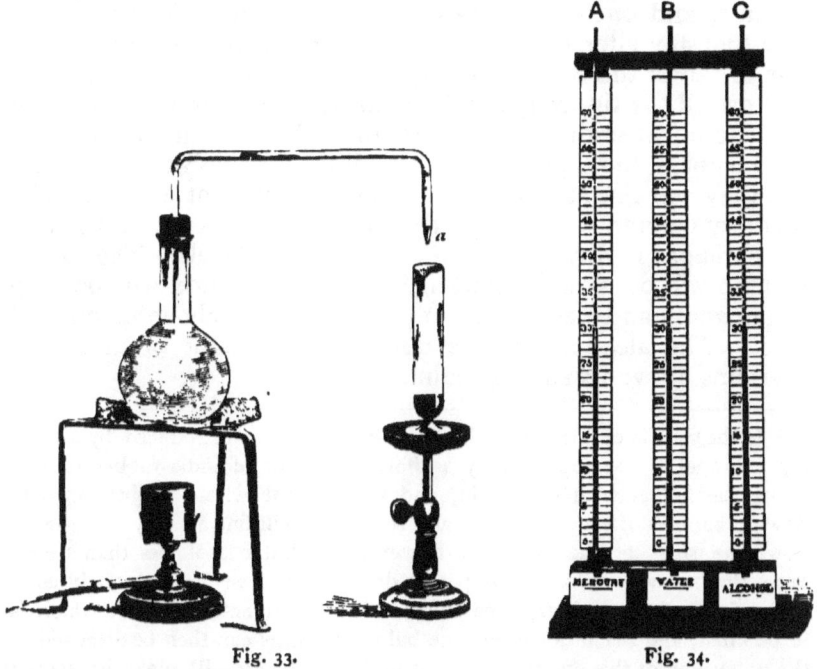

Fig. 33.   Fig. 34.

a particular coefficient of expansion, differing from that of other liquids? To investigate this point the following experiment may be made :—

Expt. 19. Arrange an apparatus as shown in fig. 34. A, B, and C are three similar tubes about 60 cm. long, and 2 mm. in internal diameter, ending in bulbs about 5 cm. in diameter. (The exact size is immaterial, but all the three tubes and bulbs should be as nearly identical in size as can be obtained.) These tubes are supported vertically in a wooden frame with the bulbs resting in a tin trough. Attached to each tube is a scale of cardboard divided into centimetres or other conveniently small and equal parts. Before use the bulb and tube A should be filled with mercury up to, say, the 30th division on the scale; B should be filled up to the same level with water (a few drops of indigo or ink being added to render it visible); and C should be filled up to the same level with alcohol (common methylated spirit will do), coloured by the addition of a little tincture of turmeric.[1] We have thus portions of three different liquids, equal in volume, and enclosed in vessels of similar material, in such a way that any alteration of volume will make itself evident by the rise or fall of the columns of liquid in the respective tubes.

Now fill the tin trough with moderately hot[2] water, stirring it so as to equalise the temperature throughout. An alteration in the height of the liquid columns in all three tubes will be noticed, but they will rise at different rates and to different extents. The mercury column will rise pretty quickly and steadily to a height of perhaps 1 division above its former level. The water column will be slow at starting,[3] but after a time will outstrip the mercury and reach a point $1\frac{1}{2}$ or 2 divisions above its original level. The alcohol will rise quickly to a height of at least 6 divisions above its starting-point.

---

[1] If the tube is of fairly wide bore the liquid may be introduced by attaching a funnel to the top of it by a short connector of india-rubber tubing, filling the funnel with the liquid, and shaking the whole slightly up and down; bubbles of air will escape, and the liquid will find its way past the air down the sides of the tube. If, however, the latter is of less than 2 mm. bore, the best way of proceeding is to draw out a piece of moderately large tubing into a long capillary tube sufficiently small to enter the tube which is to be filled, and reach down into the bulb. A funnel can then be attached to the upper end of this and the liquid poured in; the air will make its escape through the interval between the two tubes. All soiling of the upper part of the tube is thus avoided.

[2] Water at about 60° or 70° (not hotter, or the alcohol will boil) should be used. Mix in a large jug some cold water with about $1\frac{1}{4}$ times its volume of boiling water.

[3] The reason of this difference of rate in the rise of the water and of the mercury will be better understood when the subject of specific heat has been studied.

Expansion of Liquids.

From these and similar experiments it is ascertained—

1. That liquids expand more than solids when heated.
2. That different liquids alter in size to very different extents when exposed to equal alterations in temperature, water expanding nearly twice as much as mercury, and alcohol nearly six times as much, when equally heated.
3. That the amount of expansion is, in the case of water at any rate, not the same at different temperatures; for the water-column was noticed to rise much more quickly when it got hot than when it was comparatively cool, although in the former case its temperature must have been rising more slowly as it approached the temperature of the water in the trough.

It must be remembered that the amount of expansion noticed in the above experiment is only the excess of the expansion of the liquid over that of the glass vessel. Mercury really expands about seven times as much as glass, but its apparent expansion in this case is less than this by $\frac{1}{7}$, the bulb and tube getting larger themselves, and thus affording more room for the mercury, which rises proportionately less in the tube. It is pretty easy to see, however, that if we know accurately the amount that the glass has expanded in volume, we shall merely have to add this to the apparent expansion of the liquid in order to obtain the true total amount of the expansion of the latter. Thus, supposing that we know from other experiments that the glass of which the vessel is made expands so much as to enlarge the capacity of the bulb and tube by 1 c.c., while the rise of the column indicates an apparent increase in volume of 6 c.c., then the absolute increase in volume of the mercury will be $(1+6=)$ 7 c.c.

Methods of accurately determining the Coefficient of Expansion of Liquids.

1. Regnault's method.

In this an apparatus is used of which the principle is illustrated by the flask and tube employed in Expt. 18. A cylindrical bulb, made of glass of which the coefficient of expansion is accurately known, is attached to the end of a tube drawn out to a fine jet. This, after being weighed, is completely filled with the liquid at the temperature of melting ice. It is then transferred to a vessel of hot water at a known temperature, $t°$, when the liquid expands and some of it overflows. The portion which escapes from the jet is caught, and carefully weighed. The apparatus is then allowed to cool, and the weight, A, of the liquid which remains in it is ascertained. This added to the weight of the liquid which has escaped gives the total weight, B, of the liquid which fills the apparatus at 0°.

Now, it is evident that when we are dealing with portions of the same liquid under the same conditions, the size of these portions will be proportional to their weights (2 grms. of mercury, for instance, will occupy twice the space of 1 grm.). Hence we ascertain by the above two weighings that a volume of the liquid at 0° represented by the weight A will at $t°$ expand to the volume represented by the weight B: or,

$$A : B :: \text{vol. at } 0° : \text{vol. at } t°.$$

When the increase in volume of the liquid when heated through a known number of degrees, say 30, has been thus found, $\frac{1}{30}$ of this will be its increase in volume for 1°; and thus its coefficient of apparent expansion in glass is readily calculated.

To take an example :—

In a determination of the coefficient of expansion of mercury the following data were obtained :—

Weight of glass vessel empty, . = 25·72 grms.
,, do. full of mercury at 0°, = 105·82 ,,
,, do. with the remaining mercury, after being heated to 100°, = 104·60 ,,

Hence a volume of mercury at 0° represented by (104·60 − 25·72 =) 78·88 grms. will at 100° expand to a volume represented by (105·82 − 25·72 =) 80·1 grms. Then by the proportion

$$78·88 : 80·1 :: 1 : 1·0154$$

we learn that one volume of mercury at 0° expands to 1·0154

EXPANSION OF LIQUIDS.

vol. at 100°; so that the apparent increase in volume of the liquid caused by a rise in temperature of 100° is 0·0154 of its volume at 0°. The coefficient of apparent expansion of mercury in a glass vessel will be $\frac{1}{100}$ of this, or 0·000154.

Then, to find the true amount which the mercury alone has expanded, we must add to the above fraction the coefficient of expansion of the flint-glass of which the bulb was made, which was 0·0000258. Thus the absolute coefficient of expansion of mercury will be (0·000154 + 0·0000258 =) 0·0001798.

2. *Dulong and Petit's method.*

This method has the great advantage of giving at once the absolute coefficient of expansion of a liquid without requiring any knowledge of the amount of expansion of the vessel which contains it. It depends upon the hydrostatic principle that when columns of liquids balance each other, the heights of these columns are inversely proportional to the densities of the liquids. Now, it was mentioned on p. 55 that the volumes of substances are also inversely proportional to their densities; so that if the volume of a liquid varies, the height of the column of it required to balance a definite column of another liquid will vary in the same proportion. To put it briefly,—

Height of column gets greater as density gets less,

Volume gets greater as density gets less,

∴ Height of column gets greater as volume gets greater.

In the apparatus used by MM. Dulong and Petit, two moderately large vertical tubes, A and B, fig. 35, were connected below by a much narrower horizontal tube, and surrounded by wide cylinders. The tubes were nearly filled with the liquid to be experimented on, and when this was at the same temperature

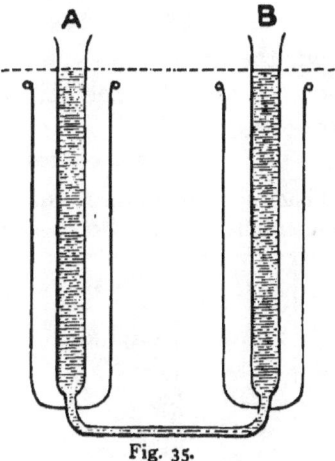

Fig. 35.

throughout, the heights of the columns of it in A and B were, of course, exactly the same; since a liquid always rises to the same level in vessels which communicate with each other. Next, the cylinder enclosing the tube A was filled with melting ice, and the other cylinder was filled with hot water or oil at a known temperature. As the liquid in B increased in temperature its density became less, as already explained, and the column in that tube rose, still, however, exactly balancing the unaltered column of the cold liquid in A. Then the heights of the two columns were again accurately measured, and the proportion between these heights gave, as above shown, the proportion between the volumes of the liquid at 0°, and at the temperature of the hot water, respectively.

For example, in one of the experiments made with the apparatus the height of the column of mercury in A, at 0°, was 544 mm., while the height of the column in B, at 100°, was 554 mm. Hence, as above explained, the density of the liquid had decreased, and therefore its volume had increased, in the proportion of 554 to 544 for a range of 100°. That is,—

Vol. at 0°.   Vol. at 100°.
544 : 554 :: 1 : 1·0184 (nearly).

Thus the increase in volume for a rise of temperature of 100° is 0·0184; and the coefficient of expansion will be $\frac{1}{100}$ of this, or 0·000184.

COEFFICIENTS OF EXPANSION OF LIQUIDS.

[The figures mean the fraction of its volume at 0° that the liquid expands in volume when heated 1° centigrade.]

Substance.	Coefficient of Expansion.
Carbon disulphide	·00114 $(=\frac{1}{880})$
Alcohol (5°—6°)	·00105 $(=\frac{1}{950})$
Do. (49°—50°)	·00122 $(=\frac{1}{820})$
Water (5°—6°)	·000022 $(=\frac{1}{45,000})$
Do. (49°—50°)	·000459 $(=\frac{1}{2,170})$
Do. (99°—100°)	·000755 $(=\frac{1}{1,320})$
Mercury	·00018 $(=\frac{1}{5,500})$

EXPANSION OF LIQUIDS.

The above are the coefficients of absolute expansion. The coefficients of apparent expansion in a glass vessel are found by subtracting from them the coefficient of expansion in volume of glass (see the table on p. 55).

Thus the coefficient of apparent expansion of mercury in glass is $(\cdot 00018 - \cdot 0000258 =) \cdot 000154$, or $\frac{1}{6,500}$, nearly.

Irregularities in the Expansion of Liquids.—Observations of the coefficients of expansion of liquids at different temperatures show that hardly any liquid expands to *exactly* the same extent when heated from, say, 49° to 50° as it does when heated from 0° to 1°; the expansion being in nearly all cases greater, and sometimes very much greater, at high temperatures than at low ones. Thus in the case of alcohol, taking its volume at 0° as the standard of comparison, it is found to expand $\frac{1}{912}$ of this volume when heated from 0° to 1°, $\frac{1}{820}$ of this volume when heated from 49° to 50°, and to a still greater extent, *viz.* $\frac{1}{710}$, at its boiling-point, 78°.

But perhaps the most singular departure from regularity of expansion is found in the case of water. This substance, when heated from 0° to 1°, actually contracts in volume instead of expanding, and continues to do so until it has been heated to 4°, after which any increase of temperature causes it to expand, slightly at first, but more and more rapidly as it approaches its boiling-point.

These variations in the coefficient of expansion are indicated in the table given on p. 70, but perhaps they can be better demonstrated in the manner shown in fig. 36.

Suppose that A represents a tube with bulb attached, like those already used for showing the expansion of liquids, and containing water which fills it up to the mark when the temperature is 0°. When the water is heated to 4°, the surface of the water column is observed to sink until it reaches the level shown in B, proving that a real contraction has taken place. If the temperature is steadily raised, it is found that the column begins to rise again in the tube, so that at 8° it has almost exactly regained its original level, as shown in C; at 20° it reaches the level shown in D, and at 40° and 60° it rises to the points shown in E and F respectively.

Now, if we draw a line through the marks which indicate the levels at which the top of the water-column stands at the different temperatures, we shall obtain a curve expressing very accurately the changes in volume which water undergoes at different temperatures. Such a curve is shown by the dotted line in fig. 36, and a diagram is given in fig. 37 which shows the same

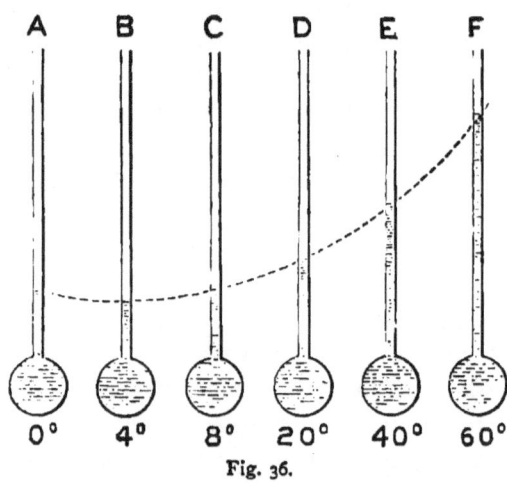

Fig. 36.

facts in a more complete and scientific way for mercury and alcohol as well as water. In this diagram the divisions on the horizontal line at the bottom indicate degrees of temperature, while the divisions on the vertical line at the left hand denote the volume occupied by the liquid. Through each of these divisions a line is drawn at right angles to the base line, and in order to express the expansion of a liquid as it is heated from 0° to 100° dots are placed on the lines in positions indicated by the results of experiments, and a line drawn through all these fixed points will show clearly and graphically the manner in which the liquid expands between the above-mentioned temperatures.

For instance, it is found by experiment that, starting with 1 c.c.

EXPANSION OF LIQUIDS.

of water at 0°, this quantity becomes 1·01 c.c. when heated to 50°: therefore on the vertical line starting upwards from 50° a dot is placed where the horizontal line starting from 1·01 crosses it. Again, at 70° the original 1 c.c. of water is found to have become 1·02 c.c. (nearly); so a dot is made where the vertical line from 70° and the horizontal line from 1·02 cross each other: and so on, until a sufficient number of positions are ascertained to enable a line to be drawn through them which shall show with

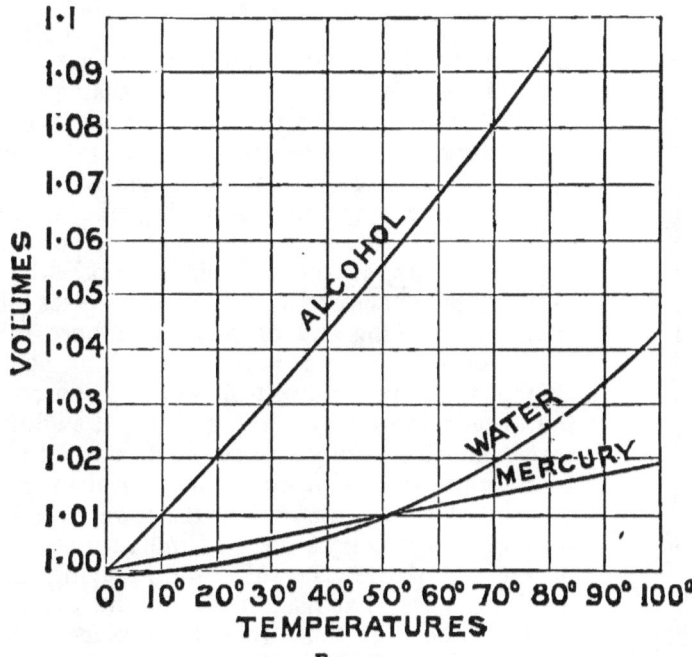

Fig. 37.

accuracy the amount of expansion of the liquid at intermediate temperatures.

It is obvious (1) that the greater the coefficient of expansion of the substance is, the more rapidly this line will slant upwards from the horizontal base line; (2) that if the liquid expands with absolute regularity, the line will be a straight one. It is very nearly so in the case of mercury, but in the case of water the line is curved, even dipping downwards slightly but perceptibly between 0° and 10°, and rising afterwards more and more rapidly as the temperature approaches 100°. In the case of alcohol the curvature is perceptible, though not very great; indicating that alcohol, like most liquids, has a greater coefficient of expansion at high temperatures than at low ones.

The irregular changes in volume which water undergoes when heated above its freezing-point may be demonstrated in the following way :—

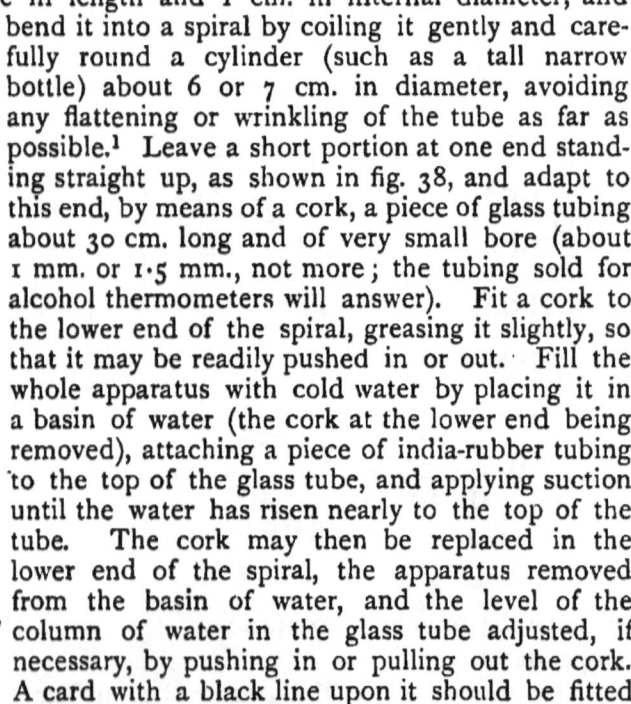

Fig. 38.

Expt. 20. Take a piece of ordinary 'composition' gas tubing, about 1 metre in length and 1 cm. in internal diameter, and bend it into a spiral by coiling it gently and carefully round a cylinder (such as a tall narrow bottle) about 6 or 7 cm. in diameter, avoiding any flattening or wrinkling of the tube as far as possible.[1] Leave a short portion at one end standing straight up, as shown in fig. 38, and adapt to this end, by means of a cork, a piece of glass tubing about 30 cm. long and of very small bore (about 1 mm. or 1·5 mm., not more; the tubing sold for alcohol thermometers will answer). Fit a cork to the lower end of the spiral, greasing it slightly, so that it may be readily pushed in or out. Fill the whole apparatus with cold water by placing it in a basin of water (the cork at the lower end being removed), attaching a piece of india-rubber tubing to the top of the glass tube, and applying suction until the water has risen nearly to the top of the tube. The cork may then be replaced in the lower end of the spiral, the apparatus removed from the basin of water, and the level of the column of water in the glass tube adjusted, if necessary, by pushing in or pulling out the cork. A card with a black line upon it should be fitted to the glass tube with wire, so that it can slide up and down.

---

[1] This can be entirely avoided by filling the tube with fine sand before proceeding to bend it, corks being fitted to each end. The sand can afterwards be shaken out, and the tube rinsed with a stream of water.

IRREGULAR EXPANSION OF WATER. 75

Now place the whole in a jar containing pounded ice (or snow) and water, in which the spiral should be completely immersed. The first effect will be a rise in the column of water in the tube, due to the contraction of the metal tube, but the level will soon sink as the water cools, and its fall should be followed by shifting the card downwards. It will, however, sink more and more slowly, and will eventually, when the water attains a temperature of $4°$, become stationary for a moment, and then begin to rise again; thus proving that water expands as it approaches the temperature of its freezing-point.

This anomalous behaviour of water is probably to be accounted for by some change in the arrangement of the molecules, which seem preparing, as it were, to arrange themselves in clusters of a definite shape; to crystallise, in fact, as they undoubtedly do when the water freezes. These crystal shapes are not such as to fit accurately and closely together like well-laid rows of bricks; and hence vacant spaces are left between them, and the whole mass takes up more room, just as the regular rows of bricks would do if simply piled together in an irregular heap. We shall see that when crystallisation actually takes place in water, there is a sudden and very great expansion in volume.

RESULTS AND PRACTICAL APPLICATIONS OF THE EXPANSION OF LIQUIDS BY HEAT.

1. A given volume, *e.g.* a litre, of a hot liquid weighs less (or, more strictly, has less mass) than the same volume of the liquid when cold.

That this must be the case is evident from the experiment already made to illustrate the fact of the expansion of liquids. (p. 64). It was found that, as the water in the flask got hot, some of it overflowed through the tube. The water which remained still filled the flask, but it must clearly weigh less by the weight of the quantity which had overflowed. In fact, 1 litre (or 1000 c.c.) of water at $4°$ weighs, of course, 1000 grms. When it is heated to $100°$ it increases in volume to 1043 c.c., and therefore a litre must weigh less by the weight of 43 c.c. of water at that temperature, *i.e.* nearly 40 grms.[1] Hence the litre of boiling water will only weigh ($1000-40=$) 960 grms., and its relative density will be only ($\frac{960}{1000}=$) 0·96.

It is easy to show that hot water is less dense than colder water in the following way :—

Expt. 21. Fill a large glass jar, or tumbler, rather more than half full of cold water. Mix in a jug some boiling water with about half its volume of cold water (to prevent any serious risk of the jar cracking); add a few drops of indigo or ink to render the liquid more visible, and pour it slowly and carefully upon the surface of the cold water in the jar, down a strip of card or sheet tin bent as shown in fig. 39, in order to break its fall and spread it horizontally over the surface of the cold water. The hot liquid will show no tendency to sink through or mix quickly with the colder water, but will remain floating as a coloured stratum above the latter.

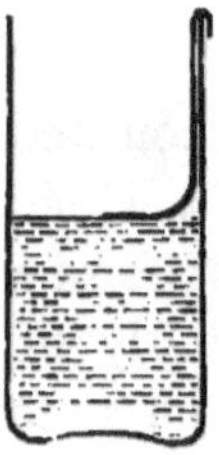

Fig. 39.

This proves that hot water is lighter, when equal bulks are compared, than colder water, just in the same way as the fact that cork floats on water shows that its relative density is less than that of water.

2. When heat is applied to a liquid at any point lower than its surface, the equilibrium of the whole mass is disturbed, and the colder, denser portions tend to displace the portions which have received the heat, and to force them upwards, just as when a cork is placed below the surface of water it is (being less dense than water) forced upwards by the preponderating pressure of the molecules below it. Thus an upward current of heated molecules commences in the liquid, necessarily accompanied by a downward movement of the colder molecules which displace the former. In fact, as long as there is any inequality of temperature between different parts of the liquid, a continual movement must, as a rule, be going on throughout the mass. The

---

[1] The method of calculating the exact weight of a liquid when its volume and temperature are known will be found in the Appendix to Chapter V.

Results of Expansion of Liquids.

results of this will be more fully considered in the Chapter on Convection of Heat.

3. It is obvious from what has been said above respecting the alteration in relative density caused by heat, that in all experiments and phenomena depending on the density of liquids strict account must be taken of the temperature of the substance which is being dealt with, otherwise scientific accuracy is not obtainable in the results. For example, in the case of the ordinary barometer, in which the pressure of the air is measured by balancing it against a column of mercury, the temperature of the mercury must be carefully noted, and the reading of the height of the column reduced to what it would have been if the mercury had been at the standard temperature of the freezing-point of water. For on a warm day the density of the mercury becomes less, and hence a longer column of it is required to balance the same air-pressure than if it was at the freezing-point. Thus a rise of the barometric column need not necessarily mean a change in the pressure of the air, but only a change in the temperature of the mercury. In scientific meteorological observations, a correction for temperature is always applied to the observed readings by tables calculated from the known coefficient of expansion of mercury.

4. The expansion of liquids has found a useful application in Graham's mercurial compensating pendulum. In this the weight or 'bob' of the pendulum is formed by a cylindrical jar containing mercury, hung in a stirrup attached to the lower end of the steel pendulum-rod, as shown in fig. 40. It is evident that, supposing a rise of temperature to occur, while the steel rod expands downwards, the mercury in the jar will expand upwards; and hence the centre of gravity of the cylindrical column, which is always at the middle of its length, will be more or less raised, so as to make the acting length of the pendulum shorter. Since the coefficient of expansion of mercury is far

Fig. 40.

greater than that of steel, a comparatively short cylinder of it (the exact required length of which can be easily calculated) will be sufficient to compensate the changes in length of the steel rod.

5. We come, lastly, to some results of the remarkably irregular changes in volume of water when its temperature is changed. It is hardly too much to say that the whole order of Nature would be utterly different from what it is, if water expanded and contracted according to the usual law. Lakes in winter would be filled up with masses of ice, rivers would be blocked up completely during a long frost, and no fish could possibly survive a single severe winter in our latitudes.

Let us consider what happens during the cooling of a mass of water, such as a lake, freely exposed to the sky. The heat is lost entirely from its surface, in ways which will be more fully explained hereafter; and as the surface water cools from a temperature of, say, 15°, it contracts, becomes relatively denser than the warmer water below, and sinks through the latter. The lower portions of the water are thus brought to the surface and get cooled in their turn, and sink in the same way. This process goes on until the whole mass of the water is brought down to a temperature of 4°, and would continue, if water obeyed the usual law, until the freezing-point was reached, and ice formed in all parts of it. But, in actual fact, as the surface-layer gets cooled below 4°, it no longer contracts and sinks, but expands and (being less dense than the warmer water below it) remains on the surface. Thus the main body of the water never becomes colder than 4°, no ice is formed in it, and fish can live in it without difficulty, if not without discomfort. As the withdrawal of heat continues during a frosty night, ice is formed on the surface only, and remains there owing to the great expansion which takes place at its formation, as will be more fully explained in Chapter VII. Section I., but the temperature of the water below remains practically unaltered.

[Problems on the expansion of liquids by heat will be found in the Appendix to Chapter V.]

Section IV.—Expansion of Gases by Heat.

Gases, in order to make them assume a definite volume, must be confined even more closely than liquids; and it is usual, in making experiments with them, to place them in a vessel closed by a movable stopper, as it were, formed of a column of liquid. We have the same three points to investigate as in the case of liquids and solids—*viz.* (1) Do gases alter in volume when their temperature is changed? (2) If so, to what extent do they alter in volume compared with liquids and solids under the same conditions? (3) Have different gases different coefficients of expansion?

The first and second questions may be easily answered by making the following experiment.

Expt. 22. Take a test-tube, the larger the better, or a thin cylindrical gas jar (a cylindrical lamp-glass with a cork tightly fitted into one end will answer), and fill it with water by immersing it in a basin of water; then, still keeping its mouth below the surface of the liquid, place it inverted on a stand (such as an inverted saucer) in the basin. Admit into it, by raising its mouth slightly, enough air to half fill it, and mark the volume of the air by an india-rubber ring placed round the tube just at the level of the surface of the water in the tube. Pour over the tube a little moderately hot water (at about 70°, *i.e.* the same temperature as the water which was used in the experiment on the expansion of liquids, Expt. 19, p. 65), and observe that the volume of the enclosed air immediately increases considerably. Next pour a little cold water over the tube: the air will shrink to its former volume; and by using water cooled by ice it may be made to contract still more.

The answer to be given to the third question may be indicated by the following experiment:—

Expt. 23. Procure two tubes, about 30 cm. in length and 3 or 4 mm. in internal diameter, ending in bulbs about 4 cm.

in diameter, and having near their upper ends short bent exit-tubes attached as shown in fig. 41. The bulbs and tubes must have *precisely* the same internal capacity,[1] which may be ascertained by filling one of them with water or mercury (by means of a funnel with a long drawn-out neck) up to the level of the exit-tube, then pouring this water into the other bulb-tube, and observing whether it fills this latter up to the same level. Support them, after carefully drying them internally (see next page), side by side in a tin trough as shown in fig. 41, in such a manner that the bent exit-tubes pass over the edges of the small glass trough shown in the figure. This trough should be nearly filled with water, and two test-tubes about 16 or 17 cm. long and 1 cm. (or rather less) in internal diameter[2] should be filled with water and supported, mouths downwards, over the ends of the delivery-tubes, so that gases may bubble up into them without loss. The open ends of the bulb-tubes should have small plugs of cork carefully fitted to them. Fill one of the bulbs with oxygen gas, passing the gas through a long, narrow, drawn-out tube reaching down into the bulb. In about two minutes the apparatus will be sufficiently full, and the narrow tube may be withdrawn and the cork plug replaced.

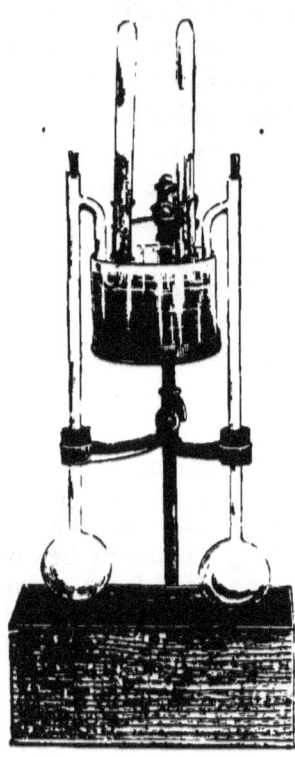

Fig. 41.

---

[1] This is of much importance, since gases expand so much that a slight difference in the original volume makes a great error in the result. If there is any difficulty in procuring such bulbs, small globular flasks may be selected from a dealer's stock, and fitted to the tubes by corks. It is worth while to take some trouble in verifying such an important law as that of the expansion of gases by heat.

[2] These receiving-tubes must have the same internal capacity; this is easily secured by selecting a piece of tubing about 34 cm. long of fairly uniform diameter, and making both of the tubes from this piece in the usual way.

The other bulb-tube should then be filled with hydrogen in a similar way.[1]

Now fill the tin trough with moderately hot water (at about 70°, as in the last experiment). Bubbles will immediately begin to issue from the ends of the delivery-tubes, showing that both the gases are expanding as they get hot; and a quantity of gas will collect in each receiving-tube, probably sufficient to half fill it. When the stream of gas has ceased, observe that the *same* volume has collected in each of the tubes, although two gases of such very different chemical characters as oxygen and hydrogen are being experimented upon.

[The cork plugs should be removed from the upper ends of the tubes before the bulbs become cool; otherwise the contraction of the gases will cause water to be forced from the trough into the bulbs. If this should inadvertently happen, the bulbs and tubes must be thoroughly dried (by warming them and drawing air through them) before being again used for the experiment.]

If similar comparative experiments are made with other gases, such as nitrogen, carbon monoxide, or ethylene, similar results will be obtained; the *same* volume of gas escaping from each delivery-tube in every case, when the rise of temperature in the gases is the same.

From the above and similar experiments we learn—

1. That gases are affected by changes of temperature in the same general way as liquids and solids, expanding when heated and contracting when cooled.

2. That, for a given change in temperature, they change in volume to a far greater extent than either liquids or solids.

3. That *all* gases, at temperatures considerably above their liquefying-points, have practically the *same* coefficient of expansion.

This last fact, first observed by Gay Lussac and Charles at

---

[1] Instead of oxygen and hydrogen the bulbs may be filled with air and coal-gas respectively, for rough demonstration; but the experiment is not then so conclusive, as both air and coal-gas are mixtures of several gases, and if some of these had a high coefficient of expansion and others a low one, the former might compensate the latter, and so produce an apparently equal expansion in both cases.

the beginning of the present century, is a very remarkable one, and a great contrast to what has been noticed in the case of liquids and solids, each of which has its own special coefficient of expansion, often differing widely from those of others. It will be worth while to examine how far such exact experiments as have been made by Regnault and others bear out the above conclusion.

The principle of one of the forms of apparatus used by Regnault will be understood from the following diagram, fig. 42.

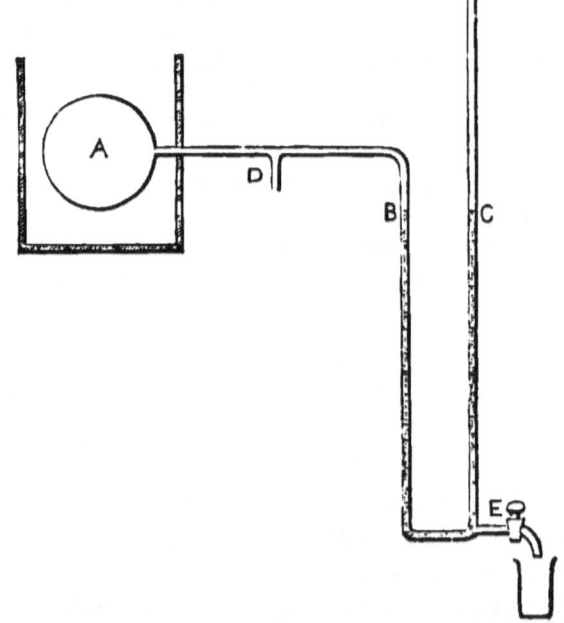

Fig. 42.

A represents a glass globe, holding about 800 c.c., attached by a narrow tube to the U-shaped measuring-tube B C. The globe is enclosed in a copper vessel into which ice may be put, or steam led, so as to maintain the mass of gas at constant known temperatures. The apparatus is filled, through the side tube, D, with the pure dry gas which is to be experimented on; and before this tube is closed the globe is surrounded by ice and mercury is poured into the open end of C until the tubes B and C are nearly full, the mercury column standing at the same height in each branch of the tube. Since the total capacity of the globe and tubes is previously determined with accuracy, we have a known volume of gas enclosed at a temperature of 0°, and under the pressure of the atmosphere. Then the ice is withdrawn and steam led into the copper vessel so as to raise the globe to a temperature of 100° or nearly so, which is observed by a thermometer. The gas, as it expands, flows into B and depresses the mercury in that branch; some of the mercury is then let out through the stopcock E until the columns in both B and C are restored to the same level, when the gas will, of course, be exposed to the same pressure as before. The increase in volume is then read off by means of the graduations on B. When corrections have been applied for the expansion of the glass vessel itself, and for the temperature of the gas contained in the tubes (which will be above 0° and below 100°) we obtain the true increase in volume of the gas for a rise of temperature from 0° to 100°; and $\frac{1}{100}$ of this will be the coefficient of expansion of the gas, *i.e.* its change in volume for 1°.

Regnault also made a second set of experiments with the same apparatus, in which, instead of keeping the pressure constant by withdrawing mercury from the tube, he poured more mercury into the open end of C so as to increase the height of the column in that branch until the gas, though heated to 100°, was compressed into its original volume by the increase of pressure. Now, from the simple law (called 'Mariotte's Law') which states the relation between the volume occupied by a gas and the pressure upon it, we learn that *the volume of a gas varies inversely with the pressure to which it is exposed:* if, for instance, we take a certain volume of a gas measured under a definite pressure and subject it to twice the pressure, the gas shrinks to one-half its original volume ($\frac{1}{2}$ being the inverse or reciprocal of 2). Hence, supposing that it was found, by measuring the difference in height of the columns of mercury in B and C, that twice the original pressure (*i.e.* the pressure of the atmosphere at the time of the experiment) was required to make the gas occupy its original volume, we should know that this volume was one-half of that which it would occupy under the original pressure, *i.e.* that if the gas had been allowed to expand freely without any change of the original pressure, it would have increased to twice its original volume.

The results of the two sets of experiments agreed very closely, and the slight differences between them can be readily accounted for. The table given below shows the corrected values of the coefficients of expansion of some of the more important gases.

COEFFICIENTS OF EXPANSION OF GASES.

[The figures mean the fraction of its volume at 0° that the gas changes in volume for a change of temperature of 1° C.]

Substance.	Coefficient of Expansion.
Oxygen	0·000367 or $\frac{1}{273}$ (nearly).
Hydrogen	0·000366 ,, $\frac{1}{273}$,,
Nitrogen	0·000367 ,, $\frac{1}{273}$,,
Carbon dioxide ('carbonic acid') .	0·000371 ,, $\frac{1}{269}$,,
Sulphur dioxide ('sulphurous acid')	0·000385 ,, $\frac{1}{260}$,,

[It must be noticed that on account of the great alteration in volume which change of temperature causes in gases (as compared with liquids and solids) it is absolutely necessary, in stating their coefficients of expansion, to specify *some particular volume* to which the volumes at other temperatures are referred; and *the volume occupied by the gas at the freezing-point of water* has been universally chosen. The numbers given in the above table are only true with reference to this standard volume; though, of course, tables might be made out in which the coefficients were fractions of some other volume, *e.g.* the volume occupied by the gas at 100°.]

It will be seen from the above table that the results of the most accurate experiments hitherto made entirely confirm the statement lately given, *viz.* that all true[1] gases have practically the same coefficient of expansion. The last two gases on the list, carbon dioxide and sulphur dioxide, have coefficients sensibly higher than the others; but this is easily accounted for, since they are (under the conditions of the experiments) much nearer their liquefying-points, and hence likely to be subject to such internal forces as cohesion, which would interfere with their normal rate of expansion.

We thus arrive at the following extremely simple law, expressing the change in volume of a gas with change of temperature, first definitely announced by Gay Lussac (A.D. 1802), and hence known as

GAY LUSSAC'S LAW.

A TRUE GAS CHANGES ITS VOLUME, FOR A CHANGE OF TEMPERATURE OF 1° C., BY $\frac{1}{273}$ OF THE VOLUME WHICH IT OCCUPIES AT 0° C.

For example, if we start with a quantity of a gas measuring 1 litre at 0°, and gradually raise its temperature, it becomes,

At	1°	$1\frac{1}{273}$ litre,	
,,	2°	$1\frac{2}{273}$,,	
,,	3°	$1\frac{3}{273}$,,	
,,	100°	$1\frac{100}{273}$,,	
,,	273°	$1\frac{273}{273}$,,	or 2 litres,
	etc.	etc.	

An interesting question arises from the consideration of the

---

[1] A true gas is one which is at a temperature and under a pressure very far removed from those conditions which would cause its conversion into a liquid. Thus oxygen, at ordinary temperatures and under the pressure of the atmosphere, is considered to be a true gas since it requires a reduction of temperature to $-120°$, together with an increase of pressure amounting to 50 times that of the atmosphere before it becomes liquid. Sulphur dioxide, on the contrary, is not a true gas in the above sense, since its liquefying-point is only 10° below zero under the ordinary pressure.

above law. Supposing that it holds good for all temperatures, low as well as high, what will be the result of lowering the temperature of a gas to a very great extent?

If we start with 1 litre of a gas at 0°, it will become—

At	$-1°$	$1-\frac{1}{273}$	litre,
,,	$-2°$	$1-\frac{2}{273}$	,,
,,	$-100°$	$1-\frac{100}{273}$	,,
,,	$-273°$	$1-\frac{273}{273}$	,, *i.e.* $1-1$, or 0.

So that if we can reduce the temperature of a gas to $-273°$ (without liquefying it), it ought to shrink to nothing,—to occupy no space at all.

Now, we have not as yet found any means of taking away so much heat energy from a body as to reduce it to this temperature; the nearest approach hitherto made to it being $-225°$, at which temperature hydrogen still obeys Gay Lussac's law. It is, moreover, probable that all gases would turn into liquids before the extremely low temperature of $-273°$ was reached. But the above considerations suggest a very convenient starting-point from which to begin the reckoning of temperatures upwards, *viz.* that represented by $-273°$ C. This point is called the 'Absolute zero of temperature,' and is often employed in investigations on the laws of heat, temperatures reckoned from it being called 'Absolute temperatures.' On this scale the purely conventional and arbitrary points called the 'freezing-point' and 'boiling-point' disappear altogether, and

0° C.	$=273°$
100° C.	$=373°$
200° C.	$=473°$
etc.	etc.

Thus, if the absolute scale be adopted, Gay Lussac's law may be put in the following very simple form:—'The volume of a gas is proportional to its absolute temperature.'

The kinetic theory of gases affords a very simple explanation of Gay Lussac's law. In fact, in order that a substance may exist in the gaseous state at all, sufficient energy must have been communicated to the molecules to overcome their cohesion entirely; and so any additional heat energy which is imparted to them has not to be spent in doing any work against internal forces such as cohesion, but is employed simply and solely in moving the molecules further apart against whatever external pressure resists their onward progress. Now, a given amount of energy transferred to a molecule will impart to it the *same* power of overcoming resistance whether that molecule be heavy or light, and whatever its chemical nature may be; just as a given charge of powder will (if all other conditions are the same) impart the same penetrative power to a bullet, whatever may be the weight and material of the latter. Thus, looking at a mass of gas as made up of a crowd of molecule-projectiles, if we impart to them a definite amount of heat energy, they will strike against obstacles with the *same* increase of force whatever be their nature; the light ones making up for their small mass by greater velocity, just as an army of light and active men may have the same effectiveness in the field as one composed of heavier and more slowly moving men.

It is, moreover, pretty certain from the researches of Avogadro and Ampère that equal volumes of different gases, whatever may be their nature, contain the same number of molecules of the substance; a litre of oxygen, for instance, contains just as many molecules of the element as a litre of hydrogen, when measured under the same conditions.

Hence it is no more than we should expect, to find that totally different gases display, when additional energy is imparted to them in the form of heat, precisely the same additional power of overcoming the pressure or resistance which opposes their progress; this increased power showing itself in expansion when the pressure or resistance is not increased as well.

Results and Applications of the Expansion of Gases.

1. A given volume of a hot gas weighs less (or rather, has less mass) than the same volume of the gas when cold.

The proof of this fact is of the same kind as that given already in treating of liquids. Supposing that a flask contains a litre of air at 0°, and that it is then heated to 100°. It will expand so much that nearly one-third will escape from the flask, if the pressure is kept constant; and so the hot air which now fills the flask will weigh one-third less than the litre of cold air.

It is for this reason that balloons filled with hot air (or, as they are commonly called, 'fire balloons') rise in the air. When the air in the balloon is heated by burning some spirits of wine under, or just within its open mouth, some of this air escapes owing to the expansion, until finally the contents of the balloon become so much lighter than an equal volume of the cold air surrounding it, that it is forced upwards like a cork in water, and will continue to rise (supposing the high temperature to be maintained) until it comes to a stratum of rarefied air possessing the same density as itself.

Expt. 24. Take a small balloon, made of the thinnest tissue-paper, not less than about 70 or 80 cm. in diameter; fold it up loosely so as to squeeze out most of the air in it, and hold it with the mouth downwards over the flame of a large Argand burner so that the stream of hot air rising from the flame may enter the mouth and fill the balloon. Care must be taken not to hold it so near the flame as to scorch or set fire to the paper. In a minute or so it will acquire sufficient buoyancy to rise to the ceiling; soon, however, descending again, as the air in it becomes cold.

2. For the same reason smoke ascends in a chimney and rises through the air from the top of it; and currents or draughts of air are set up in an unevenly heated space in precisely the same way as already explained in the case of liquids. This will be more fully treated of under 'Convection of Heat' (Chapter VII. Section II.).

3. It is obvious that the great force with which gases tend to expand when strongly heated may be utilised as a source of motive power; and the violence of the explosion of gunpowder is in a great measure due to this cause. One cubic centimetre of gunpowder yields, in the course of the chemical changes

Applications of Expansion of Gases. 89

which go on when it is 'fired,' an amount of permanent gases which would measure 300 c.c. at ordinary temperatures; but this volume is increased to about 2500 c.c. by the intense heat evolved during the chemical action; and this large quantity, when prevented from expanding by being confined in the breech of the gun, exercises an enormous pressure against the shot placed in front of it. The power of hot-air engines and (in a great measure) of steam-engines also, is due to the expansion by heat of the gas or steam used; as will be more fully explained in a later chapter.

4. Lastly, a most important application of the expansion of gases by heat is in the use of air-thermometers in the measurement of temperature; a subject which must next be more fully treated of.

[Problems on the expansion of gases by heat will be found in the Appendix to Chapter v.]

CHAPTER V.

THE MEASUREMENT OF TEMPERATURE—THERMOMETRY.

WHEN we wish to ascertain the pressure of steam in a boiler, we do so most conveniently by letting some of the steam into a 'gauge,' and observing the extent to which it compresses an elastic spring, or the height of the column of mercury which it can support in a tube.

In a somewhat analogous way, in order to ascertain the temperature of (or intensity of the heat in) a body, we usually place a small quantity of another specially selected substance in contact with the body, and as soon as the temperature of the two has been equalised we observe the change in volume which has taken place in the substance owing to the heat transferred to or from it.[1]

Instruments designed for measuring the temperatures of bodies are called 'Thermometers,' and a detailed account of some of the chief forms of them will be given as soon as the general principles involved in their construction have been explained.

In the first place, it is evident that the material to be used in the thermometer (*i.e.* the material of which the change in volume has to be measured) must be carefully selected. No one sub-

---

[1] There is, as has been already indicated (p. 5), another extremely delicate method of determining temperature, which consists in joining the ends of small rods of bismuth and antimony, applying the point of junction to the substance of which the temperature is to be measured, and observing the direction and intensity of the electric current developed by any change of temperature produced at this junction.

stance is absolutely unexceptionable, and we must take that which has the greatest balance of advantages in its favour.

The following seem to be the chief requirements in a substance which is to be used for thermometric purposes:—

1. The substance must, for a moderate change of temperature, change its volume sufficiently to be easily measured, but not too much; otherwise the scale of measurement will be inconveniently short or long.

2. It must change in volume regularly; *i.e.* its coefficient of expansion must be as nearly as possible the same at high temperatures as at low ones; otherwise the divisions on the scale will have to be varied in length in different parts.

3. It must be capable of readily and quickly taking up all the heat communicated to it, and conducting this heat rapidly through its whole mass.

4. It must not itself require much heat to raise it in temperature; for this heat has to be subtracted from the substance under examination, so that no thermometer gives strictly and truly the original temperature of the substance; and this error must be made as small as possible.

5. It must not change its state within a moderate range of temperature. The coefficients of expansion of solids, liquids, and gases are so widely different that even supposing that the same apparatus could be used for measuring the change in volume of a substance in all three states, a very different scale would be required in each case.

Now, taking some examples of substances with which experiments have been made, let us examine how far they fulfil the above requirements.

First, with regard to the solid, liquid, and gaseous states:—In which state should a substance be, in order to afford a thermometer of the most convenient kind for general purposes?

Solids, as we have seen, expand so little that rather elaborate expedients have to be resorted to in order to make their changes in volume evident at all. Gases, on the other hand, expand so much that a thermometer in which a gas is used must either be available only for a limited range of temperature or must have a large and cumbrous scale.[1]

Liquids come between the two in point of expansibility, and may certainly be selected in preference to either of the others for general purposes.

Next, out of the different liquids which have been mentioned, we must consider which fulfils the above requirements best.

Water changes its volume comparatively slightly, and does so very irregularly; it requires a great deal of heat to alter it in temperature; it is a very bad conductor of heat, and moreover a comparatively slight fall of temperature turns it into ice, and a slight rise of temperature turns it into steam. Hence it has no single point in its favour, except the readiness with which it can be obtained pure. Thus, although it is of value in obtaining certain fixed points of temperature in graduating a thermometer, no one would think of using it in the construction of one.

Alcohol has a higher coefficient of expansion than water, and requires less heat to raise it in temperature than the same weight of water, but it is a bad conductor of heat. While it boils at a temperature even lower than water, it has the valuable property of remaining liquid at the lowest temperatures which have usually to be determined; so that, in spite of its irregularity of expansion, it is extensively used in thermometers intended for measuring these low temperatures.

Mercury has a rather low coefficient of expansion, but it fulfils all the other requirements far more satisfactorily than almost any other substance. It requires nearly a red heat to turn it into vapour, and only solidifies at a temperature far below that of the coldest winter which ever occurs in our latitudes. Its coefficient of expansion throughout this wide range of temperature is very nearly the same; and since it requires the expenditure of far less heat than is required by the same weight of any other liquid to raise it in temperature to a given extent, and also conducts heat

---

[1] It must be remembered that this only refers to thermometers for ordinary practical purposes. As will be seen, gas thermometers are used in special scientific work as being the best and most accurate of all.

very readily, it indicates changes of temperature promptly and accurately. It has the further advantages of being opaque and of not adhering strongly to glass, so that its surface is always clearly marked, and when it contracts none of it remains behind on the glass tube, which would cause errors in the reading.

Of the substances already mentioned, then (and, indeed, of all others), mercury would have the preference for practical purposes; and it is, in fact, the substance selected for use in all ordinary thermometers.

Having selected the material, we must next consider how it may be used to the best advantage. The coefficient of expansion of mercury is so small that the changes in volume must be magnified in the way already explained, *viz.* by connecting a comparatively large reservoir of the liquid with a very narrow tube, into which it is free to expand. This tube must be of uniform diameter throughout, otherwise the divisions on the scale cannot be equal in length, since a given increase in volume of the liquid will fill up a shorter length of the tube in the wider parts than in the narrower ones.[1]

Construction of an ordinary Mercury Thermometer.—A properly selected glass tube has a bulb, round or cylindrical, formed at one end of it by softening the glass in a blowpipe-flame and blowing air into it until it has expanded to the proper size. This bulb has next to be filled with mercury; and here the extreme fineness of the bore of the tube (often far too small to admit a hair) causes some slight difficulty. The bulb is, of course, full of air, and before the mercury can be got in the air must be got out. There is not sufficient room for the one to pass by the other in the narrow tube, and we have recourse to

---

[1] Absolute uniformity in bore cannot be attained in practice, and all the best thermometers have their tubes examined and 'calibrated,' as it is termed; a small drop of mercury being introduced into the tube, and the length which it occupies in various parts of the tube carefully measured. The shorter this length is, the wider is the tube at that part; and all differences in diameter thus ascertained are allowed for in dividing the scale, or at any rate a record of them is kept and applied to the readings of the instrument.

the expansion of gases by heat in order to effect the filling of the bulb. The glass is heated moderately in a lamp-flame, the air within expands, and some of it escapes through the tube. Then the bulb is removed from the flame and before the remaining air has time to cool the open end of the tube is plunged into some mercury.[1] As the air in the bulb contracts, the pressure of the outside air upon the mercury forces some of it through the tube into the bulb. It is obvious that all the air cannot be thus got out of the bulb, however much it may be heated, but we have now in the bulb some of a substance, *viz.* mercury, which can be turned into a vapour occupying far more space than the liquid, and condensing again unchanged into the liquid when cooled. So the bulb is heated pretty strongly in the flame, until the mercury in it boils; the vapour soon sweeps out all the remaining air, and we have the bulb and tube filled with nothing but the pure vapour of mercury. The open end of the tube is then once more dipped into mercury, and as the vapour cools and condenses, the pressure of the air forces more mercury into the bulb until it is completely filled. Lastly, the bulb is again heated up to the highest temperature that the instrument is intended to measure, any excess of mercury is allowed to escape from the tube, and then, before the temperature falls, a blow-pipe-flame is directed upon the end of the tube until the glass melts and closes up the tube effectively and permanently. All air is thus excluded, and the mercury is protected from any chemical change.

We have now a thermometer which, even in its present state, would be useful for comparing the temperatures of different substances. Suppose, for instance, that we place it in a basin of water, A, and mark the position of the top of the mercury column in the tube; and then transfer it to another portion of water, B;— if the mercury column remains at the same level, we shall know that B is at the same temperature as A; if the mercury rises

---

[1] It was formerly usual to blow a bulb in the tube near the open end, and introduce mercury into this bulb, from which the lower bulb was afterwards filled; but this is seldom done now.

CONSTRUCTION OF A THERMOMETER.   95

higher in the tube, B must be hotter than A ; if it falls, B must be colder than A. The thermometer, in fact, will serve as a delicate sense of touch, perfectly reliable, and far more sensitive than our own nervous system.

But we may easily desire to know more than this; we may want to know *how much* hotter or colder B is than A, and we may want to convey to others an accurate idea of the differences of temperature observed in experiments. For this purpose we must fix upon certain points of temperature easily verifiable by others, which will serve as standards to which other temperatures may be referred. Now, no one has yet succeeded in taking out of a body all the heat it contains, so as to reduce it to the 'absolute zero' of temperature (p. 86), which would be the most natural point of temperature to begin from; so that we have to be content with taking some particular temperature higher up in the scale to serve as a starting-point; and the one which is universally chosen is the temperature at which pure water turns into ice. This temperature is found to be (under ordinary conditions of pressure) perfectly invariable; moreover, it remains constant during the whole process of freezing a mass of water (as will be more fully explained in the chapter on Potential Heat), so that it can be observed with deliberation and certainty.

It is usually found most convenient to observe the temperature while the reverse change of state is taking place, *viz.* while ice is turning into water; the temperature of a mass of melting ice being precisely the same as that of freezing water. So the bulb and stem of the instrument to be graduated are placed in a vessel containing a mixture of water with ice or snow, and a mark is made on the tube at the point where the top of the mercury column stands unmoved after an immersion of ten minutes or more. This mark is called 'the freezing-point.'

The thermometer is now at any rate more useful than before: by placing it in contact with anything, we can see at once whether the substance is hotter or colder than freezing water. But more than this is required; we want to know, and to be able to make clear to others, *how much* hotter or colder the substance is. Now,

CONSTRUCTION OF A THERMOMETER.

if it was easy to construct a large number of thermometers with bulbs the capacity of which bore exactly the same proportion to the internal diameter of the tubes, it would, in order to specify particular temperatures, be sufficient to observe the distance of the top of the column of liquid above or below the mark of the freezing-point. If, for instance, the column stood 1 cm., 2 cm., 3 cm., etc., above this point, it would imply that the substance was at a certain definite higher temperature than freezing water; in somewhat the same way as the measured height of the column of mercury in a steam gauge indicates the exact pressure of the steam.

But it is practically impossible to obtain absolute uniformity in the construction of thermometers. Probably there are no two thermometers in existence with bulbs and tubes bearing precisely the same proportion to one another, and hence a rise or fall of the liquid column through 1 cm. corresponds to very different changes of temperature in different instruments. Another point of temperature is therefore required in order to render the readings of various thermometers easily comparable; and the one which is universally chosen is that at which water (under a pressure equal to that of a column of mercury 76 cm. in height) boils, or turns into steam. This is, when proper precautions are taken, just as invariable and as easily observed as the freezing-point; and so the next step in graduating the thermometer will be to place it in a vessel filled with steam rising from water kept steadily boiling over a lamp. As it is necessary that the whole of the liquid in the instrument should be at the same temperature, the boiler employed for the purpose is usually constructed as shown in fig. 43. A is a metal vessel (not glass, for reasons given in Chapter VIII. Section III.) about half filled with water. At the top of this a wide tube is attached,

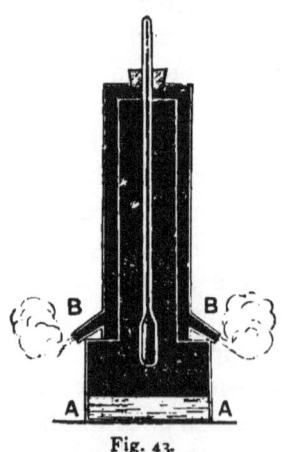

Fig. 43.

CONSTRUCTION OF A THERMOMETER.

long enough to enclose nearly the whole of the stem of the thermometer. Round this tube is placed a still wider cylinder or 'jacket,' so that the steam escaping from the top of the inner tube descends in the space between the two tubes and passes out through openings at B, B. Thus the steam in the inner tube is protected from the cooling effects of the outer air, and its temperature is maintained constant, and is found to be exactly the same as that of the water from which it was formed. When the thermometer is placed in the steam, as shown in the figure, the column of mercury will rise in the tube up to a certain point, and will remain steadily there as long as there is any water boiling in the vessel (unless the pressure of the atmosphere alters during the experiment). The instrument must be hung with its bulb just above the surface of the water in the boiler; not immersed in it, for the violent agitation of the boiling water is liable to cause some irregularity of temperature. A mark is then made on the tube where the top of the mercury column stands, and this is called the 'boiling-point.'

We have now the means of ascertaining not merely comparative but absolute temperatures. The actual distance in length between the freezing-point and boiling-point may vary to any extent in different instruments, but any given fraction of this length on one thermometer will indicate the same temperature as the same fraction of the length between the freezing-point and the boiling-point on another thermometer.

Thermometer Scales.—The position of the freezing-point and the boiling-point on a thermometer having been accurately marked, the next step will be to divide the space between them into some convenient number of parts or 'degrees'; so as to be able to specify readily the temperature of anything by stating the degree at which the top of the mercury column stands when brought into contact with it. Here, unfortunately, no general agreement has been found possible: three different systems have been devised and are in use, known as the Centigrade scale, the Fahrenheit scale, and the Reaumur scale, respectively.

1. In the *Centigrade* scale, which is the one almost always

G

used in scientific work, the space between the freezing-point and the boiling-point is divided into 100 degrees, the freezing-point being marked 0° and the boiling-point 100°. The space on the tube below the freezing-point is divided into degrees of the same length as the others (supposing that the tube is of the same internal diameter throughout), and these are distinguished from the rest by a *minus* sign attached to them. Thus if the top of the mercury column sank so low as to be opposite the 10th mark below the freezing-point, the temperature is said to be −10°. Similarly the graduation is continued above the boiling-point in divisions of the same length, and these may extend as far as 300° without any serious error, the expansion of mercury being remarkably uniform, as already explained.

2. In the *Fahrenheit* scale, which for common use is not yet superseded in this country, the space between the freezing-point and the boiling-point is divided into 180 degrees, and the freezing-point is marked 32°; the zero of the graduation being no scientifically definable temperature, but simply the point to which the mercury sank during a severe winter in Iceland.[1] Thus the number attached to the boiling-point will be 180° higher than 32°, *i.e.* 212°; and the graduation is continued above this and below the zero-point in divisions of the same length, the temperatures below zero being distinguished by a minus sign.

3. In the *Reaumur* scale, which is seldom used now except in Russia, there are only 80 divisions between the freezing-point and the boiling-point; so that each degree is longer than one on either of the other scales. The freezing-point is marked 0°, as on the Centigrade scale, and the boiling-point is therefore marked 80°.

Fig. 44 represents the scales for thermometers approximately the same in internal dimensions, graduated on the Centigrade,

---

[1] This is the account given by Boerhaave in his treatise on Chemistry, published in 1724.

According to another account Fahrenheit obtained the zero of his scale by observing the lowest temperature obtainable by a mixture of ice and salt.

Thermometer Scales.

Fahrenheit, and Reaumur systems, respectively; so as to afford a comparison of the three scales.

Of these the Centigrade scale has undoubtedly the advantage in convenience and simplicity. It is framed on a decimal system, and so is in harmony with the metric system of measures and weights, which is also decimal. Moreover, since most calculations of the volume of gases have reference to the freezing-point of water, they are much simplified by taking it as zero. The only slight disadvantages of the scale are—(1) that the length of each degree is somewhat great, so that fractions of a degree have not unusually to be dealt with; (2) that the zero-point is not so low as to avoid our having frequently to deal with *minus* quantities, which introduce a slight complication.

Conversion of one Scale to Another.—It is sometimes required, more as an arithmetical exercise than for any practical purpose, to convert a temperature expressed in degrees on one of the thermometer scales to the corresponding number of degrees on one of the others. Generally Centigrade degrees have to be translated, as it were, into Fahrenheit degrees, and *vice versâ;* and this is easily done if the following points are borne in mind.

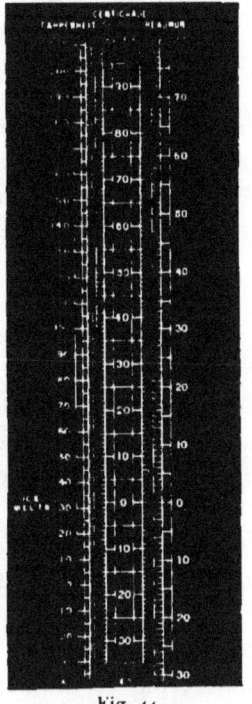

Fig. 44.

A. The same space (*i.e.* that between the freezing-point and the boiling-point of water) is divided

        On the Centigrade scale into 100 parts,
          ,,   Fahrenheit ,,   ,, 180 ,,

Hence it is clear that   100° C. = 180° F.
        or   5° C. = 9° F.
    That is,   1° C. = $\frac{9}{5}$° F.
    and   1° F. = $\frac{5}{9}$° C.

So that, if the zero of the scales marked the same point of temperature, all that would have to be done would be to take $\frac{5}{9}$ of a given number of Fahrenheit degrees as the equivalent number of Centigrade degrees; and $\frac{9}{5}$ of Centigrade degrees as the equivalent number of Fahrenheit degrees. But,

B. The point marked 0° on the Centigrade scale is marked 32° on the Fahrenheit scale.

In fact the two scales do not begin at the same level; and the Fahrenheit numbers are all higher by 32 than the corresponding Centigrade numbers. Thus

   0° C. corresponds to 0° + 32 or 32° F.
   5° C. ,, ,, 9° + 32 ,, 41° F.
   10° C. ,, ,, 18° + 32 ,, 50° F.
    etc.   etc.

We must therefore, in converting temperatures from the Centigrade to the Fahrenheit scale, after finding the corresponding value in Fahrenheit degrees by taking $\frac{9}{5}$ of the Centigrade degrees, add 32 to the result in order to bring it up to the proper level on the Fahrenheit scale.

Thus, supposing a Centigrade thermometer gave a reading of 15°; $\frac{9}{5}$ of 15° = 27°, and 27° + 32 = 59°, which is the true equivalent of 15° C. on the Fahrenheit scale.

For the same reason, in converting readings of temperature from the Fahrenheit to the Centigrade scale, we must begin by subtracting 32 from the given number of Fahrenheit degrees, in order to bring the scales to the same level, and then take $\frac{5}{9}$ of the remainder in order to get the true corresponding reading on the Centigrade scale.

Thus, if a temperature was found to be 68° F.; 68° − 32 = 36°; and $\frac{5}{9}$ of 36° = 20°, which is the equivalent reading on a Centigrade thermometer.

The following rules, then, will express the process of converting degrees of temperature on either scale into those of the other :—

A. Rule for converting Centigrade degrees to Fahrenheit degrees.

(i) Multiply the given number of degrees by $\frac{9}{5}$.
(ii) Add 32 to the product.
    (Or, more shortly, $\frac{9}{5}$ C°. + 32 = F°.)

EXAMPLE.—Alcohol boils at 78° C.
  (i) 78 × $\frac{9}{5}$ = 140·4
  (ii) 140·4 + 32 = 172·4.   *Ans.* 172·4° F.

B. Rule for converting Fahrenheit degrees to Centigrade degrees.
  (i) Subtract 32 from the given number of degrees.
  (ii) Multiply the remainder by $\frac{5}{9}$.
    (Or, more shortly, (F°. − 32) × $\frac{5}{9}$ = C°.)

EXAMPLE.—Mercury boils at 662° F.
  (i) 662 − 32 = 630.
  (ii) 630 × $\frac{5}{9}$ = 350.   *Ans.* 350° C.

In dealing with temperatures below the freezing-point on either scale, minus signs will have to be carefully attended to in working out results by the above rules.

For example—

(1) The lowest temperature observed on a particular night in winter was −10° C. By Rule *A*,
  (i) −10° × $\frac{9}{5}$ = −18°.
  (ii) −18° + 32 = +14°   *Ans.* 14° F.

(2) Mercury freezes at −38·2° F. By Rule *B*,
  (i) −38·2° − 32 = −70·2°.
  (ii) −70·2 × $\frac{5}{9}$ = −39°.   *Ans.* −39° C.[1]

Different kinds of Thermometers.—The variety of forms of thermometers is almost as great as the variety of purposes for which they are used. For such ordinary purposes as indicating the existing temperature of a room, thermometers with spherical bulbs and scales not engraved on the tube but stamped on a

---

[1] It may be noted that the freezing-point of mercury is expressed by nearly the same numerals on both scales; the subtraction of 32 from the Fahrenheit degrees compensating for the greater length and therefore smaller number of the Centigrade degrees. The exact point where the two scales coincide can be found from the equation
$$x° \text{ F.} = \tfrac{9}{5} x° \text{ C.} + 32.$$

strip of boxwood, on which the tube is fixed to protect it from injury, are considered sufficiently good. Such instruments have no pretensions to scientific accuracy, even if a special scale is made to suit each tube, which is seldom done. The wooden frames with stamped scales are made by hundreds at a time, and the thermometers are blown in the same wholesale way: the two are then fitted together without any particular care.

The form of thermometer usually employed for scientific purposes is shown in fig. 45. In this the bulb is made in the shape of a cylinder of small diameter; partly to gain extended surface, so that the whole of the mercury may quickly acquire the temperature of the substance in which it is immersed, partly in order that the instrument may be easily passed into a narrow tube or through the neck of a bottle or retort. The scale is always engraved upon the glass tube itself, and usually several points of it are ascertained by placing the ungraduated instrument side by side with a carefully made standard thermometer in water at different temperatures, and marking on the tube the different levels at which the mercury column stands. The intervening degrees are then filled in by a dividing engine.[1]

Registering Thermometers.—It is often of great interest and importance to know not merely the temperature of a body at the time of observation, but also how far it has been varied since the last observation. This is especially the case in meteorological work, records of the highest and lowest temperatures which have occurred at various intervals during the day and night being necessary in order to determine the average temperature of a place, and for other purposes. It would be practically impossible to arrange for continuous observations of the instruments, and means have been devised for making the thermometers register

Fig. 45.

---

[1] The verification of thermometers, which is one of the most useful parts of the work carried on at Kew Observatory, is conducted on a similar principle. The thermometers to be tested are grouped round a standard thermometer on

REGISTERING THERMOMETERS.

their own indications, at any rate as regards the *maximum* and *minimum* temperatures which have occurred. One or two of the simpler forms of 'self-registering thermometers,' as they are called, will be next described.

1. *Phillips's Maximum Thermometer.*—In this, as shown in fig. 46, a short portion (about 2 cm.) of the mercury column is separated from the rest by an extremely small bubble of air, just sufficient to prevent it from joining the rest of the

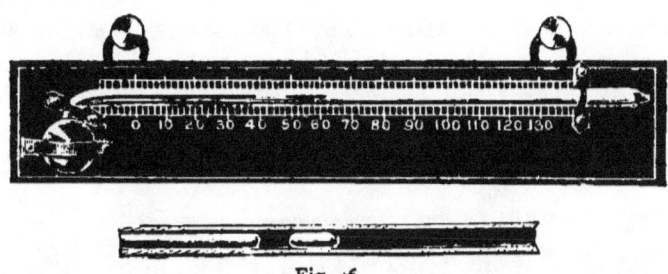

Fig. 46.

mercury. The tube is supported in a nearly horizontal position, and when the temperature rises, the short thread of mercury is pushed forward through the medium of the cushion of air as long as the main column behind it advances. When, however, the latter retreats owing to a fall in temperature, the detached portion remains where it was, since there is no adhesion between it and air; and its outer end marks the highest temperature attained by the instrument. After the reading is taken, the thermometer is 'set' for a fresh registration by simply shaking down the index-column until it rests as close as possible to the main column. This is best done by holding the instrument at arm's-length with the bulb outwards, and giving it one or two

---

a frame placed in a cistern of water. The water is stirred by a set of moving paddles, to equalise the temperature throughout, and the temperature indicated by each thermometer is recorded and compared with that of the standard instrument. Then the water is raised to some definite higher temperature, and all the thermometers are again observed. The same is done at several other temperatures, and thus a table of errors or deviations, if any, from the standard instrument is constructed and supplied with each tested thermometer.

quick swings through the air: centrifugal tendency then brings the short index-column towards the bulb. The thermometer should always be hung so that the tube is nearly horizontal, best with a slight slope downwards towards the bulb.

The only errors to which it is liable are those due to the alteration in length (1) of the bubble of air, (2) of the index-column of mercury; but these are hardly appreciable in practice.

2. *Rutherford's Minimum Thermometer.*—In this alcohol and not mercury must be used.[1] The tube, as is usual in alcohol thermometers, is rather wide in bore, and within the column of alcohol a short rod of glass is placed (as shown in fig. 47, and on an enlarged scale below it), rather smaller in dia-

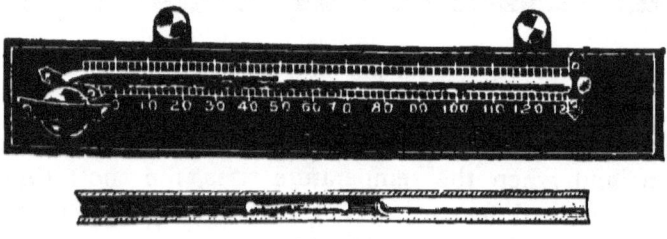

Fig. 47.

meter than the tube, so as to allow the liquid to move freely past it. Alcohol adheres pretty strongly to glass, but so long as the glass rod is wholly immersed in the liquid, this adhesion pulls it equally in all directions, forward as well as backward, and it remains in its original place. When the temperature falls, the alcohol column moves back without disturbing the index until the latter touches the surface of the liquid. Then the equilibrium of the attractive forces acting on the index is destroyed, since there is no liquid beyond the right-hand end of the index to balance the adhesion which pulls it to the left hand; and so,

---

[1] It would be far preferable to use mercury, if possible, but any index would be invisible in this liquid. No very practical form of mercurial minimum thermometer has been yet devised. Mr. Casella, of London, has invented a very ingenious one which does its work well when properly placed and carefully used, but it is too delicate and expensive for ordinary purposes.

if the column continues to retreat, the index is pulled back with it, the end of the glass keeping just level with the surface of the liquid. If after a time the temperature rises again, the alcohol column advances towards the right hand, but the index remains where it was, since the balance of attractions is restored as soon as any liquid gets beyond the end of the glass. This end, therefore, shows the lowest point to which the thermometric column has retreated since the last observation.

When the minimum temperature has been read off on the scale, in order to set the thermometer for the next observation it is only necessary to raise the bulb-end so as to bring the outer extremity of the column lower than the rest. The index will then, from its weight, run down to the end of the column, but no farther. The instrument is now restored to its horizontal position, and is ready to register the minimum temperature again.[1]

Six's Registering Thermometer.—This is represented in fig. 48, and is one of the commonest popular forms of the instrument, having the advantage of registering both maximum and minimum temperatures, although it has no pretensions to scientific accuracy. It is in reality an alcohol thermometer, although a column of mercury is seen in the lower bend of the tube ; but this column is merely an index, or, rather, serves the purpose of moving the two indices which are seen a little above its ends, one in each branch of the tube. The alcohol is contained in the long cylindrical bulb seen in the centre between the two branches, and there is also some alcohol above the mercury in the right-hand branch, so that both the indices are immersed in this liquid. These indices are rather smaller than the bore of the tube, but are kept from moving easily in it by light hair springs attached to

---

[1] It is decidedly best to hang the thermometer so that there is a slight slope downwards from the outer end of the column towards the bulb. This not only assists in moving the index back, but also prevents the accumulation of any liquid in the empty part of the tube owing to the condensation of some alcohol vapour. If such condensation should occur, the liquid at once runs down to the main column.

them which press against the sides of the tube so as to retain them in position even when the instrument is placed vertical, which is its usual position: they can, however, be moved up and down when enough force is applied to overcome the friction of the springs. They are made of steel, but are usually enclosed in thin glass tubes, to prevent their corrosion by the alcohol. When the instrument is to be used, each index is brought down until it rests upon the end of the column of mercury by holding a small magnet close to the tube, and moving it downwards, when the steel of the index will be attracted and follow it.

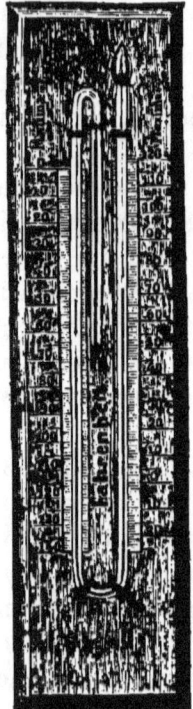

Fig. 48.

The action of this thermometer is as follows:—Suppose that a rise of temperature takes place, the alcohol in the cylindrical bulb expands, and the column of it in the left-hand branch becomes longer, flows past the index without moving it, and presses down the column of mercury in that branch. Hence the column of mercury in the other (right-hand) branch rises and pushes up the index there, since the adhesion of glass to mercury is slight, and the density of the index is much less than that of the mercury, so that it floats.

When the temperature falls, the alcohol contracts, and the pressure of the air which is always left in the little bulb at the upper end of the right-hand branch, forces down the alcohol and the mercury column below it in that branch, while the index remains where it was. Thus the lower end of this index shows the maximum temperature which has been reached. If the temperature falls below the original point, it is easy to see that the index in the left-hand branch will be pushed farther up as the alcohol contracts and the mercury column rises on that side: thus its lower end shows the minimum temperature which has occurred since the instrument was last adjusted.

AIR THERMOMETER.

Air Thermometers.—These, owing to the extreme regularity with which gases such as those which constitute air (oxygen and nitrogen) expand and contract within a very wide range of temperature, are the most accurate of all; and are, in fact, almost the only instruments by which very low temperatures such as that at which alcohol becomes solid ($-130°$ C.), or very high temperatures at and beyond the boiling-point of zinc ($940°$ C.) can be determined properly.[1]

The instrument shown in fig. 49 is rather intended to be used as a thermoscope, *i.e.* an apparatus for indicating the fact of a change of temperature in a substance, than for measuring the amount of the change. It consists of a glass bulb attached to a tube of moderately narrow bore (about 2 mm.), the lower end of which dips into some water, coloured with ink or indigo, contained in a bottle with a firm extended base or foot. The tube is maintained upright in the bottle by passing through a hole in the cork, and another small hole (or a notch) is cut in the cork to admit air. It is convenient, though not necessary, to have a small stopcock attached to the tube just below the bulb, as shown in the figure; by withdrawing air through this the column of liquid in the tube can be readily adjusted to any required height. A piece of white cardboard is fixed behind the tube, on which may be drawn a short horizontal reference line, or a scale of divisions. If there is no stopcock, the column of liquid may be raised to a convenient level by gently warming the bulb until a few bubbles of air have escaped from the lower end of the tube. Then, as the air in the bulb cools and contracts, the pressure of the external air will force up a column of liquid

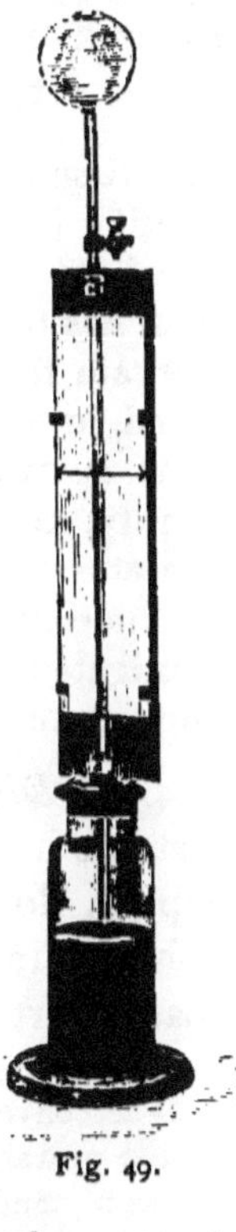

Fig. 49.

---

[1] Hydrogen, which is even more difficult to liquefy than oxygen and nitrogen, was used to determine the extremely low temperature at which liquid oxygen boils under ordinary pressure, *viz.* $-184°$ C.

in the tube, and when this has ceased to rise, the line on the cardboard may be adjusted to its level.

If, now, any substance is brought into contact with the bulb, we can easily tell whether it is hotter or colder than the air in the bulb by the motion of the column of liquid. If the latter rises, it shows that the air in the bulb has contracted, and therefore that the substance is colder. If a fall occurs, the substance must be hotter than the air in the bulb, for it has caused the latter to expand.[1] Very small differences of temperature can thus be rendered evident, since, as already explained, gases change in volume very perceptibly for even small changes of temperature.

In Regnault's air thermometer, which is adapted for delicate scientific investigations, no column of liquid is used, but the open end of the tube, while the instrument is exposed to the temperature which has to be measured, is tightly closed, and the whole is allowed to regain the ordinary temperature. Then the end of the tube is dipped into mercury and opened: the quantity of mercury which enters is carefully measured (or ascertained by weighing). This quantity, compared with the amount of mercury which fills the apparatus at $0°$, gives the change in volume which the inclosed air had undergone when exposed to the temperature which was being investigated, and from the known coefficient of expansion of air this temperature can be calculated.

Leslie's Differential Thermometer.—This instrument, which is of great use in demonstrating many of the facts and laws of heat, is simply a double air thermometer. One of the best forms of it is shown in fig. 50. Two bulb tubes, A and B, of equal capacity, are connected by swivel-joints [2] with the respective

---

[1] It is convenient to draw two arrows on the cardboard; one with point directed upward, and the word 'COLD' beside it, and the other with point downward, and the word 'HOT' beside it; indicating thus the meaning of the direction of the column's movement.

[2] These joints, made of brass, and sufficiently good to stand the slight pressure required without leaking, can be procured from any gasfitter. It is preferable to obtain small 'ball-and-socket' joints instead, since with these the horizontal distance between the bulbs can be varied within rather wide limits.

DIFFERENTIAL THERMOMETER. 109

branches of a bent U-shaped tube fixed upright on a wooden frame. The joints enable the bulbs to be turned in any direction, either downwards, when they are to be dipped into jars of liquid, or upwards, as required in many experiments on radiated heat (see fig. 97). Sufficient coloured liquid (usually a solution of indigo in sulphuric acid) is placed in the bend to fill each branch about half-way up: and a cardboard scale is fastened to the frame behind the tube by clips, so that the horizontal zero-line on it can be adjusted to the level of the surfaces of the liquid columns. A short horizontal tube with a stopcock forms a direct connection between the two branches near their upper ends, so that the pressure of the air in them can be readily equalised. In using the thermometer, this stopcock is first opened temporarily until the columns of liquid in the two branches are observed to stand at the same level, showing that the pressure of the air is equal in both bulbs. It is then closed, and the bulbs are brought into contact respectively with any two substances which are to be tested for difference of temperature. If these substances are at precisely the same temperature, whether high or low, the air in each bulb will

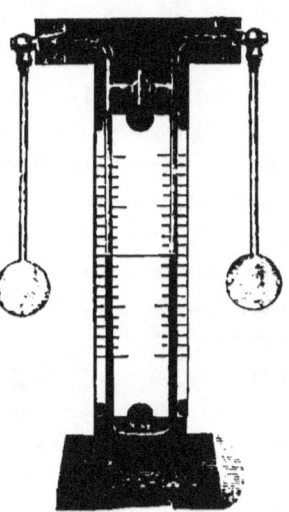

Fig. 50.

tend to change in volume to the same extent by contact with them, and the pressure upon the surface of the liquid in each branch will remain the same, so that no alteration in level will be observed. But if A, for instance, is hotter than B, it will cause the air in that bulb to expand, or tend to do so, more than that in the other bulb; and thus the pressure upon the liquid in that branch will be greater than the pressure in the other branch, and the level of the column will be depressed on the side nearest A. Thus we can always tell which of two things is the hotter (if there is any difference at all in their temperatures) by observing on which side the column of liquid is depressed.

Expt. 25. To illustrate the use of the Differential Thermometer, two beakers may be filled about two-thirds full of water at the same temperature, and the bulbs dipped, one into each beaker. The columns of liquid in the two branches will remain at the same level. A few drops of hot water may then be added to the water in one of the beakers, or the hand may be held in it for a few seconds, and the bulbs again immersed. The liquid column will now be depressed on the side of the beaker containing the slightly warmer water.

Daniell's Pyrometer.—This is one of the few instruments in which the expansion of solids by heat has been utilised for thermometric purposes. It is intended solely for the measurement of high temperatures, such as those occurring in furnaces (hence its name, from πῦρ, *fire*, and μετρέω, *I measure*); and it resembles in principle Ferguson's apparatus for demonstrating the expansion of solids, already described and illustrated on p. 46. It consists of a bar of iron or platinum enclosed in a cylindrical case of fire-clay: one end of the bar rests firmly on the bottom of the case, and a short plug of porcelain, fitting moderately tightly, is pushed down into the case until it touches the free end of the bar. Then the whole is placed in the furnace of which the temperature has to be ascertained, and as the bar expands more than the fire-clay case, the plug is pushed forward to a certain extent, and remains in that position when the case is withdrawn from the furnace and allowed to cool. The distance through which the plug has been moved affords an indication of the maximum temperature to which the instrument has been exposed, and this distance is accurately measured by applying to the top of the case a brass frame containing a lever, the short arm of which is made to rest upon the porcelain plug, while the long arm indicates the amount of expansion on a scale, as in Ferguson's instrument.

This pyrometer, however, has no pretensions to accuracy, and is seldom or never used now. High temperatures are best measured by an air thermometer with bulb of porcelain or platinum; or by observing the amount of heat given up to cold water by a ball of platinum of known weight which has been placed in the furnace. This method will be further alluded to in the chapter on 'Specific Heat.'

The Weight Thermometer. — This instrument, though not adapted for common, everyday use, has a high scientific importance and interest. Its principle will be readily understood by a reference to the account of Regnault's apparatus for determining the expansion of liquids (p. 67); for it is clear that if we know accurately from other sources the coefficients of expansion of mercury and glass, and the weight of the liquid contained in the vessel at 0°, we have only to collect and weigh the amount which overflows from the end of the tube when the whole is exposed to a higher temperature, in order to ascertain the rise of temperature which has caused the overflow. In practice, however, a slightly different method is usually adopted.

The thermometer is often made when required for an observation by closing one end of a piece of glass tubing (about 1 cm. in internal diameter) before the blowpipe flame, and drawing out the tube about 6 or 8 cm. from the closed end so as to form a bent capillary jet. But as the filling of the cylindrical bulb thus obtained (which is done in the same way as an ordinary thermometer is filled, p. 93) takes some little time, and the capillary tube is easily broken, a much preferable form is shown in fig. 51, in which the bent exit tube is separate from the bulb and fitted to the neck of the latter by grinding.

Fig. 51.

To the outer end of this tube a small deep cup is fitted by a cork having a notch cut down one side, for the purpose of catching the mercury which overflows.

In using this thermometer, the empty instrument is first

weighed and then filled completely with pure mercury; the bent tube being fitted firmly into its place, and all excess of mercury being allowed to escape into the cup, in which there should be enough mercury to cover the end of the tube. The bulb with as much of the tube as possible is then placed in a beaker of melting ice. In about ten minutes the whole mass will have acquired the temperature of the ice, *i.e.* $0°$. The cup is then completely emptied of mercury (the bulb and tube still remaining immersed in the ice), any drops which may adhere to the end of the jet being lightly brushed off. The cup being replaced, the whole apparatus is raised out of the beaker, dried, and weighed. This weight, less the weight of the empty apparatus, gives the weight of the mercury (W) which fills it at the temperature of $0°$. It is next brought into contact with the substance of which the temperature (higher than $0°$) is to be determined; some mercury will, of course, overflow into the cup, and when the overflow ceases the cup is emptied as before and replaced. Lastly the whole is weighed again; and this weight, less the weight of the empty apparatus, gives the weight of the mercury (W') which remains in it at the temperature required to be known ($t°$). Thus we have two weights,—

W representing (as explained on p. 68) the volume of mercury at $t°$
W'   „   „   „   „ $0°$

and the difference between these weights, *i.e.* W—W' represents the amount of expansion which the mercury has undergone in passing from the lower temperature ($0°$) to the higher temperature ($t°$). If, then, the coefficient of apparent expansion of mercury in glass is known, it is easy to calculate the number of degrees through which it must have been raised in temperature in order to produce the observed expansion.

For example :—

Weight of the apparatus empty	= 32·2 grms.
Do. ditto do. full of mercury at $0°$	= 154·3 ,,
Do. do. with the mercury which remained after immersion in warm water	= 153·55 ,,

Weight Thermometer.

Hence,

Weight of mercury which fills the apparatus at 0° $= (154 \cdot 3 - 32 \cdot 2 =)$ 122·1 grms.

Weight of mercury which is left after immersion in the water, $= (153 \cdot 55 - 32 \cdot 2 =)$ 121·35 ,,

Therefore the weight of mercury which overflowed in consequence of the heat imparted by the water $= (122 \cdot 1 - 121 \cdot 35 =)$ 0·75 grm.

We thus learn that a volume of mercury at 0° represented by 121·35 grms. expands to an extent represented by ·75 grm. when placed in the warm water.

Then $(\frac{\cdot 75}{121 \cdot 35} =)$ ·00618 represents the fraction of its volume at 0° which the mercury has expanded when heated to the temperature of the water.

Now, the coefficient of apparent expansion of mercury in glass is ·000154, and we have only to find how many times ·000154 is contained in ·00618 in order to learn how many degrees the mercury had been raised above 0° by the warm water.

$$\frac{\cdot 00618}{\cdot 000154} = 40 \cdot 2 \text{ (nearly).}$$

Hence the temperature of the water was 40·2°.

APPENDIX TO CHAPTER V.

Problems on the Expansion of Substances by Heat.

In working out all these, the following principles must be borne in mind :—

I. That the coefficient of expansion of a substance is (unless otherwise specified,) the fraction of its size at 0° which expresses the alteration of the substance in size when heated from 0° to 1°, or when cooled from 0° to −1°.

II. That, consequently, if a substance measures 1 unit (of length or volume) at 0°, it will measure

$1 +$ the coefficient of expansion at $1°$
$1 + 2 \times$ do.   do. $2°$
$1 + 3 \times$ do.   do. $3°$
$1 + 10 \times$ do.   do. $10°$
   etc. etc.
and $1 −$ the coefficient of expansion at $− 1°$
$1 − 2 \times$ do.   do. $− 2°$
   etc. etc.

III. That, whatever the actual measurement of the substance may be, its alteration in size will always be *in the proportion* indicated in II.

For example, the coefficient of expansion of mercury is 0·00015. Then a quantity of mercury measuring 1 volume at 0° will measure

$1 +$ ·00015 or 1·00015 vol. at $1°$
$1 + 2 \times$ ·00015 or 1·0003 vol. at $2°$
$1 − 2 \times$ ·00015 or 0·9997 vol. at $−2°$
   etc. etc.

And if the mercury measures, say, 8 c.c. at 0°, then at 2° it will measure as much more than 8 c.c. as 1·0003 is greater than 1.[1]

---

[1] Let V = the size (length or volume) of the substance at a temperature of $t°$
V′ =   ,,   ,,   ,, t'
a = the coefficient of expansion.
Then, $1 + at° : 1 + at'° :: V : V'$.

Problems on Expansion by Heat. 115

PROBLEM 1.—**Given the coefficient of expansion of a substance, to find its alteration in size for a given alteration of temperature.**

RULE.—(i) Multiply the coefficient of expansion by the original temperature, and add the product to 1. Call the result A.

(ii) Multiply the coefficient of expansion by the given changed temperature, and add the product to 1. Call the result B.

(iii) Work out the proportion sum,
$$A : B :: \text{original size} : x$$
$x=$ the altered size required.

EXAMPLES.—A copper steam pipe is 30 metres long at 0°. What will be its length when steam at 100° is passed through it?

(Coefficient of expansion of copper in length = 0·000017.)

(i) Since the original temperature is 0°, nothing is to be added to 1, and therefore A = 1.

(ii) 100 × 0·000017 = 0·0017. Hence B = 1 + ·0017 = 1·0017.

(iii) 1 : 1·0017 :: 30 metres : 30·051 metres.

Ans. 30·051 metres or 30 m. 5 cm. 1 mm.

Some mercury measuring (in glass) 500 c.c. at 50° is cooled to 10°. What will it then measure?

(Coefficient of apparent expansion of mercury = 0·00015.)

(i) 50 × 0·00015 = 0·0075. Hence A = 1 + ·0075 = 1·0075.
(ii) 10 × 0·00015 = 0·0015. Hence B = 1 + ·0015 = 1·0015.
(iii) 1·0075 : 1·0015 :: 500 c.c. : 497 c.c. (nearly).

Ans. 497 c.c. (nearly).

A quantity of air measures 140 c.c. at 21°. What will it measure when cooled to 0°?

(Coefficient of expansion of air = 0·00366).

(i) 21 × 0·00366 = 0·07686. Hence A = 1 + 0·07686 = 1·07686.
(ii) B = 1.
(iii) 1·07686 : 1 :: 140 c.c. : 130 c.c. (nearly).

Ans. 130 c.c. (nearly).

This last example is an instance of a calculation which has very frequently to be made in scientific work. In experiments on gases, the volumes have to be measured under very various conditions of temperature (and also of pressure), and in order to render the results of a series of experiments comparable with each other, the observed volumes of gases are always reduced by calculation to the volumes which the gases would occupy at the standard temperature of 0° and under the standard pressure of a column of mercury 760 mm. in height.

The following examples will show how this reduction to standard temperature and pressure is effected; and it will slightly simplify the calculation if the coefficient of expansion is taken as $\frac{1}{273}$.[1]

Some gas in a measuring-tube has a volume of 36 c.c.; the temperature of the room is 15°. What volume will it occupy at 0°?

$$\left(1+\frac{15}{273}=\right)\ \frac{288}{273}:1::\overset{c.c.}{36}:\overset{c.c.}{34\cdot 125}.$$

Ans. 34·125 c.c.

It is easy to see that, since the second term is always 1, the process of reduction to standard temperature consists simply in multiplying the observed volume by 273, and dividing the product by (273 + the observed temperature).

In the next examples the correction for pressure will be also applied.

By Mariotte's Law, p. 83, the volume of a gas alters inversely with the alteration in the pressure upon it. Hence the volume of a gas under the standard pressure of 760 mm. height of mercury-column will be

    (*a*) Greater than the observed volume if the observed pressure is more than 760 mm.

    (*b*) Less than the observed volume if the observed pressure is less than 760 mm.

And in order to calculate the precise volume we have the proportion—

$$\frac{mm.}{760}\ :\ \begin{array}{c}\text{observed}\\ \text{pressure}\end{array}\ ::\ \begin{array}{c}\text{observed}\\ \text{volume}\end{array}\ :\ \begin{array}{c}\text{volume under}\\ \text{pressure of 760 mm.}\end{array}$$

EXAMPLES.—A gas is found to measure 190 c.c. under a pressure of 800 mm. What will it measure under the standard pressure (temperature remaining unaltered)?

$$\overset{mm.}{760}:\overset{mm.}{800}::\overset{c.c.}{190}:\overset{c.c.}{200}$$

Ans. 200 c.c.

---

[1] A still simpler number is $\frac{1}{3000}$, since the multiplier in the calculation is an easy figure: and this may be used instead of $\frac{1}{273}$ in working out examples.

Problems on Expansion by Heat.   117

The observed volume of a gas, at a temperature of 27° and under a pressure of 700 mm., was 120 c.c. What will it measure at 0° and under a pressure of 760 mm?

(i) Correction for temperature—
$$\left(\frac{273}{273+27}=\right)\frac{273}{300}\times 120\overset{\text{c.c.}}{=}109\cdot2\text{ c.c.}$$

(ii) Correction for pressure—
$$\overset{\text{mm.}}{760}:\overset{\text{mm.}}{700}::\overset{\text{c.c.}}{109\cdot2}:\overset{\text{c.c.}}{100\cdot58}\text{ (nearly)}.$$
$$\textit{Ans. } 100\cdot58\text{ c.c. (nearly)}.$$

When the principles of the calculation are fully understood, the following formula will be useful in reducing gas-volumes to standard temperature and pressure:—

Let V = observed volume of a gas.
P = observed pressure upon it.
$t°$ = observed temperature of the gas.
a = coefficient of expansion.
V' = volume at 0° and under 760 mm. pressure.

Then $V'=V\dfrac{P}{760(1+at°)}$.

PROBLEM 2.—**Given the alteration in size of a substance originally at 0°, and its coefficient of expansion, to find the alteration in temperature to which it has been subjected.**

This is the problem which has to be solved in using Regnault's air thermometer and the weight thermometer (p. 112).

RULE.—(i) Divide the alteration in size by the original size of the substance. The quotient will be the fraction of its size by which the substance has been altered.

(ii) Divide this quotient by the given coefficient of expansion. The resulting quotient will be the number of degrees which the substance has been altered in temperature.

EXAMPLE.—50 c.c. of air at 0° were heated until the volume became 70 c.c. What increase in temperature had taken place?

(i) The increase in volume of the air was (70 − 50 =) 20 c.c.
$$\frac{\text{Increase in vol.}}{\text{Original vol.}}=\frac{20}{50}=\frac{2}{5}=0\cdot4.$$

Hence the volume of the air had increased by 0·4 or $\frac{2}{5}$ of its original amount.

Now, the increase in volume for 1° is 0·00366,

Hence, $\dfrac{0\cdot 4}{0\cdot 00366} = 109\cdot 29$ (nearly).

Therefore the air had been heated from 0° to 109·29°.

PROBLEM 3.—**Given the coefficient of expansion of a substance, and the weight of a unit volume of it (say 1 c.c.) at a certain temperature, to find the weight of a given volume of it at some other specified temperature.**

In making this calculation it must be remembered that the density of a substance varies inversely with the volume it occupies (p. 55): and hence if we know the alteration in volume which a substance has undergone, the alteration in density can be found from the proportion—

Altered . Original .. Original . Density at the
volume ⋅ volume ∷ density ⋅ given temperature.

And therefore, since the densities above mentioned denote the comparative weights (or, more strictly, masses) of equal volumes of the substance, we have—

Altered . Original .. Wt. of unit Wt. of the same
volume ⋅ volume ∷ volume at : volume at the
orig. temp. given temp.

from which the weight of the given volume of the substance is readily found.

EXAMPLES.—The mean coefficient of expansion of alcohol may be taken as 0·0011. The weight of 1 c.c. of alcohol at 0° is 0·8 grm. What will be the weight of 100 c.c. of alcohol at 60°?

Vol. at 60°.   Vol. at 0°.   Wt. of unit vol. at 0°.   Wt. of unit vol. at 60°.

$1 + (60 \times 0\cdot 0011) : 1 :: 0\cdot 8$ grm. $: 0\cdot 75$ grm. (nearly).

Hence, since 1 c.c. of alcohol at 60° weighs 0·75 grm., 100 c.c. will weigh $(100 \times 0\cdot 75 =)$ 75 grms.

Ans. 75 grms.

The weight of 1 litre of carbon dioxide at 0° is 2 grms. (very nearly). What will be the weight of 8 litres of the gas at 91°?

Vol. at 91°.   Vol. at 0°.   Wt. of 1 litre at 0°.   Wt. of 1 litre at 91°.

$1 + \dfrac{91}{273} : 1 :: 2$ grms. $: 1\cdot 5$

$\dfrac{273}{364} \times 2$ grms. $= 1\cdot 5$ grm.

Problems on Expansion by Heat. 119

The weight of 1 litre, then, at 91° is 1·5 grm., and the weight of 8 litres will be $(8 \times 1\cdot5 =)$ 12 grms.   *Ans.* 12 grms.

PROBLEM 4.—**Given the weight of a unit volume of a gas at the standard temperature and pressure, to find the weight of a certain volume of the gas at any other specified temperature and pressure.**

This resembles the last Problem, but the effect of pressure as well as temperature in altering the volume (and therefore the density) of the gas has to be taken into account.

From Gay Lussac's and Mariotte's Laws it appears that the volume of a gas varies directly with the absolute temperature (p. 86), and inversely with the pressure. Hence, if V = vol. at $t°$ and under a pressure of P mm., and V' = vol. at 0° and 760 mm.,—

$$273 : 273 + t :: V' : V, \text{ or } V = V' \frac{273+t}{273}.$$

Also $\qquad P : 760 :: V' : V, \text{ or } V = V' \frac{760}{P}.$

Combining the two proportions, we get

Vol. at given temp. and pressure = (vol. at 0° and 760 mm.) $\times \dfrac{760\,(273+t)}{273\,P}.$

Then, as in the last Problem,

Vol. at given temp. and pressure : Vol. at 0° and 760 mm. :: Wt. of unit vol. at 0° and 760 mm. : Wt. of unit vol. at given temp. and pressure.

EXAMPLE.—The weight of 1 c.c. of air at 0° and 760 mm. is 0·001293 grm. What is the weight of 200 c.c. of air at 27° and 500 mm.?

1 c.c. of air at 0° and 760 mm. becomes, at 27° and 500 mm.

$$\left(1 \times \frac{760 \times 300}{273 \times 500} = \right) 1\cdot67 \text{ c.c., nearly.}$$

Then 1·67 c.c. : 1 c.c. :: 0·001293 grm. : 0·00077 grm. ...

Hence 1 c.c. of the air weighs 0·00077 grm., and 200 c.c. will weigh 0·154 grm.

The following formula will be useful in solving similar problems :—

Let D = density of a gas relatively to air.
V = vol. of the gas in c.c. at $t°$ and P mm.
W = weight of the volume V.

Then $\qquad W = 0\cdot001293\ DV\ \dfrac{760\,(273+t)}{273\,P}\ $ grms.

CHAPTER VI.

THE MEASUREMENT OF QUANTITY OF HEAT—CALORIMETRY.

SECTION I.—**The Relative Capacities of Substances for Heat.**

WE have hitherto been considering only the *intensity* of the heat which exists in substances, and its effects in changing their size and state. But this is only one part of the subject; we must also examine the *quantity* of heat (or, more strictly, of the energy which is convertible into heat) which substances contain, before we can be said to have a full, complete view of their condition as regards heat; just as an engineer in estimating the value of a steam-boiler has to inquire not only into the pressure of the steam but also into the quantity of the steam which it contains.

Now, it has not hitherto been found possible to take out all the heat from a body, and thus observe the absolute amount of heat which it contains at any particular temperature; all that we can do is to add to, or subtract from, the heat which is already contained in it. We must content ourselves, therefore, with endeavouring to find out to what extent different substances have the power of associating heat energy with their molecules,—to find out what may be called their 'capacities for heat'; just as an engineer might set himself to find out whether two boilers, similar in external size, had the same internal capacity for steam. One of them, for instance, might be filled with tubes which would lessen its internal capacity in a way not easily ascertained by mere inspection.

The first thing to be observed is that in these investigations the ordinary thermometer cannot, except in a few cases, give us any direct information as to the quantity of heat in a body; any more than the height of the column of mercury in a barometer shows the number of litres of air in a room, or a steam-gauge can tell us how much steam there is in a boiler. The thermometer, as has been mentioned already, simply indicates the intensity of heat in a substance, just as a barometer indicates the intensity of the pressure of the atmosphere. It is true that if we take equal masses [1] of the same substance at the same temperature, and raise one of them 1° in temperature and the other 2°, we may be justified in saying that twice the quantity of heat has entered the body in the latter case as in the former. It is true also that if we take 1 grm. of a substance and communicate heat to it until its temperature is raised from 0° to 100°, and also take 100 grms. of the same substance in the same state and heat it until its temperature is raised from 0° to 1°, we shall not be far wrong in assuming that the *same quantity* of heat has entered these different weights of the substance in the two cases. In fact, it may be stated as a general law sufficiently correct for ordinary purposes (when the range of temperature is not very extended) that, when the same substance is being dealt with, *the rise in temperature due to a given quantity of heat being imparted to it is inversely proportional to the weight taken.* For example, to heat 5 kilogs. of water 4° requires as much heat as would raise 10 kilogs. of water 2°. Twice the mass is only raised one-half as much in temperature, three times the mass only one-third as much, and so on, by a given quantity of heat. Thus we arrive at the convenient conclusion that in experiments relating to quantities of heat we need not necessarily take equal weights of the substances under examination, or raise them to equal temperatures. If we are careful to multiply the weight of each substance by the number of degrees of its change in temperature, the products will denote definite, comparable quantities of heat.

---

[1] We must take equal masses and not equal volumes of the substance, because masses are related to quantity of matter, and heat is a motion of matter.

Suppose, for example, that one gas-burner was found to heat 500 grms. of water through 30° in the same time that another burner heated 800 grms. through 20°, we might compare the heating powers of the burners in the following way:—

$$500 \text{ grms.} \times 30 = 15{,}000.$$
$$800 \quad ,, \quad \times 20 = 16{,}000.$$

The products 15,000 and 16,000 represent the relative quantities of heat afforded by the two burners.

Methods of Determining the Relative Capacities for Heat of Different Substances.

Suppose that it was necessary to find out the relative capacities of several boilers under circumstances where direct measurement was impossible. This might be done in several such ways as the following:—

1. The boilers might be fitted with pressure gauges and steam got up in them until the gauges showed the same high pressure in each. Then some of this steam might be allowed to escape into empty reservoirs, of the same size, also fitted with gauges, until the pressure became equal in both boiler and reservoir. Obviously that boiler the steam from which raised the pressure in the reservoir connected with it to the greatest height must have the largest capacity.

2. The boilers might be filled with steam up to the same high pressure as before, and some of the steam from each might, through the medium of an engine, be made to do some definite mechanical work such as grinding corn, until the pressure fell to the same extent in each boiler. The boiler which gave the highest results in such work would certainly be the one of largest capacity.

The above illustrations show the principles of the two chief methods of finding out the relative capacities for heat (or 'specific heats,' as they are called) of different kinds of matter. The pressure gauges represent thermometers, and the steam, or other work-producing fluid, represents the heat energy which can be communicated to the molecules of bodies.

RELATIVE CAPACITY FOR HEAT.

1. *First Method of determining Capacity for Heat.*—We may take known masses of different substances at the same high temperature, and make them give up some of their heat to known masses of cold water, and then observe how much each of them in cooling raises the water in temperature. For example, we may compare the relative capacities for heat of water and iron in the following way:—

Expt. 26. Take a small flat block of iron about 4 or 5 cm. in diameter and 6 or 8 mm. in thickness (an iron pulley-sheave, which can easily be procured from an ironmonger, answers well), and tie a piece of thin string to it, to lift it up by. Counterpoise a beaker, wide enough to allow the iron to lie in it, and holding about 150 c.c., in one scale of a balance: place the piece of iron in the scale containing the counterpoise, and pour into the beaker enough water to balance the iron. We have now equal masses of water and iron. Put the iron into the water in the beaker, and heat the whole on a sand-bath to about 70° or 75°. While this is being done, take two beakers wide enough to allow the bulbs of the differential thermometer to go easily into them, put into each an equal weight of cold water, *e.g.* 150 grms. (or 150 c.c.), and arrange them on blocks so that the bulbs of the thermometer may dip simultaneously into them and remain wholly immersed in the water. Close the stopcock in the cross tube, and, having noted that the water in the two beakers is at the same temperature, take the thermometer bulbs out of the beakers and set the instrument aside. When the water and iron have attained a temperature of about 75°, take out the iron, letting it drain for a moment only, and quickly pour the hot water into one of the beakers containing cold water, and put the iron into the other beaker, moving it up and down once or twice and then letting it rest at the bottom of the beaker. Now place the bulbs of the differential thermometer again in the beakers, and observe that the column of liquid in it is greatly depressed on the side where the beaker is placed to which the hot water was added; proving that the water in this beaker had received more heat than that in which the iron was placed.

We thus learn that a definite weight of hot water gives out much more heat in sinking through a certain number of degrees than an equal weight of equally hot iron; that water has, in fact, a much greater capacity for heat than the same mass of iron.

Expt. 27. Dry the counterpoised beaker used in the last experiment, replace it on the scale of the balance, and re-adjust the counterpoise if necessary. Dry also the piece of iron and place it in the other scale: then pour into the beaker enough mercury to balance the iron, put the iron into the beaker with the mercury, and heat the whole to 100°, or nearly so, in a vessel of boiling water, such as a saucepan.[1] A loose plug of cotton wool or tow should be put into the beaker to prevent loss of heat, and about ten minutes should be allowed for the iron and mercury to attain the same temperature. Meanwhile arrange the differential thermometer and beakers containing equal quantities of cold water, as in the last experiment. Place the heated iron in one of the beakers and pour the equally hot mercury into the other as quickly as possible: then immerse the bulbs of the thermometer. Observe that the column of liquid is depressed on the side of the beaker into which the iron was put. This proves, of course, that the iron has given up to the water more heat than the mercury; *i.e.* that iron has a greater capacity for heat than the same mass of mercury.

A circumstance which demonstrates the relative capacities for heat of water and mercury will have been incidentally noticed already in making the experiment to illustrate the expansion of liquids (p. 66). The column of mercury began to rise in the tube and attained its highest point much sooner than that of the water, though heat was being communicated equally to both; showing that water required much more heat in order to raise it to the temperature of the hot water in the cistern, even though its weight was far less than that of the mercury. Hence the capacity of water for heat must be greater than that of mercury; a fact further proved by the results of the two experiments which have just been made.

2. *Second Method of determining Capacity for Heat.*—We may take definite masses of different substances, heat them to the

---

[1] A glue-pot forms a very convenient water bath for this and similar purposes.

same high temperature, place them in contact with some easily fusible substance such as wax or ice, and observe how much of it is melted by each of the substances. That which has the greatest capacity for heat will melt most wax.

Expt. 28. Procure or make three metal balls, one of lead, one of tin, and one of iron, of the same size, about 2·5 cm. in diameter,[1] with small brass rings screwed into them, to which pieces of thin string should be attached to suspend them by. Place the balls in a vessel of water kept gently boiling for eight or ten minutes at least. Support a plate of soft wax,[2] about 17 or 18 cm. long, 10 cm. broad, and 7 mm. thick, on wooden blocks or in a tin frame as shown in fig. 52, laying under it a folded piece of blotting-paper to catch the melted wax. Take the lead ball out of the boiling water, and quickly place it on the slab of wax, of which it will soon melt enough to embed itself. Place the tin ball in like manner upon another part of the wax, and then the iron ball in the same way. Observe the different amounts of wax melted by each metal. The lead ball will hardly melt enough to embed itself up to the centre in the slab. The tin ball will bury itself much more deeply, though it will probably hardly melt enough wax to pass quite through. The iron ball will quickly melt its way right through the slab and drop upon the folded

Fig. 52.

---

[1] The lead and tin balls may be easily cast in a large bullet-mould (these can be obtained up to 2·5 cm. or one inch in diameter). The iron ball will probably have to be bought, but a ball of zinc, which will answer nearly as well, can be cast in the same mould as the others. Strictly speaking, of course, the balls ought to be of the same weight, but the densities of the metals do not differ so widely as to make equal-sized balls unsuitable for the above illustrative experiment.

[2] This should be made of a mixture of wax and lard : directions for making it are given in the Appendix.

paper underneath,[1] or at any rate as far as the string allows it. This experiment clearly shows that iron has a greater capacity for heat than tin, and tin than lead.

Unit Quantity of Heat: Calorie.—An extensive series of experiments have shown the remarkable fact that water has a capacity for heat far exceeding that of the same mass of any other ordinary substance, and it has on this account been selected as the standard to which the capacities of other bodies for heat are referred. Moreover, it has been found convenient to take the quantity of heat energy which is required to raise the temperature of 1 grm. of water from 0° to 1° as the unit or definite quantity, in terms of which other quantities of heat are expressed: just as a gramme, a grain, or a pound is a unit of mass, and a litre or a gallon is a unit of volume. To this unit-quantity the name '**calorie**' has been given, from the Latin word 'calor' meaning 'heat.'

The '**calorie**' then, or 'Heat-unit' may be defined as—

'THAT AMOUNT OF HEAT ENERGY WHICH, IF APPLIED TO ONE GRAMME OF WATER AT 0°, WILL RAISE IT 1° C. IN TEMPERATURE.'

This is the strict definition of the term: but since the heat capacity of water does not vary much between its freezing-point and boiling-point, we may without serious error consider a calorie to be the amount of heat required to raise 1 grm. of water through 1° C., whatever its temperature at starting may be.

Thus it becomes easy to express concisely in calories the amount of heat produced or transferred in any action between substances. We have merely to communicate this heat to a known number of grammes of water, and to observe how much the water is raised in temperature. The weight of the water in grammes, multiplied by its rise of temperature in degrees, will give the number of calories of heat imparted to the water, and

---

[1] In the figure a fourth ball, made of copper, is shown: this will be found to melt nearly as much wax as the iron.

Specific Heat.

this expresses the quantity of heat which is concerned in the action.

For example, in a trial of one of the Bunsen burners described on p. 32, having a tube 1 cm. in internal diameter, and burning 200 litres (7 cubic feet) of gas per hour, 1000 grms. of water in a copper kettle were raised from 20° to 70° (*i.e.* through 50°) in five minutes. Now, to raise 1 grm. of water through 50° would require 50 calories; and the 1000 grms. of water will of course require 1000 times as much heat, or 50,000 calories. Further, it was known that 900 calories were expended in heating the copper vessel itself 50° in temperature. Hence the burner was giving out 50,000+900, or 50,900 calories in five minutes, *i.e.* ($\frac{1}{5}$ of 50,900=) 10,180 calories per minute.

Section II.—Specific Heat.

As mentioned above, the capacities for heat of almost all substances are less than that of water, so that (with hardly any exceptions) less than a calorie of heat is required to raise 1 grm. of a substance 1° in temperature. The number called the '**Specific Heat**' of a substance is the fraction of a calorie which is required to heat 1 grm. of it 1° C. in temperature.

For example, the specific heat of iron is said to be 0·1 (nearly). This means that only one-tenth of a calorie is sufficient to raise 1 grm. of iron 1° in temperature. Of course the ratio 1 : 0·1 expresses generally the proportion between the amounts of heat required to raise any equal weights of water and of iron through any equal number of degrees, except that the specific heats of substances vary slightly at different temperatures.

[A table of specific heats is given on p. 135.]

Exact Methods of Determining Specific Heats.

These are based generally on the two principles explained and illustrated already.

First Method.   Method of Mixture.—The process which was

devised by Regnault,[1] and which is most usually employed is called the 'Method of Mixture.' It consists in heating a known weight of the substance to a definite temperature, generally about 100°, placing it in a known weight of cold water at a definite temperature, and observing accurately the temperature of the mixture. From these data the specific heat of the substance can be calculated in a manner which may be rendered clear by the following practical examples.

Expt. 29. Coil a strip of sheet copper about 2 cm. broad, and 30 cm. long, into a flat spiral, about 5 cm. in diameter, tie a piece of strong thread to its centre (to take it up by), weigh it, and place it in a vessel of boiling water for ten minutes at least, keeping the water gently boiling all the time. Meanwhile select a beaker holding about 200 c.c., and wide enough to let the spiral rest on its bottom; weigh it, and wrap it in several folds of thick flannel or wadding, covering the bottom as well as the sides, to prevent any loss of heat (loose wool being an extremely bad conductor of heat). Place the beaker thus enveloped in a larger beaker (or a tin canister) so that the whole of the interval between the two is filled up by the wool, and put into it 100 grms. (or 100 c.c.) of cold water. Take the temperature of the water by a good thermometer, reading it to half a degree, or even a tenth if possible, and stirring the water with the thermometer just before taking the final reading, which should be recorded at once in a note-book. When the copper has gained the temperature of the boiling water (which may be assumed to be 100°, unless the barometer column is very much below or above 76 cm., *i.e.* 30 inches),[2] take it out by the thread and quickly transfer it to the cold water in the beaker, moving it up and down in the water so that it may rapidly part with its heat: then without delay put in the thermometer and read the temperature of the mixture, stirring with the thermometer just before reading it, and noting the high-

---

[1] *Annales de Chimie et de Physique*, lxxiii. 5.
[2] If the barometer column stands as low as 73.3 cm. (29 inches nearly), the temperature of the boiling water will be 99°. If, on the contrary the column is 78.7 cm. (31 inches nearly) the temperature of the water will be 101° (see Chapter VII. Section III.). From these data an approximate correction may be made, if necessary.

est point attained by the mercury column (it will soon begin to fall owing to unavoidable loss of heat).

We have now the data necessary for the calculation of the specific heat of copper. Thus, neglecting for the moment the small amount of heat absorbed by the glass beaker (or 'calorimeter,' as it is termed), in an experiment made as above, the following facts were observed:—

Weight of water in beaker	200 grms.
Temperature of water	10°.
Weight of copper	50 grms.
Temperature of copper	100°.
Temperature of mixture	12°.

We have here a certain amount of heat transferred from the copper to the water. This quantity has raised the temperature of 200 grms. of water from 10° to 12°, *i.e.* through 2°, and must therefore (see p. 121) be (200 × 2 =) 400 calories. Now, the temperature of the copper has fallen from 100° to 12°, *i.e.* through 88°; and hence, as no heat is lost in the transfer, 50 grms. of copper, in falling through 88°, have given out 400 calories of heat. One gramme of copper would, of course, have given out $\frac{1}{50}$ of this, *i.e.* ($\frac{1}{50}$ of $\frac{400}{1}$ =) 8 calories. Supposing that a change of temperature in the reverse direction had taken place, the copper gaining instead of losing heat, it is clear that 8 calories would raise the temperature of 1 grm. of copper 88°. Knowing this fact, we can find out how much heat is required to raise 1 grm. of copper 1° in temperature, *i.e.* the specific heat of copper, by the proportion sum,—

$$\text{Calories.} \quad \text{Calorie.}$$
$$88 \; : \; 1 \; :: \; 8 \; : \; 0\cdot09 \text{ (nearly).}$$

Hence the specific heat of copper is 0·09.

Strictly speaking, the heat given out by the copper has raised not only the water but also the glass 'calorimeter' 2° in temperature, so that rather more than 400 calories must have been given out by the hot metal, and hence its specific heat must be really rather higher than 0·09. It is easy to make the necessary correction for this if we know from other sources the specific heat of the material of the calorimeter. We have only to multiply the weight of the vessel in grammes by its specific heat to obtain the number of calories of heat required to raise its temperature 1°; and the number thus obtained, which is called the 'water value' of the calorimeter, is to be added to the weight of the water which is placed in the vessel. Thus in the above experiment the weight of the glass beaker was 30 grms., and the specific heat of glass is 0·2 (nearly): therefore the water value of the calorimeter is 6 calories, *i.e.* as much heat is spent in raising it 1° as is required to raise 6 grms. of water 1°. We may consider, then, the whole mass of water to which the copper gave up its heat as (200 + 6 =) 206 grms.; and this quantity multiplied by the rise in temperature, *viz.* 2°, will give (206 × 2 =) 412 calories as the true quantity of heat yielded up by the copper. The calculation should be worked out with this corrected value.

It will be well to make a few other experiments of a similar kind with about the same weight of mercury, glass, lead, sulphur, turpentine,[1] etc., and to compare the results with those given in the table on p. 135. The numbers obtained will almost certainly be lower than those given in the table, chiefly owing to unavoidable loss of heat in transferring the heated substance from the boiling water to the calorimeter: in fact, the temperature of the substance when it reaches the cold water may be pretty safely assumed to be 98° instead of 100°. Regnault avoided this source of error by placing the calorimeter immediately under the steam bath in which the substance was heated, so that the latter dropped directly into the cold water. A description of his apparatus will be found in Ganot's *Physics*, and should be consulted if possible.

The calculation of specific heats from the data obtained by the method of mixtures as above described may be put in a shorter form from the following considerations.

Let M = weight of substance.
$T°$ = temperature to which it is raised.

---

[1] The temperature of this should not be raised above 60°.

SPECIFIC HEAT.

w = weight of cold water (including 'water-value' of the calorimeter).
$h°$ = temperature of water in calorimeter.
$\theta°$ = temperature of mixture.
x = specific heat of substance.
Then $T°-\theta°$ = fall of temperature of substance.
$\theta°-h°$ = rise of temperature of water in calorimeter.

And since no heat is lost in being transferred from the substance to the water (which follows from the Law of the Conservation of Energy),

$$x M (T-\theta) = w (\theta-h)$$
$$\text{Hence, } x = \frac{w (\theta-h)}{M (T-\theta)}$$

Or, putting it in plain words,

Specific heat of substance = $\dfrac{\text{Wt. of water} \times \text{its gain in temperature}}{\text{Wt. of substance} \times \text{its loss in temperature}}$

and this rule, when the principle of the method has been clearly understood, may be advantageously used in calculating specific heats.

Second Method of determining Specific Heats. Lavoisier and Laplace's Calorimeter.—This method consists in observing the weight of ice melted by a known weight of a substance, heated to a known high temperature. The form of apparatus used by Lavoisier and Laplace (in 1780)[1] is shown in fig. 53, and consists essentially of three metallic vessels, nearly similar in shape and of such sizes as to fit, one within the other, with an interval of about 5 cm. between them. The smallest and innermost, A, about 18 cm. in diameter, is made of iron wire gauze, and serves to contain the substance which is being experimented on. The next outer one, B, is a cylinder of thin tinned iron, ending below in a cone or funnel with exit-tube and stopcock, D. This vessel is filled with

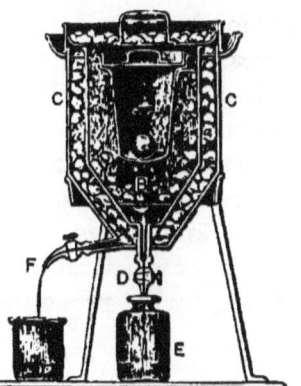

Fig. 53.

---

[1] *Œuvres de Lavoisier*, vol. ii. p. 283.

small fragments of ice, which are kept by the wire gauze of the inner vessel from actual contact with the substance, but which readily take up heat from it and melt: the water thus formed passes through the tube D, into a jar, E, placed below. The outermost vessel, C, is of the same shape as B, and is filled with fragments of ice merely for the purpose of preventing the heat of the external air from reaching the ice contained in the inner vessel. Any water formed by the melting of the ice in C passes off through the side-pipe, F, into a separate jar. A cover, on which ice is placed, is fitted on this vessel, and thus the inner vessel and its contents are kept constantly at the temperature of melting ice, *i.e.* 0°, and can only receive heat from the substance placed in A.

In using this calorimeter, a piece of any substance [1] of which the specific heat is to be determined is weighed and heated to a temperature of about 80° in hot water. Its exact temperature having been noted, it is transferred quickly to the vessel A, where it soon parts with its heat to the ice in B, sinking itself to the temperature of 0°. Some of the ice is thus melted, and the water formed runs down into the jar E, the weight of which has been previously taken. This jar is afterwards again weighed, and the increase in weight shows the amount of ice which has been melted by the heated substance. Comparing this with the amount of ice melted by an equal weight of equally hot water, we learn the relative specific heats of the substance and of water. For example, a piece of iron the weight of which was equivalent to 3·5 kilogs. was heated to 79° and placed in the calorimeter. The water which ran into the jar E was found to weigh 386 grms. Hence, by the proportion,—

Kilog.	Kilog.	Grms.	Grms.
3·5	: 1	:: 386	: 110,

we learn that 1 kilog. of iron in sinking from 79° to 0° would melt 110 grms. of ice. Now, it is found (see Chapter VII. Section 1.)

---

[1] Liquids were put into glass flasks.

that 1 kilog. of water at 79° will, in sinking to 0°, melt 1000 grms. of ice. Hence,—

Weight of ice melted by water	Weight of ice melted by the same weight of iron	Sp. heat of water	Sp. heat of iron
1000 :	110	:: 1	: 0·11.

This method, though correct in principle, does not in practice give such accurate results as Regnault's method, mainly because some water is liable to remain entangled in the fragments of ice and never reaches the vessel E at all. Lavoisier and Laplace, by employing large quantities (3 or 4 kilogrammes) of the substances experimented on, greatly lessened this source of error; but a much better ice-calorimeter has been lately invented by Professor Bunsen, in which the quantity of ice melted is estimated by observing how much the volume of the mixture of ice and water differs from that of the original ice. For, as will be more fully explained in the next chapter, ice in turning into water contracts to the extent of nearly one-tenth of its volume.

Specific Heat of Gases.—The above forms of calorimeter are only adapted for experiments with liquids and solids. The determination of the specific heats of gases is a much more difficult matter, owing to the small mass which even a large volume of a gas possesses. A litre of air, for example, does not weigh much more than a gramme and a quarter, and in dealing with such small weights the unavoidable errors may amount to a large percentage in the final results.

The most reliable experiments on the specific heat of gases have been made by Regnault,[1] whose method consisted in passing a measured volume of a gas through two long spiral tubes, one immersed in hot water, the other in a closely adjacent vessel of cold water. The gas was left free to expand and contract during its passage (in other words, the pressure on the gas was kept constant throughout), and the volume of it, which was in some experiments 100 litres or more, was accurately measured: from this the weight of the gas which was actually dealt with can be easily calculated. The gas was first passed through the heated

---

[1] *Mémoires de l'Académie Royale des Sciences*, xxvi. 1.

spiral, and thus raised to a certain known temperature; then it passed at once into the spiral immersed in the cold water, where it parted with some of its heat to the latter. The rise in temperature of the water was observed, and thus Regnault determined the number of calories given out by the known mass of the gas in sinking through a known number of degrees of temperature: from which the specific heat of the gas was deduced in the usual way.

But it must be noted that experiments made under these conditions do not give the true amount of heat expended solely in raising the temperature of a gas. Besides heating the gas which was passing through the spiral tube, the imparted heat is causing it to expand, *i.e.* to overcome the resistance due to the pressure upon it; and to do this some heat-energy must be expended. The specific heat, then, of a gas under constant pressure shows the total amount of heat required to produce *both* these effects, and the true amount which is expended in simply raising the temperature of the gas must be less than this. It might seem easy to determine the latter by confining the gas in a strong closed vessel so as to prevent it from expanding at all (by keeping, in fact, its volume constant), and then determining its specific heat in a calorimeter of the usual kind. But the comparatively small mass of any manageable volume of gases has prevented this method from being carried out successfully in practice; and the specific heat of a gas at constant volume has been usually deduced from considerations of the velocity with which sound passes through gases at different temperatures, in a manner which can hardly be entered into here.[1] The important fact has

---

[1] Perhaps the following outline may give an idea of the principle of the mode of calculation. Sound is transmitted through gases by a wave-motion consisting of impulses of alternate expansion and compression like those which pass along a spiral spring, when struck or pulled at one end in the direction of its length. If this extension and compression of the gas could go on without changing its temperature, the velocity of transmission of sound would not be altered during its passage, and would be that which was calculated by Newton for a gas which maintains a constant volume. But heat is really developed along the line of waves, and this increases the elastic force

thus been made out, that when a gas is allowed to expand under constant pressure, about 1·4 times as much heat is required to raise it 1° in temperature as when its volume is kept constant and the heat is solely employed in raising its temperature.

TABLE OF SPECIFIC HEATS.

[The figures denote the number of calories required to raise 1 gramme of the substance from 0° to 1°.]

SOLIDS.		LIQUIDS.		GASES.*		
					Equal weights, *i.e.* 1 grm.	Equal volumes, *i.e.* 1 litre.
Ice	0·504	Water	1·000	Hydrogen	3·409	0·305
Sulphur	0·202	Alcohol	0·615	Steam	0·480	0·385
Glass	0·200	Turpentine	0·426	Air	0·237	0·306
Iron	0·114	Mercury	0·033	Oxygen	0·217	0·310
Copper	0·095			Carbon dioxide	0·217	0·319
Tin	0·056					
Lead	0·031			Chlorine	0·121	0·383

* Under constant pressure.

SECTION III.—Explanation of the Differences in the Specific Heats of Solid and Liquid Substances.

On the old theory, according to which heat was an actual fluid, the fact that different bodies take in very different amounts of heat in rising 1° in temperature, was explained by supposing that they afforded very different amounts of space between their

---

of the gas, and consequently the velocity of the sound-wave. Hence it may be seen that the difference between the observed velocity of sound in a gas, such as air, and its theoretical velocity will afford a measure of the amount of heat-energy employed in raising the temperature of the gas when its volume remains unaltered. In fact, since the energy imparted to a given mass of a substance is indicated by the square of the velocity it produces, we have the ratio,—

$$\left(\frac{\text{Observed velocity}}{\text{of sound in air}}\right)^2 : \left(\frac{\text{Theoretical velocity}}{\text{of sound in do.}}\right)^2 :: \frac{\text{Specific heat of air at}}{\text{constant pressure}} : \frac{\text{Sp. heat of do. at}}{\text{constant volume}}$$

$$333^2 \text{ metres} : 279^2 \text{ metres} :: 0.2375 : 0.1684$$

molecules for the fluid to occupy; just as different sponges absorb very different amounts of water.

On the dynamical theory an explanation quite as conclusive, though not quite so obvious, is given. Perhaps it will be best understood by considering the way in which heat is employed when it enters a substance. It imparts increased velocity to the molecules, and thus increases their energy or power of doing work. In this way four effects are produced: (1) their inertia is overcome; (2) their internal cohesion is partly or entirely counteracted, leading to a change of state from solid to liquid, and liquid to gas; (3) they gain increased power of overcoming external resistance, and thus tend to enlarge their orbits of motion, so that the whole mass increases in volume or tends to do so; (4) they acquire an increased disposition to part with their energy to substances possessing less of it: in other words the temperature of the whole mass is raised. It is only this last effect which is directly observed in determining specific heats; and since the strength of cohesion varies very greatly in different substances, and in different states of the same substance, we should quite expect to find that varying amounts of heat-energy are spent in overcoming it.

Part, then, at any rate, of the difference in the specific heats of bodies is explained as due to the different amounts of internal work which the heat has to do. But a further reason has to be sought, to account for the wide difference between the specific heats of equal weights of such very analogous substances as tin and lead: and it is readily found when we consider the very different number of molecules which equal weights of these elements contain. It is pretty certain that a gramme of tin contains nearly twice as many molecules as a gramme of lead does; and since heat-energy, when imparted, is employed in setting in motion each single molecule,[1] it is easy to see that

---

[1] It will not matter, of course, whether the molecule has great mass or little mass: a given amount of energy will impart the *same* power of overcoming resistance to a light molecule as to a heavy one, the light one making up for its small mass by a greater velocity; just as a given charge of powder

Specific Heat.

nearly twice as much of it would be wanted in the case of tin as for an equal weight of lead, to raise it 1° in temperature. As another example we may take iron. There are good grounds for believing that 1 grm. of iron contains nearly four times as many molecules as the same weight of lead, and it is obvious that this fourfold number of molecules will require four times as much heat to make the mass 1° hotter as is required by the much smaller number of lead-molecules in the gramme of lead: so that the specific heat of iron will be nearly four times that of lead, as experiments prove it to be.

We obtain an estimate of the relative number of molecules in the same mass, *e.g.* 1 grm., of different elements by finding out the comparative weights of their single molecules. Thus the molecule of iron is, in all probability, about $\frac{1}{4}$ (strictly speaking $\frac{56}{207}$) the weight of the molecule of lead,[1] and hence nearly four times as many of these lighter molecules will be wanted to make up a gramme of iron, as are wanted to make up the same mass of lead.

An interesting experiment may be made as follows, to illustrate what has been said above.

Expt. 30. Weigh out 56 grms. of iron and 207 grms. of lead (in the form of coiled-up strips, as usual). These weights are in the same proportion as those of the molecules of iron and lead respectively, and so we may assume, without serious error, that these masses of iron and lead contain the *same* number of molecules of the elements. Heat them to the same temperature in boiling water, and arrange the differential thermometer with bulbs dipping into beakers containing 150 grms. of cold water, as in previous experiments. Plunge the equally hot metals one into each beaker, and after stirring the mixture, immerse the bulbs of the thermometer. If the experiment is carefully done, no differ-

---

will (if other conditions are equal) make a bullet penetrate into a bank as deeply as a much heavier cannon-ball.

[1] It is, at any rate, certain (from chemical reasons) that the weight of the iron-atom is to that of the lead-atom as 56 : 207, and the same proportion must exist between the weights of the molecules of the two elements, if they are similarly constituted, as they probably are.

ence in the levels of the columns of liquid in the thermometer will be perceptible. We infer, of course, that the iron and lead have given up *equal* amounts of heat to the water; and since there are the same number of molecules in each mass, we may conclude that a single molecule of each element would do the same.

The above considerations will also point to the reason why the specific heats of equal volumes of all elements in the state of gas are the same. There are many grounds for believing that all gases have the same physical constitution, whatever the chemical nature of the substance may be. It has been seen already (p. 85) that all gases change in volume to the same extent for a given change in temperature (Gay Lussac's Law); and it is also found that a given change in the pressure on a gas causes precisely the same change in volume, whatever the gas is (Mariotte's Law, p. 83). Many chemical reasons, also, lead us to a belief in the important statement known as Avogadro's Law, *viz.* that 'EQUAL VOLUMES OF DIFFERENT GASES CONTAIN, IN ALL CASES, THE SAME NUMBER OF MOLECULES.'

Now, if this is true, it is no more than we should expect, to find that equal volumes of all elements in the state of gas have the same specific heat. There are the same number of molecules in each of these equal volumes, and since in gases cohesion is entirely overcome, there is no internal work for which heat would have to be spent. Therefore a given amount of heat must be able to impart to these equal numbers of molecules an equal power of doing external work and an equal rise of temperature.

The dynamical theory of heat, then, accounts satisfactorily for all the facts of specific heat, as ascertained by experiment; the great variations in the specific heats of equal weights of different substances being mainly due to our not taking Nature's unit, the molecule, in stating the results of experiments, but an entirely different and arbitrary one, the unit of mass. So far, at any rate, as relates to the elements, there is every reason for concluding that the molecules of all elements, when existing under the same conditions, have the same specific heat.

SPECIFIC HEAT.

Dulong and Petit's Law.—Since, as has been mentioned, the number of molecules in a given mass of a substance depends on the weight of the molecule, more of them being present in proportion as the weight of each molecule is less; and since the specific heats of equal masses of solid and liquid elements vary with the number of molecules present in the mass, it follows that the less the molecule weighs, the greater will be the specific heat of the substance. What is true of the molecule is also true of the chemical atom, if the weight of the latter bears a fixed relation to the weight of the molecule. Hence we naturally deduce the general statement known as Dulong and Petit's Law, *viz.*—

'THE SPECIFIC HEATS OF EQUAL MASSES OF SOLID AND LIQUID ELEMENTS VARY INVERSELY WITH THE WEIGHTS OF THE ATOMS OF THE ELEMENTS.'

From this it follows, of course, that the product obtained by multiplying the specific heat by the atomic weight of an element is a constant quantity, which, adopting the usual units, *viz.* the calorie and the hydrogen atom, is nearly 6·4. This will be plain from the following examples:—

Name of Element.	Specific Heat.	Weight of Atom.	Product.[1]
	(Water=1)	(Hydrogen-atom=1)	(Sp. heat × wt. of atom).
Iron	0·114	56	6·38
Sulphur	0·202	32	6·48
Lead	0·031	207	6·41
Mercury	0·033	200	6·6

The use of Dulong and Petit's Law, connecting the specific heat of an element with the weight of its atom, is obvious. Given a correct value of the specific heat, the weight of the atom must be some number which, multiplied by the former, gives a product approximating to 6·4. This method is often resorted to by

---

[1] The slight variation in these figures may be sufficiently explained as due to (1) unavoidable errors in determining the specific heats; (2) differences in the amount of internal work done by the heat (p. 136).

chemists in order to decide between two different values for the weight of an atom.

It must be noted that Dulong and Petit's Law only applies to elements. In the case of compounds the conditions are complicated by varying chemical affinities and strains among the constituent atoms, and no such simple connection between the weight of the molecule and its capacity for heat is observable.[1]

SECTION IV.—Results and Applications of Specific Heat.

1. The very high specific heat of water is of the greatest importance in Nature. About three-fourths of the earth's surface is covered with water, and this is constantly either receiving heat from the sun or giving it off into space. If water changed quickly in temperature with the heat it receives or loses, the climate of islands and places on the sea-coast, and indeed of the earth generally, would be constantly fluctuating between almost unbearable heat and equally unpleasant cold. A single frosty night would reduce all the water in lakes, ponds, and streams nearly to the freezing-point, and a single summer day might raise them to a temperature prejudicial or fatal to the inhabitants of the water. But, as a matter of fact, water acts as a remarkably efficient equaliser of climate. It can take in a great deal of heat without much rising in temperature, and it can part with a great deal of heat without becoming much colder. This uniformity of temperature is, of course, shared by the air above it; and hence all places in the vicinity of water enjoy a very equable climate as compared with regions in the interior of a continent. Taking Great Britain as an instance, we find, on comparing its climate with that of districts in Russia having the same latitude, that the range of temperature at Edinburgh is rarely so great as 45° (*minimum*, −15°, *maximum*, 30°), while at Moscow the range is commonly 70° (*minimum*, −38°, *maximum*, 32°).

---

[1] It is found, however, that closely related classes of compounds, such as the carbonates, the chlorides, the sulphates, etc., the chemical structure of which is presumably similar, show the same specific heat, or nearly so.

So, again, water acts as an equaliser of the temperature of the human body. About two-thirds of the weight of an animal consists of water in one form or another, and the uniformity in temperature of this is shared by the tissues which enclose it and by the whole organism. The cooling effect of water as a drink depends upon the same property; it takes in a large amount of heat from the overheated body without becoming much hotter itself. Its use in foot-warmers and in the apparatus for heating buildings is explained in the same way; apart from its cheapness and abundance, nothing else would store up so much heat in so small a mass. It must be observed, however, that this high specific heat is a disadvantage attending the use of water in engine boilers; a large amount of fuel being consumed in raising it to the boiling-point.

2. The low specific heat of mercury supplies an additional reason for using it in thermometers. It quickly assumes the temperature of any substance placed in contact with it, and comparatively little heat is spent in obtaining an equilibrium of temperature between the two.

3. Lastly, the high specific heat of iron increases the facility with which it is worked. Many things must be wrought into shape while the iron is at a red heat, and so much heat is stored up in the metal that it retains the high temperature necessary for dealing with it much longer than platinum, for instance, would do.

APPENDIX TO CHAPTER VI.

Problems Relating to Specific Heat.

PROBLEM 1.—Given the specific heat of a substance, its weight, and the rise of temperature it causes when placed in cold water, to find the original temperature of the substance.

This illustrates the use of the 'Method of Mixture' in determining high temperatures, *e.g.* the temperature of a furnace.

It is clear that if the specific heat of a substance is known we can readily find out the temperature to which it has been raised by observing how many calories of heat it gives out when immersed in cold water. For example, suppose that a piece of platinum weighing 100 grms. is heated in a furnace and plunged into 200 grms. of water at 12°, with the result of raising the temperature of the water to 24°, *i.e.* through 12°. Then we know that (200 × 12 =) 2400 calories of heat must have been given by the platinum in sinking from the temperature of the furnace to 24°. The specific heat of platinum is 0·032, *i.e.* 1 grm. of platinum requires 0·032 calorie to raise it 1°. Hence the 100 grms. of platinum will require (100 × ·032 =) 3·2 calories to raise the mass 1°, or, conversely, will give out 3·2 calories in sinking 1°. Then from the proportion,

$$3\cdot 2 \; : \; 2400 \; :: \; 1° \; : \; 750°,$$

we learn that in order to give out 2400 calories the heated platinum must have sunk through 750°, *i.e.* that the temperature to which it had been exposed must have been 750° above 24° (the temperature finally attained by the platinum). Hence the temperature of the furnace must have been (750° + 24° =) 774°.

Problems on Specific Heat. 143

PROBLEM 2.—**Given the specific heat of a substance, its weight, and its temperature, to find the rise of temperature produced when it is put into a known weight of colder water.**

This is sometimes required to be solved, more in order to answer questions in an examination paper than for any scientific or practical purpose. It may be done on the following principles. Consider that the heated substance cools down to the original temperature of the water, and calculate the number of calories which it would in that case have given out. Then regard this quantity of heat as imparted to the mixture of water and the substance, and find out the temperature to which this mixture will be raised.

For example, suppose that 50 grms. of sulphur (specific heat, 0·2) at 80° were put into 40 grms. of water at 14°. The number of calories which would be given out by the sulphur in sinking from 80° to 14°, *i.e.* through 66°, is (0·2 × 50 × 66° =) 660. Now, the heat required to raise 40 grms. of water 1° is, of course, 40 calories; and that required to raise 50 grms. of sulphur 1° is (0·2 × 50 =) 10 calories. Therefore the heat required to raise the mixture of water and sulphur 1° is (40 + 10 =) 50 calories. Then, if the whole quantity of heat given out by the sulphur was 660 calories, we learn from the proportion,

$$50 : 660 :: 1° : 13·2°,$$

that the mixture would be raised by the 660 calories 13·2°, *i.e.* from 14° to 27·2°.

The above problems may also be easily solved from the equation given on p. 131, from which T or θ can be found when the other quantities are known.

CHAPTER VII.

PHENOMENA CONNECTED WITH CHANGE OF STATE.

HEAT, as we have seen (p. 40), owing to its power of overcoming cohesion, is able to change a solid substance to the liquid condition, and a liquid into a gas. These general facts have been illustrated already, and it will now be advisable to examine more closely the phenomena attending such changes of state, and consider

1. *The Precise Temperature at which they occur.*
2. *The Changes of Size which take place.*
3. *The Amount of Heat-energy which is spent in producing them.*

SECTION I.—Conversion of a Solid into a Liquid, and *vice versâ.* Fusion and Solidification.

1. *Temperature at which the Change occurs.*—Liquefaction in most cases takes place at a definite point of temperature, called the 'melting-point' of the substance, which is, under ordinary conditions, perfectly constant for the same substance. Thus the melting-point of ice is 0°, and is so constant that it is taken as a reliable point of reference in graduating thermometers.

The reverse change of condition, *viz.* the transformation of a liquid into a solid, takes place at exactly the same temperature, which is in such a case called the 'freezing-point' or 'solidifying-point.' It is a curious fact, however, that several substances, of which water and sulphur are examples, can be cooled down several degrees below their true solidifying-points and will yet remain liquid if kept quite undisturbed. But in these cases, when the liquid is slightly shaken or stirred, solidification at once sets in, and the temperature rises to the usual solidifying-point.

FUSION AND SOLIDIFICATION.

A table of the melting-points of a few important substances is given below:—

TABLE OF MELTING-POINTS.

Substance.	Melting-point.	Substance.	Melting-point.
Alcohol	$-130°$	Sulphur	$114°$
Mercury	$-38.5°$	Lead	$330°$
Ice	$0°$	Zinc	$420°$
Wax	$65°$	Copper	$1050°$
Iodine	$114°$	Cast-iron	$1200°$
		Platinum	$2000°$

Some solids when heated get soft and 'viscous' before their actual melting-point is reached; the cohesion between their molecules being gradually overcome. Sealing-wax, glass, and iron are excellent examples of this class.

Expt. 31. Heat the middle of a piece of glass tubing, about the size and length of a pencil, in the flame of a Bunsen burner (or, better, a blowpipe), turning it constantly round and bringing it *gradually* into the hottest part of the flame. Observe that, when it attains a red heat, the glass becomes as pliable as a piece of string, but yet does not fall down in drops as a liquid would. When it is thoroughly soft, take it out of the flame and pull the ends a metre or more apart. The glass will be drawn out into a fine thread which is still a perfect tube,[1] and not a rod, as it would be if the glass had really liquefied.

This experiment illustrates an important application of the viscosity produced in some substances by heat. Glass tubing is made exactly in the above manner, and glass cups, bottles, etc., are manufactured, not by casting in moulds, but by heating the material until it becomes plastic, and then fashioning it into the required shape. Similarly wrought iron, though its melting-point is extremely high, becomes soft when merely heated to redness, and can then be hammered or bent into shape almost as easily

---

[1] This may be easily proved by breaking off a piece and dipping the end of it into ink. The thin column of ink which rises in the capillary tube is readily seen when it is held against a piece of white paper.

as if it were lead. Moreover, two pieces of iron can be most firmly joined by heating the ends white-hot, and then hammering them together: the molecules will cohere as strongly as in any other part of the metal. This process is called 'welding,' and is applied to platinum and glass, as well as iron.

2. *Change of Size during Liquefaction.*—In nearly all cases a sudden and decided increase in volume takes place in the process of liquefaction. Thus 100 c.c. of solid wax become 104 c.c. of liquid wax at the same temperature. The alteration in volume is perhaps more easily observed when the reverse change of state takes place: 100 c.c. of liquid wax contract, on solidifying, into 96·1 c.c.

Expt. 32. Melt a quantity of beeswax in a saucer (best, by placing the vessel on a saucepan in which water is kept boiling), and leave it to cool and harden. Cracks will be noticed in the material, when solid; and often the whole will shrink away from the sides of the saucer as it contracts. Similarly if some wax is melted in a test-tube, the surface will, as it solidifies, be depressed at the centre, forming a cavity. This is still more strikingly shown when some melted lead is poured into a deep narrow iron crucible, or a large bullet-mould: the parts in contact with the iron solidify first, and as the rest cools, a hollow, often extending half through the block, is formed in the centre.

A practical consequence of this shrinkage during solidification is that sharp castings cannot be made in pure wax,[1] or in such metals as silver, copper, or lead, since the substance shrinks away from the sides of the mould and fails to fill it, when solid. Hence ornaments, etc., of silver are generally hammered into shape; coins are stamped out of flat plates, and not cast.

Ice is a remarkable exception to the above law. When it is liquefied it contracts suddenly and considerably, 100 c.c. of ice becoming only 91·8 c.c. of water.

---

[1] In order to obtain delicate casts of medals in wax, some infusible solid, such as white lead, must be added to the melted material: this keeps the molecules stretched, as it were, when they solidify.

Change of Size in Liquefaction. 147

Expt. 33. Fit a flask or bottle (holding about 250 or 300 c.c.) with a cork through which passes a piece of glass tube about 30 or 40 cm. in length. Fill the flask with lumps of ice, and add water (which may be coloured with ink to render it visible) until it overflows, then press the cork firmly into its place, so that a column of liquid rises in the tube. Place the flask in tepid water, and observe the continuous contraction which takes place; the water sinking quickly in the tube. The latter may be filled again with coloured water through a small funnel, but the contraction will go on as long as any ice remains unmelted.

The converse of this is, perhaps, more frequently noticed; *viz.* that water expands greatly in freezing; 100 c.c. of water at 0° form 110 c.c. of ice, the expansion being thus equal to one-tenth of its bulk.

Expt. 34. Place the flask filled with coloured water, used in the last experiment, in a mixture of ice with about half its weight of salt (this, as will be shortly explained, effects a reduction of temperature much below 0°). Add a little more coloured water, if necessary, so that the column stands in the tube several centimetres above the cork. In a short time freezing will set in, and the column of coloured water will rise rapidly in the tube, showing the great expansion which takes place as the ice forms.

This exceptional behaviour is probably due to the fact that ice is crystalline in structure: its molecules being arranged in clusters of a definite shape, *viz.* six-sided prisms. Thus they must necessarily take up more room than when they lie loosely side by side as in the uncrystallised liquid.[1]

When resistance is opposed to this increase in volume;—when, for instance, water is frozen in a tightly closed vessel which it completely fills, the force with which the expansion tends to take place is found to be enormous. It is, indeed, exactly equivalent to the force required to compress 11 c.c. of water into 10 c.c., and this is at least 2000 kilogrammes per square centimetre.

The following experiment may be made to illustrate the immense amount of energy involved in this change of state.

---

[1] Thus a given number of bricks would take up more room if arranged in patterns than if packed in layers.

Expt. 35. Obtain a small bottle of cast-iron about 9 or 10 cm. in length, 4 cm. in diameter, and with sides 1 cm. or more in thickness, fitted with a stopper which can be tightly screwed in.[1] Fill it completely with cold water, brushing out any air bubbles from the interior with a feather, and screw the stopper (previously greased) tightly into its place. Then put the bottle into a wooden or iron bowl, filled with a mixture of ice and salt (p. 147), cover it with a piece of board (on which it may be well to place a heavy weight) and leave it for a time. In a few minutes the bottle will burst more or less violently, even before the whole of the water in it has frozen. The same result will, of course, take place if the bottle is left out of doors on a frosty night; but it should be put in some place where the flying fragments can do no damage.

The reason will now be plain why jugs of water are frequently broken when the water in them is frozen, and why pipes conveying water often burst during a frost. The damage is usually not noticed until the thaw comes and a house or a street is flooded by the leakage at the crack, but it is really done at an early period of the frost, and any pipes which are exposed to cold weather should either be emptied of water or else thoroughly protected by a thick covering of straw or felt.

Much of the destruction of rocks, or 'denudation,' which is taking place on the earth's surface is due to the same quiet but intensely powerful action of freezing water. Rain sinks into the cracks and pores which all rocks are liable to contain, and when it freezes there, the crack is inevitably widened or the structure of the rock loosened. Thus room is made for more water, which acts in the same way when it freezes; and so by degrees immense masses of rock and earth are loosened from the mountain-side, nor does the action end until the material is reduced to the finest soil. Farmers fully appreciate the aid which frost renders them in loosening and crumbling up the stiffest soil, ready for the plough in spring.

---

[1] Such a bottle can be easily procured from an optician. Instead of it any gasfitter will supply a piece of iron pipe with 'blank caps' screwed on at each end. Or a soda-water bottle may be used, the cork being strongly wired down.

Change of Size in Liquefaction. 149

Another result follows from the fact that ice is larger in size than the water from which it is formed. Since this increase in volume takes place without any change in weight, it is clear that (as already explained and illustrated in reference to the expansion of solids, p. 55) the density of ice must be less than that of water. Hence, on well-known hydrostatic principles, ice floats on water with only so much of its bulk submerged as displaces a quantity of water equal to the whole weight of the ice. Thus the ice formed from 100 c.c. of water measures 110 c.c., but only weighs, like the water, 100 grms.: hence only 100 c.c. of it require to be submerged in order to displace an amount of water equal to it in weight (100 grms.), and then it will be wholly supported by the pressure of the surrounding water with the remaining 10 c.c. projecting above the surface. Hence it is that icebergs float with about one-tenth of their whole bulk visible above the sea. Hence also ice, when formed at the top of a mass of water, such as a lake, remains there instead of sinking, as it would be liable to do if it obeyed the usual law and contracted in volume (which would imply increase of density) in being formed. We have here, in fact, another reason, besides the one given on p. 78, why masses of water such as lakes do not become solid masses of ice. Not only does the colder water always remain at the surface, but the ice which is formed also remains there, buoyed up by the water below, and forms a layer through which heat passes out very slowly, so that it is some time before even a layer thick enough for skaters is formed. It is extremely rare to find ice formed below the surface of water, although loose, spongy masses of 'ground-ice' sometimes collect round the stems and roots of water-plants in running streams, probably from the action of eddies and currents in carrying down the freezing surface-water; the ice which is forming in it then attaches itself to any solid support which it meets with, and collects there.

Besides ice, a few metals, *viz.* iron, bismuth, and antimony (all of them, it may be noted, decidedly crystalline in structure) contract in becoming liquid; and hence, of course, these metals when melted expand in solidifying. Sharp castings can therefore be obtained by pouring the liquid metal into moulds; since when it cools and becomes solid the expansion forces it into every crevice of the mould. Antimony and bismuth communicate this property to their alloys: type-metal, for instance, is an alloy of antimony with lead, and yields sharp impressions of the most delicately cut letters used in printing.

Effect of Pressure in altering the Melting-point of a Substance.—In the case of the great majority of substances, *viz.* all those which expand on becoming liquid, increase of pressure, since it resists this expansion, prevents the solid from melting until a higher temperature than usual has been reached: in other words, it raises the melting-point. Thus wax, which under ordinary conditions melts at 65°, when subjected to a pressure of about 520 kilogrammes per square centimetre does not melt until heated to 75°.

An important result follows from this. It has been mentioned already (p. 14) that the interior of the earth is at a temperature far above the ordinary melting-point of the substances of which it is composed. Yet it is probable, or rather, almost certain that, owing to the enormous pressure due to the weight of the rocks near the surface, a great part of the interior is solid in spite of its high temperature, at any rate to a depth of 1300 kilometres.

In the case of those exceptional substances, such as ice, which contract in becoming liquid, an increase of pressure, since it assists in producing that reduction in volume which would naturally occur, enables the substance to melt at a lower temperature than usual. This has been proved by the following experiment made by M. Mousson.[1] A strong hollow steel cylinder, closed at each end by a screw, was nearly filled with water, a small piece of copper was dropped in, and the whole was cooled until all the water was frozen. The open end was then closed by its screw-plug, and the cylinder was inverted, so that the copper was at the top of the column of ice within, the latter (being a solid) supporting it in that position. The cylinder was kept at a temperature of $-18°$, while the screw-plug was turned round so as to apply a

---

[1] *Annales de Chimie et de Physique* (3d series), vol. lvi. p. 252.

pressure estimated at about 1300 kilogs. per square centimetre to the ice. The pressure was then removed, and the position of the piece of copper was examined. It was found to have fallen to the other end of the cylinder, which it could not have done unless the ice had become liquid. Hence we may infer that under a pressure of 1300 kilogs. per square centimetre the melting-point of ice is lower than $-18°$. This influence of pressure upon the melting-point of ice has been thought to account for the curious fact that two pieces of ice will stick firmly together, and that ice will freeze to flannel, even though the surrounding air may be comparatively warm and the ice itself not below $0°$. Some amount of pressure must exist between the surfaces in contact, and this enables liquefaction to occur there. The film of water formed is slightly below $0°$, owing to the heat necessarily absorbed during liquefaction (p. 152); but as it is squeezed outwards it escapes from pressure, and then its low temperature causes it to solidify again and cement the surfaces together. This property of ice is called 'Regelation,' and renders it easy to compress snow into a firm, hard snowball when the air is above the freezing-point. The phenomenon, however, is perhaps better accounted for by supposing that ice at $0°$ possesses a slight amount of softness or viscosity, just before it actually becomes liquid, as has been shown to be the case with glass and iron (p. 145).

3. *Disappearance of Heat during Liquefaction.*—In order that a substance may pass from the solid to the liquid state, the force of cohesion must be in a great measure overcome. To effect this some energy must necessarily be expended, and this energy is supplied by the heat which is communicated to the substance. Obviously, then, a quantity of heat proportional to the work done upon the molecules must disappear during the process of change of state, being converted into some other form of energy which exists in the liquefied mass in a statical condition. The whole process may be likened to what occurs when balls, held together by a spring, are pulled a little distance apart. In doing this the force of cohesion of the spring is to a certain extent over-

come, and energy is expended, or, more strictly, transferred to and stored up in the separate balls in a statical condition; not destroyed, for if the balls are let go, they come together with a force which is the exact equivalent of that which was employed in pulling them apart.

The fact that during the liquefaction of a solid a large amount of heat disappears, so that its presence cannot be shown by any of the usual methods, such as the application of a thermometer, may easily be shown in the following way:—

Expt. 36. Pour a little water on some finely crushed ice and leave it for a few minutes to attain the same temperature as the ice. When the ice and the water are both at $0°$, pour off the water into another beaker, place thermometers in both the water and the ice, and leave the beakers for a short time near each other in a warm room or standing in slightly warm water. Observe that the water gradually rises in temperature, while the ice, although just as much heat must be entering it, remains at $0°$: nor does it alter in temperature as long as any of it remains unmelted.

The same disappearance of heat occurs whenever a solid substance is dissolved in a liquid, as, for instance, salt in water; for here the substance is as truly liquefied as in the ordinary process of fusion by heat; the cohesion of its molecules being overcome in consequence of their attraction for, and tendency to diffuse themselves through the mass of the solvent. To enable this to take place, heat is absorbed from every available source; the solid, the liquid, and the vessel which contains them becoming decidedly colder.

Expt. 37. Arrange the differential thermometer with its bulbs dipping into beakers as in previous experiments. Place in each beaker 150 c.c. of water, and after proving that these portions of water are at the same temperature, put into one of the beakers about a teaspoonful of powdered sodium sulphate; stir it up in the water, and replace the bulbs of the thermometer in the beakers. The water in which the salt is dissolving will be observed to have become decidedly colder than that in the other beaker, and will continue so as long as any salt is being dissolved.

The heat which is thus shown to be absorbed by a melting solid is called 'the Potential Heat of Liquefaction'; the term 'potential' implying that, although there are no signs of its presence in the body as heat, it is there as some power (Lat. *potentia*) or energy which can be regained in the form of heat. Formerly it was known as 'Latent Heat,' from an idea that the heat-fluid lay hid (Lat. *latens*, lying hid) in some inexplicable way within the substance; but this misleading term should be dropped altogether, as wholly out of harmony with our present ideas as to the nature of heat.

4. *Evolution of Heat during Solidification.*—In order that a liquid substance may become solid, cohesion must be free to act between the molecules to its full extent; and hence the molecules must be deprived of all that energy which, by counteracting cohesion, kept the substance in the liquid state. This energy can be most readily withdrawn in the form of heat, and thus it is found that precisely the same amount of heat makes its appearance when a liquid solidifies as was spent in liquefying the solid. We do not usually notice in a direct way this evolution of heat, because the colder surrounding bodies absorb it, so that the solidifying substance does not become hotter; but the fact can be easily proved by the following experiment, in which solid crystals are caused to form in a solution without contact with any colder substance.

Expt. 38. Dissolve 200 grms. of sodium sulphate in 150 c.c. of hot (but not boiling)[1] water. When the whole has dissolved, filter the hot liquid at once through thin white blotting-paper into a *perfectly clean* flask, and close the neck with a pellet of cotton wool or tow (not a cork) to prevent the entrance of dust. Leave the flask absolutely undisturbed for several hours at least, until the solution has become quite cold. If the interior surface of the flask

---

[1] At 33° water dissolves more sodium sulphate than at any other temperature. If a solution saturated at this temperature is heated to boiling, some of the salt separates out in small crystals.

is quite clean and free from roughness, and if dust is excluded, none of the salt will crystallise out although the quantity of water taken is only sufficient to dissolve 60 grms. of sodium sulphate at the ordinary temperature of 15°. Under such circumstances the solution is said to be 'supersaturated,' and is in an unstable condition, ready to deposit some of the solid salt if shaken or stirred, or otherwise encouraged to do so. Take out the plug of cotton-wool and drop in a small crystal of sodium sulphate, placing also a thermometer in the liquid. The crystal will serve as a nucleus or centre of attraction on which crystals will be observed to form rapidly; and soon the whole of the solution will be filled with crystals, becoming, in fact, nearly solid. Observe that, while this solidification is going on, a great deal of heat is evolved; the thermometer rising 10 degrees or more, and the flask becoming perceptibly warm to the touch.

Sodium thiosulphate ('hyposulphite of soda'), a salt which can be procured easily and cheaply, being manufactured in large quantities for photographic and other purposes, shows the phenomenon still more strikingly. About 100 grms. of it may be melted in a clean flask (preferably by supporting the flask in a saucepan of boiling water, if a regular water-bath is not at hand) and treated in the same way as the sodium sulphate.[1] Enough heat is evolved during its solidification to boil ether in a small test-tube placed in the flask.

Methods of Determining the Exact Amount of Heat which Disappears during Liquefaction.

1. The earliest accurate observations on this subject were made by Dr. Black, Professor of Chemistry at Glasgow, in 1762, by a method of which the principle is illustrated by Expt. 36, already made. He took equal weights (about 140 grms.) of water and of ice, both at 0°, and placed them in jars in a warm room which was kept at a uniform temperature during the whole time of the experiment. Thermometers were placed in each jar, and every

---

[1] If the liquid requires filtration (and it is always best to filter it), the filter must be placed in a funnel kept hot by surrounding it with boiling water, as may be done by supporting it in a larger tin funnel, and filling the interval between the two with water kept hot by a lamp-flame applied to the exterior.

half-hour the temperatures indicated by the thermometers were observed and recorded. At the end of the first half-hour the thermometer in the water had risen from 0° to 4°, while that in the ice was still at 0°, though some of the ice had melted. He could therefore infer that the quantity of heat which was entering the ice every half-hour was such as would raise the same weight of water from 0° to 4°. At the end of twenty half-hours the whole of the ice was just melted, the temperature of the resulting water being 0°; and at the end of half an hour more this water had risen to 4°, from which it might be concluded that the same quantity of heat had been imparted to the ice during every half-hour throughout the experiment. Since this quantity was sufficient to raise an equal weight of water 4°, and since the ice took twenty half-hours to melt, Dr. Black inferred that the total amount of heat spent in melting a certain weight of ice was such as would raise the same weight of water ($20 \times 4° =$) 80°. Thus if 1 grm. of ice at 0° is taken, the heat spent in its liquefaction was, according to Dr. Black's experiments, sufficient to raise 1 grm. of water from 0° to 80°: and this may be simply stated as 80 calories.

The above method, though historically interesting, can only give approximately correct results, and is, moreover, applicable to few substances besides ice.

2. More recent investigators, such as Person[1] and Regnault, have always followed a plan resembling in principle the 'method of mixture' already explained (p. 128) as applied to the determination of specific heats. In this (which is illustrated by Expt. 37, already made) a known weight of the solid is liquefied by placing it in a known quantity of water, and the heat absorbed is determined by observing the number of degrees which the water is reduced in temperature.

Expt. 39. Take the beaker wrapped in cotton-wool and fitted into a larger jar (the calorimeter, in fact), which was used in the determination of specific heats (Expt. 29, p. 128); counterpoise the whole on one scale of a balance by bits of lead or shot placed in

---

[1] *Annales de Chimie et de Physique* (3d series), vol. xxvii. p. 250.

the other scale: then fill it with finely crushed ice[1] and leave it for a few minutes to cool to 0°. Meanwhile mix some boiling water with a little cold water in a jug, so as to reduce its temperature to a little above 80°. Empty all the ice out of the beaker, replace the latter on the scale (the counterpoise being retained in the other scale) and quickly weigh into it 200 grms. of the powdered ice, as free from water as possible. Then bring the water in the jug to a temperature of 79° by adding a little cold or hot water as required, place another 200 grm. weight in the scale containing the counterpoise and the 200 grms. used in weighing the ice, and pour into the beaker containing the ice enough of the hot water to restore equilibrium. We have then a mixture of 200 grms. of ice at 0° with 200 grms. of water at 79°. Stir the mixture, taking the temperature at intervals. If the experiment is carefully done, the final result will be that the ice will be all liquefied, and the temperature of the contents of the beaker will be approximately 0°.[2]

We infer from this that the quantity of heat spent in liquefying 200 grms. of ice is that which is given out by 200 grms. of water in sinking from 79° to 0°. If 1 grm. of ice had been taken, and 1 grm. of water at 79° had been added, it is evident that the result would be the same. But 1 grm. of water gives out, in sinking from 79° to 0°, 79 calories. Hence the potential heat of liquefaction of ice may be stated as 79 calories.

In accurate determinations of potential heat by this method it is preferable to take water at a lower temperature than that which was just now employed, and a proportionately smaller quantity of ice, so that the final temperature of the mixture may be nearly that of the room. In this way errors arising from the passage of heat to or from surrounding substances are greatly lessened.

Expt. 40. Take the counterpoised calorimeter used in the last

---

[1] Ice is most easily crushed finely by wrapping a large lump in a piece of strong canvas and hammering it on a board with a wooden mallet or any block of wood.

[2] It may perhaps be slightly above this, since (1) the ice weighed out may not be all ice, but contain some entangled water, (2) the heat of the room can hardly be prevented from melting some of the ice and raising the temperature of the mixture.

experiment, and weigh into it 150 grms. of water at a temperature of about 40°. After observing the exact temperature of the water, put a 30 grm. weight into the other scale, and add finely crushed dry ice to the water in the beaker until equilibrium is restored. Stir up the ice with the water (most conveniently with the thermometer itself) so as to melt it as quickly as possible, and read the thermometer from time to time, noting the lowest temperature reached.

Then we have all the data required to determine the potential heat of liquefaction of ice. Suppose that the results obtained in the above experiment are as follow :—

Weight of water in beaker . . .	150 grms.
Water-value (see p. 130) of beaker	10 grms.
Temperature of water .	41°.
Weight of ice added . .	30 grms.
Temperature of ice . . .	0°.
Lowest temperature reached .	22°.

Here (150 + 10 =) 160 grms. of water have been reduced from 41° to 22°, *i.e.* 19°, and must therefore have given up (160 × 19 =) 3040 calories. This quantity of heat has done two things :—

1. It has converted 30 grms. of ice at 0° into water at 0°.
2. It has raised this weight of water from 0° to 22°. For this latter purpose (30 × 22 =) 660 calories are required, and the remainder, (3040 − 660 =) 2380 calories, must have been spent in simply melting the ice. Then if 30 grms. of ice require 2380 calories to melt them, 1 grm. will require $\frac{1}{30}$ of this, *i.e.* ($\frac{2380}{30}$ =) 79·3 calories for the purpose. Hence the potential heat of liquefaction of ice is 79·3.

If the solid has a high melting-point, such as sulphur or tin, it is more convenient to determine the amount of heat given out by the liquid in becoming solid: for this, as above explained, is precisely equivalent to that required to liquefy the solid. For this purpose some of the melted substance is allowed to cool slowly, and when it has just begun to solidify, the liquid portion is poured into a weighed quantity of cold water in a calorimeter, and the rise in temperature of the water is noted; from which the number of calories which have entered it can be calculated in the usual way. The increase in weight of the vessel and its contents gives the weight of the substance which has been used. The rest of the calculation is made on the same principle as in the experiment last performed.

TABLE OF POTENTIAL HEATS OF LIQUEFACTION.

[The figures denote the number of calories which are spent in changing 1 grm. of the substance from the solid to the liquid state.]

Substance.	Potential Heat.	Substance.	Potential Heat.
	Calories.		Calories.
Ice,	79	Sulphur,	9
Sodium Sulphate, dissolving in water,	49	Common Salt, dissolving in water,	9
Tin,	14	Lead,	5.4

RESULTS AND APPLICATIONS OF THE POTENTIAL HEAT OF LIQUEFACTION.

1. The high potential heat of liquefaction of ice produces quite as important results in Nature as the high specific heat of water. Before even a small mass of ice or patch of snow can melt, so large a quantity of heat has to be taken in that the melting proceeds with great slowness, and when a thaw sets in we have no sudden, unmanageable, and dangerous floods. For a similar reason lakes and rivers take a long time to freeze, and seldom or never become solid masses of ice. So also there is a decided decrease in the coldness of the weather just before a fall of snow occurs, since the small particles of water which form mist and cloud must, in freezing into snow-flakes, give out all the potential heat which kept them in a liquid state.

2. Many of the most effective methods for obtaining a great reduction of temperature depend upon the fact that a solid, in order to become a liquid, will absorb a large quantity of heat.

POTENTIAL HEAT OF LIQUEFACTION. 159

Thus, if we encourage a solid to liquefy under conditions in which it would usually remain solid, it takes heat from its own mass and from surrounding bodies, so as to make them much colder. One of the simplest examples of this has been given already in Expt. 37, p. 152, where sodium sulphate was dissolved in water, with the result of making the solution decidedly colder. But a much greater reduction of temperature can be obtained by dissolving in water certain mixtures of salts, and such mixtures are called 'Freezing mixtures.' For instance, if equal weights of ammonium chloride (sal ammoniac) and potassium nitrate (saltpetre) are powdered together and dissolved in twice the weight of water (on a small scale, 25 grms. of each salt and 100 c.c. of water may be taken) a thermometer placed in the mixture will fall about 20°. So again, if 5 parts of strong hydrogen chloride (hydrochloric acid) are added to 8 parts of powdered sodium sulphate (10 c.c. of the acid and 20 grms. of the salt may be taken to illustrate this) the temperature falls nearly 30°. But the readiest and cheapest freezing mixture, when ice or snow is obtainable, is made by mixing 1 part of common salt with 2 parts of either material. The salt has a strong attraction for water, and hence causes the ice to liquefy, and dissolves in the water produced. Thus from two solids a liquid is obtained which remains in that state at a temperature much below 0°, and a great absorption of heat takes place.

Expt. 41. Weigh out 200 grms. of finely crushed ice or snow, and 100 grms. of salt, stir them together in a basin, and put the mixture into a tin canister standing on a plate containing a little water. Put a thermometer into the canister, and observe that the temperature falls to about −20°. The water in the plate will soon freeze, so that when the canister is lifted up the plate will be lifted up with it, the two being cemented together by the ice.

Such a freezing mixture is used practically on a large scale by confectioners in making 'cream-ices,' etc. It is also formed unintentionally when salt is sprinkled on snow in order to melt it; a very common but unadvisable practice. The snow is certainly melted, but the ground or pavement is rendered extremely cold and liable to cause chills in those who walk on it, and to freeze any water employed to cleanse the surface into dangerously slippery ice.

Section II.—Conversion of a Liquid into a Gas, and vice versâ. Vaporisation and Condensation.

1. *Temperature at which the change takes place.*—When a substance is in the liquid state, its molecules, owing to their rapid motion, are able so nearly to overcome the constraining influence of cohesion that we might naturally expect that, if there is a clear space above its surface, those molecules which are at or near the surface would show a tendency to shoot upwards into that space, and remain there in the comparatively free condition of a gas. This is actually found to occur: liquids form vapour at almost all temperatures in any space above their surfaces which is unoccupied, or only occupied by a gas. We cannot, as in the case of liquefaction, assign any particular point of temperature at which the formation of vapour begins, although there certainly is one definite point at which it is formed with great rapidity. This is called the 'boiling-point' of the liquid, and will be considered later.

Expt. 42. Pour a few drops of common alcohol upon a glass plate, and incline the plate so as to distribute the liquid over the surface. Observe that it soon disappears entirely, having turned into vapour even at the ordinary temperature of the room.

Expt. 43. Pour a few drops of ether into a beaker, and (having closed the bottle and removed it to some distance) bring a lighted match within the mouth of the beaker. A blue flame will appear, showing that the ether has been converted into vapour sufficient to fill the beaker.

Water, in the same way, evaporates at ordinary temperatures, though not so quickly as ether or alcohol. Ponds and ditches dry up during a drought: wet clothes are soon dried by hanging them out in the air; indeed, nearly all the enormous amount of water-vapour existing round the earth was formed by quiet, ceaseless evaporation from the surface of masses of water which never even approached their boiling-point. Mercury also gives off a perceptible amount of vapour at ordinary temperatures, as may be proved by hanging a strip of glass covered with gold leaf[1] in a bottle containing a little mercury, taking care that the gold does not actually touch the liquid. In a few months of moderately warm weather the gold will become white, owing to the vapour of mercury having combined with it to form an amalgam. The crews of vessels carrying cargoes of mercury are liable to symptoms of poisoning from inhaling the vapour from mercury which has accidentally leaked from bottles.

There are even some solids which give off a perceptible amount of vapour at temperatures much below their melting-points. Ice and snow do this, and disappear slowly during a long frost, when the air is never above the freezing-point. The smell of such substances as camphor sufficiently proves that they are giving off vapour, and very perfect crystals are often found near the neck of a bottle containing camphor or iodine.

2. *Change of Size during Vaporisation.*—When a liquid turns into a gas, if the pressure upon it is preserved unchanged, there is, in all cases, a sudden and very great increase of volume. Thus 1 c.c. of water at 100°, when converted into vapour at the same temperature and pressure, becomes 1630 c.c. of steam : 1 c.c. of alcohol at 78° yields 528 c.c. of vapour, and so on for other liquids.

The general fact of expansion is easily observed by boiling some water in a flask fitted with a cork having a short piece of glass tubing passed through it. The flask will be filled with steam, as proved by the latter escaping from the exit-tube, without any perceptible diminution in volume of the water. But

---

[1] This may be done by laying the gold leaf loosely on the glass, and then introducing a little water underneath it from a pipette. The leaf will float smoothly on the water, losing all wrinkles, and the water may then be poured off very gently at one corner, leaving the gold spread flat on the plate, which should be dried by a gentle heat.

L

perhaps the great disproportion between the volume of water and that of the steam it forms is better proved by the reverse experiment, *viz.* by condensing a certain volume of steam into water, and observing the space occupied by the water.

Expt. 44. Fit a good cork to a globular[1] flask about 10 cm. in diameter: bore a hole in the cork and fit into it a short bit of glass tubing drawn out at one end to a long point and sealed. Pour into the flask about 6 or 8 c.c. of water, and add a small teaspoonful of sand or iron filings to prevent breakage, which is otherwise liable to occur as the last drops of water are being evaporated. Fit the cork into its place, the glass tube being withdrawn for the present, and heat the flask on a small sandbath or on wire-gauze,[2] covering the neck and upper part of it with a cone of sheet tin, fitting over it like a cape, to prevent loss of heat. Continue the heat until the water has all boiled away, and as soon as the current of steam issuing from the hole in the cork begins to slacken, fit the wide end of the glass tube firmly into the cork (the sealed point being outwards), and immediately remove the flask from the hot sand. We have now the flask full of steam, and this will soon condense into water as the temperature falls: the cooling may be hastened by blowing air from bellows upon the flask, and finally by pouring a little cold water over it. Observe that the whole of the steam collapses into a mere slight film of dew, hardly amounting to a drop of water. Plunge the mouth of the flask rather deeply into some water in a basin, and break off the sealed point of the tube. Notice the force with which the water is pressed in by the external air: if all the air in the flask has been displaced by steam, a nearly complete vacuum will have been formed, and the water will entirely fill the vessel.

If this expansion during vaporisation is resisted by enclosing the liquid in a vessel not much larger than is sufficient to contain it, and heating it in this condition, a very intense pressure is

---

[1] The flask must not have a flat bottom, which would be forced inwards and broken by the pressure of the air.

[2] An air-bath, large enough to contain the body of the flask, is much preferable, if at hand. A tin canister (with seamed, not soldered joints) may be used; a hole being cut in the cover, to allow the neck of the flask to project. No sand or iron filings need then be placed in the flask.

Change of Size in Vaporisation. 163

exerted by the vapour against the sides of the vessel; a pressure equivalent, in fact, to that which is required to compress the vapour from the volume it would occupy under ordinary conditions into the volume of the vessel. This can be easily illustrated as follows:—

Expt. 45. Place one of the small glass bulbs (7 or 8 mm. in diameter) nearly filled with water, and sealed, known as 'candle-bombs,' on a piece of wire-gauze: cover it with another small piece of gauze, and support it over the flame of a Bunsen burner. In a minute or so the bulb will burst with considerable violence, owing to the pressure of the vapour within.

Results and Applications of the Expansion which occurs during Vaporisation.

1. A comparatively small supply of water is sufficient for the boilers of steam-engines, as the liquid yields so large a volume of steam. For this reason, apart from others, there is no substance so suitable as water for use in engines, since no substance expands, in vaporising, to at all the same extent.

2. An approximate vacuum can be readily obtained by displacing the air in a vessel by steam, and then condensing the steam by cooling it: since (as illustrated in the last experiment but one) the vapour shrinks into a very small, practically negligible bulk, leaving nothing else in the vessel but the slight amount of vapour which can exist at ordinary temperatures. This principle was taken advantage of in several of the early forms of steam-engine (for instance, Newcomen's atmospheric engine, of which an account will be given later), and still contributes greatly to the efficiency of modern 'condensing' engines, as they are called.

Disappearance of Heat during Vaporisation.

In order that a liquid may assume the state of gas, enough energy must be supplied to it not only to overcome the internal cohesion entirely, but also to enable the freely moving molecules to force back obstacles and enlarge the area of their movements very considerably indeed. As already stated, a gas may be regarded as consisting of a vast crowd of separated molecules moving with enormous velocity in straight lines, and constantly striking against one another, and against anything else which comes in their way. In order to give them this great velocity, which in the case of steam is at least 600 metres per second, it is obvious that a large amount of heat energy must be expended; in other words, the potential heat of vaporisation might be expected to be very great. An experiment like that made already (p. 152) to illustrate liquefaction will readily prove that such is the case.

Expt. 46. Place a flask containing some water on wire-gauze over an Argand burner. Fit a cork to the flask and bore two holes in it: through one of these pass the stem of a thermometer, arranging it so that the bulb is just above the surface of the water in the flask. The other hole is intended to allow steam to escape freely. Heat the water to boiling for some time, and observe that, although heat is continuously supplied to it, the temperature of the steam does not rise above 100°.[1] Hence heat must be disappearing during the formation of the steam.

Expt. 47. Wrap a bit of muslin round one of the bulbs of the differential thermometer, and pour a few drops of alcohol upon it. The alcohol will at once begin to turn into vapour, and the thermometer will show that there is a considerable fall of temperature: heat being withdrawn from the liquid, the muslin, and the bulb, and disappearing in the formation of the vapour.

The same fact may be noticed in a simpler way by pouring a little alcohol or eau de Cologne (which is merely a solution of aromatic oils in alcohol) upon the back of the hand. The sensation of cold will be very decided, and still more so if ether is used.

Methods of Determining the Potential Heat of Vaporisation.

Dr. Black was the first to estimate with fair accuracy the amount of heat which is spent in converting water into steam

---

[1] If the reading of the barometer is much above 76 cm. the temperature will be slightly higher, for a reason to be explained shortly (p. 191).

POTENTIAL HEAT OF VAPORISATION.

without altering its temperature. His experiments were made in 1762, about the same time as those on the potential heat of liquefaction, and on a very similar principle. He took a flask containing about 300 grms. of water at 10°, placed it upon a heated iron plate, maintained at a uniform high temperature during the whole of the experiment, and observed (i) the time which elapsed before the water began to boil, (ii) the time which subsequently elapsed before it had all boiled away. In one of his experiments the water began to boil in four minutes, and twenty minutes more elapsed before it had entirely disappeared in the state of steam. Assuming that the same amount of heat had been communicated to the water during each minute that the experiment lasted, he inferred that, since 20 is five times 4 minutes, it took five times as much heat to turn a certain quantity of water into steam as to raise the same weight from 10° to 100°, *i.e.* through 90° of temperature. Now, to heat 1 grm. of water 90° requires, of course, 90 calories: hence five times 90 or 450 calories are spent in changing the state of 1 grm. of water from liquid to gas.

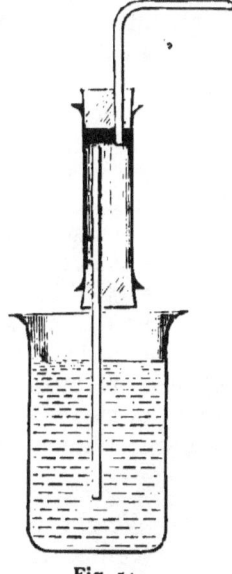

Fig. 54.

The above number is below the true value; indeed, Dr. Black's method is hardly capable of giving very exact results. The principle of the method by which Regnault and others have accurately determined the potential heats of vaporisation of many substances consists in condensing the vapour of the substance in cold water in a calorimeter, and observing the number of calories given out by a known weight of it in passing into the liquid state. The following experiment may be made in illustration of it :—

Expt. 48. Fit to a moderate-sized flask a tube bent twice at right angles like that shown in fig. 33, p. 65, but not drawn out at the end *a*. To the outer end of this tube adapt the arrangement of tubes shown in fig. 54, consisting of a short piece of tubing about

8 cm. long and 2 cm. in diameter, with well-fitting corks at each end, through the upper one of which the tube leading from the flask is passed, while into the lower cork is fitted a straight tube about 13 or 14 cm. long. The holes for these tubes should be bored near the sides of the corks, as shown, so that the end of one tube may pass by that of the other, to avoid the chance of any water dropping direct into the lower tube. The object of this 'water-trap' is to prevent the water, which is sure to condense in the tube leading from the flask, from passing into the water in which the lower tube is dipped; thus insuring that nothing but pure steam shall enter the water in the calorimeter. Fill the flask about one-third full of water, add two or three small bits of clean coke or cinders to make the water boil regularly (the reason is explained on p. 187), and place the flask on wire-gauze over a lighted lamp, removing the water-trap for the present. While the water is being heated, counterpoise the beaker-calorimeter and weigh into it 150 grms. of cold water, the temperature of which should be taken by a thermometer. When steam issues freely from the end of the bent tube, fit on the water-trap and dip the lower end of the exit-tube deeply into the water in the calorimeter, supporting the latter on blocks so that the end of the tube may reach nearly to the bottom of it. The issuing steam will be at once condensed by the water, and the temperature of the latter will rise owing to the heat given out. Stir the water constantly with the thermometer, noting the rise in temperature; and when it has reached about 30° or 35° withdraw the tube, and note once more the exact temperature of the water. Finally replace the calorimeter in the scale, and weigh the whole. The difference between its present weight and its original weight denotes the weight of steam which has been condensed. Thus the following data are obtained (the numbers are taken from an actual experiment):—

Weight of water in calorimeter	150 grms.
Temperature of do.,	10°.
Water-value of calorimeter	6 grms.
Weight of steam condensed	6 grms.
Temperature of the mixture	33°.

We have here 156 grms. of water raised from 10° to 33°, *i.e.* through 23°. Hence ($156 \times 23 =$) 3588 calories must have entered it. This quantity of heat has been given out—

POTENTIAL HEAT OF VAPORISATION. 167

(*a*) By the 6 grms. of steam at 100° in condensing into water at 100°.

(*b*) By the 6 grms. of water thus formed in sinking from 100° to 33°, *i.e.* through 67°.

This latter source of heat (*b*) will account for (6 × 67 =) 402 calories; and the remainder, *viz.* (3588 − 402 =) 3186 calories, must have been given out by the steam in simply being converted into water without change of temperature. Then by the proportion,

6 grms. : 1 grm. : : 3186 calories : 531 calories,

we learn that 1 grm. of steam gives out, in becoming water, 531 calories. And since the same quantity of heat must obviously be spent in producing the reverse change of state, the result of the experiment shows that the potential heat of vaporisation of water is approximately 531 calories.[1]

TABLE OF POTENTIAL HEATS OF VAPORISATION.

[The figures denote the number of calories which are spent in changing 1 grm. of the substance from the liquid to the gaseous state.]

Substance.	Potential Heat.	Substance.	Potential Heat.
	Calories.		Calories.
Water	536	Sulphur	362
Alcohol	209	Mercury	62
Ether	90	Iodine	24

RESULTS AND APPLICATIONS OF THE POTENTIAL HEAT OF VAPORISATION.

1. Some of the most effective methods of obtaining a low temperature depend on the fact that a liquid must obtain heat

---

[1] This number is slightly too low, since no allowance has been made for the unavoidable loss of heat by radiation, etc., which must occur as soon as the water rises above the temperature of the air. This source of error might be nearly avoided by taking the water to begin with at a temperature as far below that of the air as it will be above that temperature at the end of the experiment. Then the heat it gains from the air in the first part of the experiment will be equal to that which it loses in the last part.

from some source or other in order to turn into vapour. Thus water, when sprinkled on a hot road or on the floor of a room, evaporates with sufficient rapidity to cause a refreshing coolness. In hot countries water may be cooled considerably below the temperature of the air by placing it in jars made of porous, unglazed earthenware: the water soaks through the sides of the jar and evaporates as it comes to the outer surface, thus cooling the jar and the remaining water. The rate of evaporation is increased by placing the vessel in a current of air, since the vapour is then carried off as soon as it is formed. Water-coolers of the same kind are often used in England, although in temperate climates the action is not so decided.

Our bodies are similarly cooled externally by the evaporation of the perspiration which is constantly coming to the surface of the skin, and internally by the formation of water-vapour in the lungs, which is emitted at every breath. Rather more than a litre of water is, on an average, carried off thus by evaporation from the skin, and half a litre more from the lungs in the course of every twenty-four hours: so that above 800,000 calories must be withdrawn from the body every day from this cause alone. If it were not for this extensive withdrawal of heat it would be almost impossible to endure the high temperatures of tropical climates or of a Turkish bath-room, and quite impossible to bear exposure to the furnaces in iron-foundries and glass manufactories, or to work even for an hour in the stoke-holes of steam-ships. The men who are employed in such places drink copious draughts of water, and are encouraged to do so: perspiration is thus increased, and evaporation also, so that the temperature of the body is kept down, and great heat can be borne without danger.

It is easy, indeed, to reduce the temperature of the skin much below its normal point by evaporation; as happens when a 'chill' is caught by standing about after bathing or by exposure to a draught. The surface of the body is then liable to become so cold that the healthy action of the vessels of the skin is checked and a 'cold in the head' comes on.

2. The refreshing coolness of effervescing draughts is in a great

measure due to the production of a large quantity of gas in the liquid: the latter necessarily losing the heat required for the formation of the gas.

Expt. 49. Powder and mix together in a mortar 5 grms. of bicarbonate of soda and 5 grms. of tartaric acid.[1] Arrange the differential thermometer as in previous experiments, with its bulbs dipping into 150 c.c. of water in beakers, and put the mixture of salts into the water in one of the beakers. A quantity of carbon dioxide ('carbonic acid gas') will be evolved, and the thermometer will show that the water in that beaker has become decidedly colder.

Of course we have, in the above experiment, another cause contributing to reduce the temperature of the mixture, *viz.* the solution of the two solid salts in the water, which is (as has been explained on p. 152) equivalent to their liquefaction.

3. The immunity of the contents of fireproof safes is secured in the following way. The safe is made of two iron cases, one inside the other, with an interval of 6 or 8 cm. between them. This interval is loosely filled with a mixture of salts which when moderately heated either volatilise entirely (such as ammonium chloride or 'sal ammoniac') or give off a large quantity of water-vapour (such as alum). When the safe is exposed to a fire, the heat is all expended in volatilising or forming steam from these salts, and not enough of it reaches the interior of the safe to damage its contents.

4. A much greater reduction of temperature can be obtained by causing very volatile liquids to evaporate quickly, as may be done by making them expose a large surface from which evaporation may take place. This is the principle of Richardson's 'ether-spray' apparatus, which is shown in fig. 55. It consists of a tube, A, about 3 mm. in internal diameter, the lower end of which passes through a cork fitted into the neck of a bottle containing ether. The tube ends above in a fine jet, and within it is fixed a much smaller tube, B, as shown in the figure, open at both ends, the lower end dipping into the ether in the bottle,

---

[1] If any of the so-called 'citrate of magnesia' (which is only a mixture of the above two substances) is at hand, it will do very well for the experiment.

and the upper end reaching nearly to the jet. A side tube, C, is attached to the larger tube, through which a strong stream of air is forced by means of bellows or a simple form of indiarubber blower usually supplied with the instrument. This stream of air divides when it enters the vertical tube; one portion of it passes downwards into the bottle, and presses on the surface of the ether, forcing some of it up the narrow tube to the jet. The rest of the blast of air rushes upwards and issues from the jet, carrying the ether with it. The sudden expansion of the air when it gets out of the jet separates the liquid ether into a cloud of fine particles, or spray: and each particle being surrounded by air evaporates almost instantaneously, taking the requisite heat from the air, or anything it comes in contact with. Thus any substance immersed in the spray becomes quickly cooled to a temperature much below the freezing-point of water.

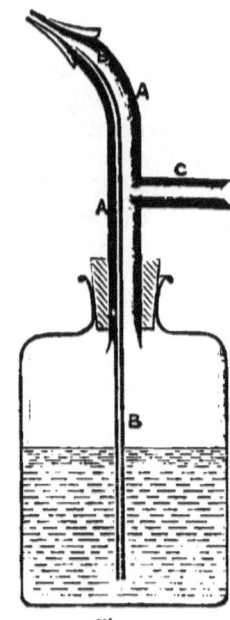

Fig. 55.

Expt. 50. If an ether-spray apparatus is at hand[1] its effects may be shown by holding a small thin test-tube, about one-third filled with water, in the stream of spray about 2 or 3 cm. (not more) from the jet. Some of the water will be quickly frozen; and without waiting for the whole to become solid, the tube of ice may be detached from the test-tube by clasping the latter in the warm hand for a moment, and shaken out upon a plate.

The purpose for which the instrument was originally designed, *viz.* to benumb the flesh by intense cold, and thus render it insensitive to the pain produced by surgical operations, may be illustrated by holding the back of the hand in the spray, rather

---

[1] The ether used should not be the ordinary 'sulphuric ether,' but that of sp. gr. 0·718 (usually made from methylated spirit) specially prepared for use with this instrument.

POTENTIAL HEAT OF VAPORISATION. 171

near the jet. A white patch of benumbed, insensitive flesh will very shortly appear on the hand.

Expt. 51. If an ether-spray apparatus is not procurable, the following simpler experiment will serve the same purpose. Place five or six drops (not more) of water in a watch-glass laid on a folded cloth. Pour about 2 c.c. of ether into a test-tube, and support the tube upright with its lower end dipping into the water in the watch-glass. Force a rapid stream of air through a glass tube dipping into the ether, and connected by an india-rubber tube with a pair of bellows of the kind used for a gas blowpipe. The ether will quickly evaporate, and absorb so much heat that the water in the watch-glass will be frozen in a minute or so, and will cement the glass to the test-tube, as may be proved by raising the latter.

5. The lowest temperatures hitherto obtained have been produced by the evaporation of substances much more volatile than even ether, *viz.* the liquid and solid forms of bodies which we only know as gases under ordinary conditions, such as carbon dioxide, oxygen, and nitrogen (the two latter being the chief constituents of common air). The methods of liquefying such gases will be explained later; in the meanwhile, it may be noted that they are only retained in the liquid state by enclosing them in very strong vessels so that they cannot expand into gas. Liquid carbon dioxide, for instance, is kept in thick wrought-iron bottles under a pressure of at least 50 times that of the atmosphere, and when some of the liquid is allowed to escape from a jet attached to the bottle, the moment that it is released from pressure it flashes into gas, absorbing so much heat in doing so that a portion of it is cooled below its melting-point and converted into a white solid resembling snow in appearance. This solid carbon dioxide can be collected in a receiver attached to the jet, and preserved for some time; its temperature being so low that it does not evaporate quickly. When, however, it is mixed with ether, the evaporation is so much hastened that a thermometer placed in the mixture sinks to at least $-90°$. If some of the semi-fluid mass is mixed with mercury, the latter is at once frozen into a malleable solid resembling lead; and if any of it is allowed to

touch the skin it produces an inflamed sore as painful as that caused by red-hot iron. When solidified nitrogen (which is much more volatile than even liquid carbon dioxide) is caused to evaporate very quickly by reducing the pressure upon it, a temperature of $-225°$ C. has been produced, which is the lowest point yet attained.[1]

We may now turn to some instances in which the potential heat of vaporisation is taken advantage of for purposes of heating, instead of cooling.

Steam, in consequence of the very large amount of heat which is produced in its condensation, is extremely effective and useful as a heating agent. Manufactories and other buildings are warmed by leading steam from a boiler into a series of pipes distributed throughout the building, provision being, of course, made for catching the condensed water in cisterns so that it may not choke up the pipes. Wherever a constant heat is required which shall not under any circumstances rise above 100° (or a few degrees higher), steam affords the best means of obtaining it. Thus gunpowder is dried without risk of explosion by placing it on trays in a chamber heated by steam-pipes supplied from a boiler placed at a safe distance. A 'steam cupboard,' constructed with double walls between which steam is made to circulate, is constantly used in laboratories for drying substances which would be altered or decomposed if strongly heated. In sugar refineries, the solution of sugar is evaporated until concentrated enough to deposit crystals by the heat from steam led through coils of pipe placed in the evaporating-pans. In distilleries the spirit of wine is vaporised by a similar arrangement of steam-pipes in the boiler or 'still,' and thus the serious risk of the vapour catching fire is avoided. Sometimes the steam is led directly into the liquid to be heated; as in the case of the water used for supplying an engine-boiler, into which the waste steam from the engine is often allowed to flow, thus saving a great deal of fuel.

Scalds from exposure to steam are extremely severe from a

---

[1] *Comptes Rendus*, vol. ci. p. 238.

similar reason; far more so than those produced by boiling water, although the steam is really no hotter than the water.

Section III.—Evaporation at different Temperatures.

Evaporation may be defined as the process of formation of vapour from the free surface of a substance. As has been already remarked, it does not begin or cease at so definite a point of temperature as liquefaction, although the rapidity with which it takes place is greatly dependent on certain conditions which will now be considered more minutely.

Tension of Vapour.—Every vapour or gas confined in a vessel is observed to exert some pressure against the sides of the vessel; or, indeed, against anything which resists its diffusion on all sides. This outward pressure is called the 'tension' of the vapour, and it is very simply and completely explained by the kinetic theory of gases (p. 10) as due to the rapid motion of the molecules in straight lines, which causes them to knock, like small, swiftly flying cannon-balls, against anything which opposes their onward progress.

The amount of tension which a vapour exerts is usually ascertained in the same way as the pressure of the air is measured in an ordinary barometer, *viz.* by observing the height of the column of mercury which the vapour can support in a tube. Thus it is found that the vapour given off from water at 15° exerts a pressure sufficient to raise and support a column of mercury 12·7 millimetres high in a tube: hence the tension of water-vapour at 15° is said to be 12·7 mm. The principle of the method may be illustrated by the following experiment, which will also show that the tension of alcohol-vapour at ordinary temperatures is greater than that of water-vapour mentioned above.

Expt. 52. Prepare a bent glass tube about 5 or 6 mm. in internal diameter, of the form shown at A B in fig. 56. The upturned branch of the tube A should be about 30 cm. in length, and the other branch, B, about half as long. Prepare also a short piece of tubing rather wider in bore, about 6 cm. long, and after slightly rounding the ends in the lamp-flame, fit into each end a short plug of cork,[1] fitting so accurately as to be air-tight and yet admit of being pushed along the tube by a piece of thick wire or a wooden rod. Adapt the two tubes by a well-fitting cork to a globular flask holding about 300 c.c., as shown in fig. 56, supported in a retort-stand or otherwise.

Withdraw the tube A B from the cork and pour into it sufficient mercury to fill it nearly to the level of the horizontal branch; then fit it again into its place in the cork. This tube will then serve as a gauge to indicate any change of pressure which may take place within the flask. The next step will be to introduce some alcohol into the flask without opening any communication with the external air. To do this, take out the upper plug from the short tube and fill the latter with alcohol; the lower plug will serve to prevent any of the liquid running into the flask. Now replace the upper plug and push it down upon the alcohol. As the latter is compressed it will force out the lower plug and drop down into the flask, where, in spite of the presence of the air, it will soon form as much vapour as can exist in the space. This vapour immediately begins to exert a tension, and as it cannot escape through the short tube in which the cork plug still remains (it need not be pushed more than half-way down) it will press upon the surface of the mercury in the gauge, and drive up a column of it in the branch A until the height of this column above the level of the mercury in the branch B is just sufficient to balance the tension. Measure the difference in level of the columns in the two branches; this difference (which may be from 20 to 30 mm. according to the temperature) will, as above explained, express the tension of alcohol vapour at the temperature of the flask.

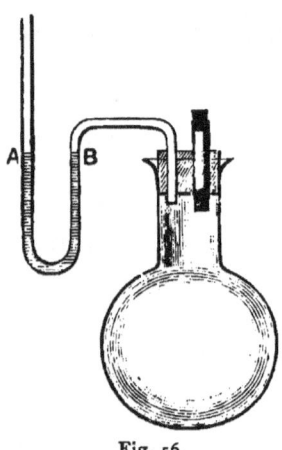

Fig. 56.

It will, moreover, be easy to prove that the tension of a vapour increases with increase of temperature, by pouring over the flask some warm water (not hotter than 50°). The difference in level

---

[1] The plugs cut out by the usual cork-borers in making holes through a cork will do very well, if smoothly cut.

Tension of Vapour.

of the mercury columns will increase [1] and may be measured as before.

If any ice or snow is at hand some of it may be mixed with water and poured over the flask; the tension of the vapour will be found to diminish.

Another method, which was that originally employed by Dalton in investigating the subject, and in which the tension of the vapour, instead of being measured directly, is estimated by observing how much of the total pressure of the atmosphere it will balance, may be illustrated by the following experiment (if sufficient mercury is at hand).

Expt. 53. Prepare a tube such as is used for an ordinary barometer, about 90 cm. long and 7 or 8 mm. in internal diameter, with one end drawn out, closed, and rounded in the blow-pipe flame. After cleaning and drying it thoroughly, fill it with mercury to within 1 cm. of the open end: then close it firmly with the finger and raise the other end so as to cause the enclosed air to pass as a large bubble along the tube and collect all small bubbles which may have adhered to the glass. When these are got rid of, fill up the tube completely with mercury, place the finger tightly on the open end, and immerse this end in a small trough of mercury (a strong porcelain mortar will answer the purpose), keeping the tube slanting until the end is well below the surface of the mercury. Now withdraw the finger and slowly raise the closed end of the tube until it is upright, in which position it should be supported by a firm, steady holder, such as a Bunsen holder with a weight placed on the foot. The mercury will sink from the upper end of the tube, since the pressure of the external air on the surface of the mercury in the cistern is not enough to support a column of mercury higher than about 760 mm. in the tube, and a practically empty space will be left above the mercury, into which vapour may be introduced. Measure the exact height of the column, starting from the surface of the mercury in the cistern, to the highest point of its curved surface in the tube; this will express the amount of the pressure of the external air at the time of experiment. Fill a small short test-tube with alcohol, close the tube with the finger, and bring it

---

[1] Of course, however, a part of this increase will be due to the expansion by heat of the air in the flask independently of the vapour; for it is a law, first discovered by Dalton, and known as 'Dalton's Law,' that every vapour acts quite independently of others which may be present.

mouth downwards under the surface of the mercury: then pass up a few drops (not more) of the alcohol into the barometer tube. As soon as the liquid reaches the top of the mercury some of it will evaporate into the empty space, and the tension of the vapour will depress the column of mercury in the tube: the final height of the column should then be measured as before. The difference between this height and the original height of the column will express the amount to which the pressure of the vapour has acted against the pressure of the atmosphere, *i.e.* the tension of the vapour.

If, after emptying, cleaning, and drying the tube it is refilled with mercury and a few drops of ether are passed up into it, a much greater depression of the mercury will take place: and the amount of this depression, measured as before, will indicate the tension of the vapour of ether at the temperature of the room.

If a cloth dipped in hot water is placed round the upper part of the tube, the mercury column will be much further depressed, showing that the tension of the vapour increases with the temperature.

Experiments similar to the above made with water, ether, and other liquids would show that at the same temperature the tension of the vapours they form is very different indeed: as will be seen from the following table, compiled mainly from the results of Regnault's experiments.[1]

THE TENSION OF VAPOURS.

[The figures express the height in millimetres of the column of mercury which balances the pressure of the vapour.]

Temperature.	Ammonia.	Sulphur Dioxide.	Ether.	Alcohol.	Water.
	mm.	mm.	mm.	mm.	mm.
$-36°$	760	200	...	...	...
$-10°$	2150	760	115	6	2.1
$0°$	3162	1165	184	13	4.6
$10°$	4612	1720	287	24	9.1
$20°$	6384	2462	433	44	17.4
$35°$	...	...	760	103	41.8
$78°$	...	...	2800	760	327
$100°$	...	...	4953	1695	760

[In this table, and in general, the tension of the vapour means its *maximum* tension, as explained on p. 179.]

---

[1] *Annales de Chimie et de Physique* (3d series), xi. 273.

It will be seen from the above table that the tension of a vapour increases at a much greater rate than the temperature. Thus the increase of the tension of water-vapour for the 10° between 0° and 10° is (9·1−4·6=) 4·5 mm.; while for the next 10°, *i.e.* between 10° and 20°, it is (17·4−9·1=) 8·2 mm.

Distinction between a Vapour and a Permanent Gas.—It will also be observed in the table, that, assuming the average pressure of the atmosphere to be 760 mm., the tension of the vapours of some substances, *e.g.* ammonia and sulphur dioxide, is greater than this, even at 0°, while the tension of the vapours of other substances, *e.g.* ether, alcohol, and water, is less than this at 0°. This will enable us to understand the distinction conventionally made between a 'vapour' and a 'permanent gas.' There is no essential difference between the two, in so far as both are names for the gaseous condition of a substance: but the sense usually attached to the words is the following :—

A '**vapour**' is the gaseous form of any substance which at 0° C. exerts a tension less than 760 mm.

A '**permanent gas**' is the gaseous form of any substance which at 0° C. exerts a tension greater than 760 mm.

Thus we talk of the 'vapours' of what we know as volatile liquids, *e.g.* ether, while we call sulphur dioxide a 'permanent gas': although at temperatures of 40° and higher (such as occur in the rooms of Turkish baths, and occasionally in the Tropics) ether would exist as an easily condensable gas, while in the Arctic regions sulphur dioxide would be known as a very volatile liquid.

In stricter scientific language the term 'gas' is confined to the vapours of substances which are existing under conditions so far removed from those under which they become liquid, that they obey approximately the laws of Gay Lussac (p. 85), and Mariotte (p. 83), when the temperature and pressure are changed.

CIRCUMSTANCES WHICH AFFECT THE FORMATION OF VAPOUR.

Although evaporation is always going on in most liquids, yet the rate at which it proceeds may be increased or diminished very greatly by varying certain conditions, of which the following are the chief:—

1. The extent of surface afforded by the liquid.

Obviously the larger the surface is, the more of the liquid will evaporate in a given time. An illustration of the effect of this is given by the ether-spray apparatus already described (p. 170), the effects of which are mainly due to the extremely rapid evaporation from the surfaces of the multitude of minute drops into which the mass of the ether is divided. When solutions of sugar or of salts, such as alum, are to be concentrated by driving off the water until the substance separates out in crystals, they are always placed in broad pans or troughs so as to expose a large area of surface at which evaporation may take place. So also in Spain and other hot countries salt is obtained from sea-water by admitting the latter into large shallow pools where the water is soon evaporated away by the heat of the sun alone. In Hungary the same quick evaporation is secured by allowing the water from brine-springs to trickle down loose bundles of twigs or strings: it is thus spread over a large surface and so quickly concentrated that masses of salt form on the strings.

2. The quantity of vapour already existing in the space round the liquid.

The molecules which form the vapour are rapidly moving in all directions through this space, and some of them of course strike upon the surface of the liquid and dive into it. Thus a constant exchange is going on, some molecules issuing from, others returning to the liquid; and the amount of the liquid lost by evaporation is evidently determined by the excess in number of the former over the latter. If the space is allowed to become more and more crowded with molecules, a time arrives when just as many of them return to the liquid as issue from it; and then of course there is no further increase in the amount of vapour present in the space round the liquid. This space is then said to be 'saturated' with vapour, and although evaporation does not stop (as it is commonly said to do), its effect in lessening the quantity of the liquid is reduced to nothing.

Maximum Tension of Vapour. 179

Maximum Tension of a Vapour.—When a space is 'saturated' with vapour, *i.e.* when the greatest number of molecules are present in it that can be present under the existing conditions, the vapour always exerts the greatest resistance to compression and liquefaction that it can do, and is then said to have its 'maximum tension.' This it always has when an excess of liquid is present; and if the pressure upon it is increased, or the temperature lowered, some of it becomes liquid, the rest showing the maximum tension corresponding to the new conditions. In fact, whenever the tension of a vapour is referred to, its *maximum* tension is always meant unless the contrary is specified.

It is clear that if the vapour is removed from the space above the liquid as fast as it is formed, the molecules have no chance given them of regaining their places in the liquid, and the latter is quickly diminished in quantity. This principle is practically and usefully applied in many ways. Thus wet clothes and roads dry much more quickly on a windy day than on a calm one. The porous water-coolers mentioned on p. 168, act much more effectively when placed in a current of air. The chilling effects of a draught, even of summer air, are well known, and tea, soup, etc. are cooled by blowing over the surface, and thus removing the vapour as fast as it rises.

A good illustration of the effects of thus quickening evaporation is given by a instrument invented by Dr. Wollaston, and called (rather fantastically) a 'cryophorus' (κρύος φορὸς, frost-carrying). It consists of a glass tube bent as shown in fig. 57, having a bulb at each end, and containing a little water. Before the apparatus is finally sealed up the water is boiled for some time until its vapour has driven out all the air (as was done in Expt. 44, p. 162), and then the tube is quickly closed before the blow-pipe flame. Thus it contains nothing but water and water-vapour, the space above the liquid being saturated as explained just above. Now, if all the water is collected in the upper bulb, and the lower bulb is cooled below the freezing-point, so much heat is withdrawn

Fig. 57.

that all the vapour in the bulb is converted first into water and then into ice; and not only the vapour in the bulb but also (owing to the quick diffusion of gases into an empty space) the vapour from the tube and the other bulb is very soon condensed almost entirely. Then the water in the upper bulb quickly evaporates, and so much heat is spent in the formation of this vapour, that the remaining liquid freezes in a short time into a mass of ice.

Expt. 54. If a cryophorus is at hand, its action may be shown in the manner indicated above, a freezing mixture of ice and salt being used. The upper bulb should not be more than half filled with water (otherwise it may be cracked by the expansion of the ice at the moment of its formation), and it should be surrounded with cotton-wool, or flannel, to prevent the heat of the room from reaching it. In twenty minutes or so a fair amount of ice will be formed.

In chemical work it is often requisite to dry substances completely, which cannot be heated much above the ordinary temperature without decomposition. This is readily done by placing them, spread out in a dish, under a bell-jar in which is also enclosed some substance, such as quicklime or sulphuric acid, which has a strong affinity for water; this absorbs all the water-vapour as fast as it is given off from the substance, and the latter is rapidly dried. One form of such a 'desiccator,' as it is called, is shown at AB in fig. 58.

3. **The amount of other vapours or gases present in the space round the liquid.**

Even though these have no chemical action on, or tendency to dissolve in the liquid, yet by their mere presence they crowd the space and thus impede the movements of the molecules of the liquid in departing from the surface. Hence evaporation is rendered slower, although it is found that practically the same amount of vapour is eventually formed in the space as if no other substance was there. This is true even if many liquids in succession (which have no chemical action on one another) are allowed to evaporate into the same space: each forms, if sufficient time is allowed, as much vapour as if none of the others were present, and exerts its own tension quite independently of the rest, so that the total pressure upon the sides of the containing vessel and upon the surface of the liquids is the sum of the separate tensions due to each vapour. This law was discovered by Dalton in 1802, and is known as 'Dalton's Law.'

Fig. 58.

The rapidity with which evaporation takes place into a nearly empty space is well shown by an experiment due to Leslie, the inventor of the differential thermometer and other apparatus. If a watch-glass containing a little water is supported over a dish of strong sulphuric acid placed under the receiver of an air-pump (fig. 58), and the air is then pumped out of the receiver as completely as possible, the water quickly gives off vapour which is at once absorbed by the sulphuric acid. Thus, as in the cryophorus already described, so much heat is absorbed from the remaining water that the latter freezes. Ice-making machines on this principle have been constructed on a large scale, but the cost of working them is great, and the yield of ice comparatively small.

4. The pressure upon the liquid.

A gas, as has been explained, p. 161, always occupies a much greater space than the liquid from which it is formed, and any pressure which resists this increase in volume impedes the formation of vapour. On the other hand, if the pressure on a liquid is diminished the molecules start from its surface with greater freedom: in other words, evaporation is encouraged and increased. The influence of pressure will be further alluded to presently.

5. The temperature of the liquid.

This is undoubtedly the most important factor in modifying evaporation, since heat must be supplied in order to give the molecules of the liquid sufficient energy to enable them to exist as gas, and rise of temperature implies the communication of energy in the form of heat, in sufficient quantity not only for the above purpose but also to give an increasing tension to the vapour formed. The latter is thus enabled more readily to overcome obstacles such as are due to pressure and to the presence of crowds of other molecules, and is therefore formed more quickly as well as in greater quantity. On the other hand, if the temperature of a substance is reduced to a low point, cohesion asserts its power so strongly that the molecules remain in the liquid or solid condition and little or no vapour is formed. Thus we may state it as a general fact, that

    Increase of temperature } promote vaporisation.
    Decrease of pressure

    Decrease of temperature } hinder vaporisation.
    Increase of pressure

BOILING, OR EBULLITION.

We may next consider the effect of gradually heating a cold, not very volatile liquid, the pressure upon its surface being maintained constant, as is the case when the liquid is in a vessel open to the air. The pressure on its surface is then, of course, that due to the air, indicated by the height of the mercury column in a barometer, and may be assumed not to vary during the process.

As the temperature of the liquid rises, the tension of the vapour which is formed increases, and hence vaporisation proceeds more and more quickly, but only from the surface of the liquid; since the pressure of the air on the surface, added to the pressure of the liquid itself, prevents the formation of vapour in other parts. If, however, the communication of heat is continued, the tension caused by it eventually becomes so great that it overcomes the whole pressure; and vapour is then observed to form in all parts of the liquid (especially in those parts which are nearest to the source of heat). This vapour rises to the surface in bubbles, which burst and discharge their contents into the space above. Then the liquid is said to 'boil.'

Expt. 55. Half fill a flask or beaker with clean water (distilled water, if it is at hand), support it on wire-gauze over a lamp, and heat it gradually till it boils. Note all that happens: first, the disengagement of small bubbles of air which all water contains; then the clouds of condensed vapour, which hover over the surface and increase in quantity as the water becomes hot; then the formation of small bubbles of vapour at the most strongly heated parts of the vessel, which collapse almost immediately when they enter the cooler parts of the liquid; and finally the development of large bubbles which rise through the liquid without contracting, and burst at the top, throwing the whole surface into commotion.

'Boiling,' then, or 'ebullition' (*bulla*, a bubble) may be defined as the condition of a liquid in which vapour is formed in bubbles in all parts of it, and not at the surface only, as in evaporation.

It can, obviously, only occur when the tension of the vapour is at least equal to the pressure at the place where it is being formed. If the application of heat is continued, no further rise in temperature or increase of tension takes place as long as the pressure is unchanged; the whole of the communicated heat being spent in the mere formation of fresh vapour and the maintenance of the existing tension.

Definition of 'the boiling-point.'—The particular temperature called the 'boiling-point' of a substance may now be exactly defined as

> 'That point of temperature at which the tension of the vapour of the liquid is equal to the pressure upon the place where the vapour is being formed.'

A reference to the table of vapour-tensions on p. 176 will illustrate this definition. It will there be seen that there is a particular point of temperature for each of the substances mentioned, at which the tension of its vapour is equal to 760 mm. of mercury column. If 760 mm. be taken as expressing the average pressure of the air, then the above-mentioned temperature will be the approximate boiling-point of the substance when heated in an open vessel under ordinary conditions. Thus the boiling-point of sulphur dioxide will be $-10°$, and that of alcohol $78°$; and so on.

The best method of determining with accuracy the boiling-point of a liquid has been indicated already in treating of the graduation of thermometers, p. 96. It is advisable to support the thermometer so that its bulb may be immersed in the vapour just above the surface of the liquid and not in the liquid itself, because the latter is liable to be rather uneven in temperature owing to the currents which are set up in it. The temperature of the vapour, however, just above the liquid is found to be very constant, and may always be taken as the true boiling-point of the liquid under the existing conditions. The height of the barometer at the time must be carefully noted, and a correction made if it differs materially from 760 mm., as will be more fully explained directly.

Although in very accurate determinations it is essential that the whole of the mercury-column in the thermometer should be exposed to the same temperature as the vapour, as it will be in the apparatus shown in fig. 43, p. 96, yet for ordinary purposes the simpler apparatus described in the following experiment will give good results.

Expt. 56. Fit a cork to a moderately large test-tube and bore a hole in its centre, through which a thermometer is to be passed. Into another hole in the cork fit a small tube bent as shown in fig. 59, to carry off the vapour. Fill about one-fourth of the tube with ordinary alcohol, and put in a bit of crumpled tinfoil (or, better, platinum foil) to promote regularity of boiling (p. 187). Fit the cork into its place, adjusting the thermometer so that its stem passes down the middle of the test-tube and its bulb is just clear of the surface of the liquid. Support the tube in a beaker of water heated over a lamp, and when the alcohol begins to boil note the reading of the thermometer at short intervals until it remains steady. The temperature now indicated will be the boiling-point of the specimen of alcohol used, under the pressure indicated by the barometer, which should be observed.

Fig. 59.

The boiling-point of ether may be taken in a similar way: care being taken that no lighted lamp is near, or the vapour may catch fire.

THE BOILING-POINTS OF DIFFERENT SUBSTANCES
(under a pressure of 760 mm. of mercury-column).

Substance.	Boiling-point.	Substance.	Boiling-point.
Zinc	$930°$	Alcohol	$78°$
Sulphur	$448°$	Ether	$35°$
Mercury	$357°$	Sulphur dioxide	$-10°$
Sulphuric acid	$325°$	Carbon dioxide	$-78°$
Water	$100°$	Oxygen	$-184°$

It may be worth while to explain here the cause of the rather imperfect musical note which is often heard when water is being heated in a kettle or other metallic vessel, and is commonly called the 'singing' of the kettle. A little observation will show that this singing only occurs when a metallic vessel, and not one of glass or earthenware is being used; that it begins at a particular point in the heating of the liquid, rather below its boiling-point;[1] and that it ceases entirely when the liquid boils. Any true explanation of it must account for all these facts. Now, in the experiments on boiling lately made it will have been noticed that bubbles of vapour form in contact with the heated surface of the vessel some time before the actual boiling-point is reached, but that these collapse almost immediately by contact with the colder mass of liquid. It is this quick formation and condensation of small bubbles of vapour which cause the sound: the bubbles, as they are formed, knock against the surface of the vessel, and when they suddenly collapse there is a quick movement of the liquid to fill their places. Thus the liquid as well as the vessel is thrown into a state of vibration which is communicated to the surrounding air, and when these regular vibrations or waves succeed each other more rapidly than about 40 times per second, they produce, as is well known, the sensation of a musical note when they strike upon the ear. It is plain that the quick vibratory movement will only begin when the liquid is sufficiently hot to admit of vapour-bubbles being momentarily formed, and that it cannot occur during actual boiling, because then large bubbles are formed which do not collapse at all. We can understand also why singing would not occur in a vessel made of glass or any bad conductor of heat, since in such vessels the heat is not conveyed from the source to the water with sufficient readiness to form bubbles *quickly*, and hence a rapid vibration cannot be set up.

---

[1] In the case of water, singing commences at about 68° in a copper vessel, and about 80° in one of tinned iron.

Boiling of Liquids.

Circumstances which modify the Boiling of Liquids.

1. The nature of the surface of the vessel in which the liquid is heated.

In order that a bubble of vapour may be formed elsewhere than at the free surface of the liquid, the cohesion of its molecules or their adhesion to the surface of the vessel has to be overcome: they must be thrust apart in order to give the vapour a space to expand into. The work of doing this is much easier when there is any irregularity in the surface of the vessel, or when it is made of a material to which the liquid has only a slight adhesion. Thus, all other conditions being the same, water will boil in a metal vessel at a temperature at least $1°$ lower than that at which it boils in a glass vessel: and if the surface of the glass is perfectly clean and smooth the temperature may sometimes be raised $4°$ or $5°$ above the true boiling-point before ebullition begins.

The effect of projecting points, or sharp edges, or even a slight roughness of surface, in aiding the formation of bubbles is very marked. If, while the water is gently boiling in the flask used in Expt. 55, p. 183, a small bit of coke or a crumpled piece of platinum foil is dropped into it, quantities of bubbles will be seen to form at the projecting points and the rapidity of boiling will be increased. In fact, whenever a steady, uniform rate of boiling is required, it is best to put into the liquid one or two small bits of coke, which from their exceeding roughness of surface seem to answer better than anything else. The effect of such substances is due to their destroying the continuity and evenness of the liquid surface in contact with them, and thus rendering it more easily broken up by the tension of the vapour as it struggles to form: just as a piece of scratched or furrowed glass, or of rusty, corroded iron is easily broken.

2. The presence of air, or other similar gases, dissolved in the liquid.

These greatly facilitate boiling, since their presence between the molecules of the liquid breaks the continuity of the whole mass, and weakens its power of resisting disruption (much in the same way as a sheet of postage stamps is most easily torn along the lines of perforations). When water is freed from air by long continued heating or otherwise, the process of boiling becomes very irregular, and the temperature may sometimes be raised many degrees above the true boiling-point without the formation of any bubbles at all; then an almost explosive burst of vapour takes place, and the temperature sinks to the ordinary boiling-point. It is, indeed, doubtful, whether the boiling of a liquid free from all traces of dissolved gases has ever been witnessed; the steam from water has invariably been found, when condensed, to leave a small bubble of uncondensable gas.[1]

Some liquids, such as alcohol and sulphuric acid, which dissolve but little air, boil in a manner rather difficult to control; they boil with 'bumping,' as it is called, *i.e.* large bubbles of vapour form at irregular intervals and so suddenly and explosively that the vessel, if of glass, may be broken by the concussion. This inconvenience is much lessened by putting into the liquid a few bits of coke (or of platinum foil, if coke is acted on by the substance) which regulate the boiling as above explained.

3. The presence of salts dissolved in the liquid.

It seems likely that there is always some sort of chemical combination between the salt and the liquid which dissolves it; and since some energy must be spent in decomposing this compound before the vapour can be set free, we should expect to find that solutions of salts do not boil until a higher temperature is reached than the boiling-point of the pure liquid. This is found to be the case: a saturated solution of common salt in water does not boil until heated to 107°.

Expt. 57. Dissolve 35 grms. of salt in 100 c.c. of water, and heat the solution in a beaker placed on wire-gauze over a lamp, supporting a thermometer with its bulb dipping into the liquid. Place a small test-tube containing a little water in the solution, and observe that the water in the tube boils before the solution

---

[1] Grove; *Journal of the Chemical Society*, vol. xvi. 263.

BOILING OF LIQUIDS.   189

of salt shows any signs of doing so. Note the temperature of the liquid when boiling commences; it will be found to be nearly 107°.

Another substance, calcium chloride, when added in large quantity to water forms a solution which can be raised to 179° before it boils; and this, as well as the solution of salt above mentioned, is often used in chemical work for heating substances and maintaining them at a definite temperature, above that of boiling water.

4. The pressure upon the liquid.

From the definition of the boiling-point given already, p. 184, it is clear that we can make a liquid boil at almost any temperature we like by exposing it to different amounts of pressure. If the pressure is small, the vapour need only possess a slight tension in order to overcome it and form in bubbles, and thus a comparatively low temperature is sufficient to cause boiling. If the pressure is great, then no bubbles of vapour can be formed until the temperature is considerably raised, so as to give the vapour the high tension requisite to overcome the resistance to its formation. The effects of variation of pressure may be illustrated as follows:—

A. Reduction of pressure lowers the boiling-point.

Expt. 58. If an air-pump is at hand, the following experiment may be made. Heat some water in a beaker (which should not be more than one-third full) to a temperature of about 80°, not higher. Make a loose pad of 6 or 8 folds of coarse thick flannel, thoroughly dried (this is to absorb the steam and prevent its injuring the valves, etc. of the pump), and lay it on the plate of the pump. Place the beaker of hot water on the flannel, cover it with the usual bell-shaped receiver, and work the pump. The pressure of the air upon the liquid is thus lessened, and the water will soon begin to boil violently. If some air is admitted into the receiver by turning the screw plug of the pump, the boiling will cease, but it will recommence if the air is again exhausted.

Ether may in a similar way be made to boil at ordinary temperatures; but the experiment should not be tried with a good air-pump, as the ether damages the valves and packing of the pistons.

Expt. 59. Fit a good cork to a globular flask, about 400 c.c. in capacity; fill the flask about half full of water, and heat it to boiling, laying the cork loosely on the mouth of the flask to allow the steam to escape and to check the inward diffusion of air. When the boiling has continued briskly for three or four minutes, so as to ensure the complete expulsion of air by the vapour, remove the lamp and immediately press the cork firmly into its place, and invert the flask, resting it, neck downwards, upon a ring of the retort-stand. It now contains water at 1° or 2° below 100°, subject to the pressure of its own vapour, which, at the moment the flask was closed, was equal to that of the atmosphere, and is sufficient to prevent boiling from taking place at the existing temperature of the water. Pour a little cold water (or, better, slightly warm water, to avoid risk of cracking the glass) over the flask; this will cause the condensation of some of the vapour and reduce the pressure within the flask, and although the temperature of the water is reduced at the same time, boiling will commence again, and may be maintained for some time by pouring more cold water upon the flask so as to condense the vapour as fast as it is formed.

An important application of this principle is made on a large scale in sugar refineries. The solution of raw sugar is comparatively weak, and some of the water must be removed by evaporation in order that crystals of the pure substance may be deposited from the liquid as it cools. If the evaporation is carried on at 100°, or nearly so, a large portion of the sugar is found to be converted by the high temperature into an uncrystallisable substance resembling ordinary treacle; and a great loss of material was formerly occasioned by this fact, only about 30 per cent. of the sugar present in the juice being obtainable in a pure, crystallised form. In 1812, however, a manufacturer named Howard introduced the system of evaporating the solution of sugar in large pans, having close-fitting dome-shaped covers, from which the air and vapour were rapidly withdrawn by pumps worked by a steam-engine. The boiling-point of the solution was thus brought down to 63° or less, at which temperature little or no deteriorating change in the sugar takes place.

B. Increase of pressure raises the boiling-point.

In order to increase the pressure upon a liquid, and yet leave a free space above it into which vapour may diffuse, we must of course increase the quantity of gas or vapour in the space. This may be done either by pumping in more gas, or, more simply, by closing the vessel so that the vapour which is formed as the temperature is raised cannot escape, but must exercise an increasing pressure upon the liquid.

Expt. 60. If sufficient mercury is at hand (about a kilogramme is wanted), the following experiment may be made. Adapt to a globular flask, such as was used in the last experiment (it must be of fairly strong glass), a tube bent at right angles in two places, like that shown in fig. 33, p. 65, but having the right-hand branch about 30 cm. long. Fit also into the same cork a thermometer, with its bulb reaching down to the surface of the water, with which the flask should be about half filled. The cork should be tied down with wire or string, to prevent its being forced out by the pressure of the steam. Dip the open end of the tube deeply into a tall, narrow jar filled with mercury,[1] so that the end of the tube is about 20 cm. below the surface of the mercury. The jar should be placed on a tray or large plate, to catch any mercury which may be spilled accidentally. Heat the flask carefully on wire-gauze or a sand-bath (since the pressure inside will rise decidedly, though not dangerously, above that of the atmosphere), and observe that the thermometer does not cease rising when 100° is reached, but goes up to 106° or so before regular boiling goes on. If, for instance, the column of mercury in the jar is 180 mm. above the end of the immersed tube, and if the barometer stands at 760 mm., the total pressure to be overcome before vapour can escape is $(760 + 180 =)$ 940 mm, and at this pressure the boiling-point of water is very nearly 106°.

---

[1] One of the jars on feet sold for use with hydrometers is the most convenient for the purpose; but a long wide test-tube of stout glass, or a narrow cylindrical lamp glass, closed at one end with a tight-fitting cork, will answer.

This experiment illustrates the conditions which exist in an ordinary high-pressure steam boiler. The pressure of the steam accumulated in it prevents the water from boiling until a temperature much above 100° is reached. Thus if steam is supplied to the engine at a pressure of 4 kilogs. per square centimetre (58 lbs. per square inch), the boiling-point of the water will be 143°; and it must be maintained at this temperature in order to afford a continuous supply of steam. In the boilers of locomotive engines a pressure of 10 kilogs. per sq. cm. (about 140 lbs. per sq. in.) is the usual one, and the water in such boilers must be maintained at a temperature not less than 180°, which is not much below the melting-point of soft solder (an alloy of tin and lead).

TABLE OF THE BOILING-POINTS OF WATER UNDER DIFFERENT PRESSURES.

Pressure (in mm. of mercury column).	Boiling-point (approximate).	Pressure (in mm. of mercury column).	Boiling-point (approximate).
100	52°	900	105°
300	76°	1000	108°
500	89°	3000	144°
700	98°	5000	163°
760	100°	7000	177°

The possibility of thus keeping water in the liquid state at such high temperatures as would cause its immediate conversion into steam under ordinary conditions is of great importance in several manufactures, for water has its power of dissolving most substances greatly increased by raising its temperature. The earliest practical application was made by Papin in what he called a 'digester,' a drawing of which (from the paper which he communicated to the Royal Society in 1681) is given in fig. 60. It consists of a strong jar of iron, with a cover held firmly down by screw clamps. In the cover is a conical hole, closed by an accurately fitting plug, which is pressed down by a lever, on the long arm of which a weight is arranged to slide, so as to give varying amounts of pressure according to its position on the lever. This contrivance, which appears in every modern steam boiler under the name of 'the safety valve,' is intended to prevent the pressure of the steam rising to a dangerous extent, for as soon as the force is sufficient to lift the valve, the steam escapes through the opening, and no further increase of pressure or temperature is possible.

A more modern form of digester, fitted with a pressure gauge to indicate the actual tension of the vapour, is shown in fig. 61.

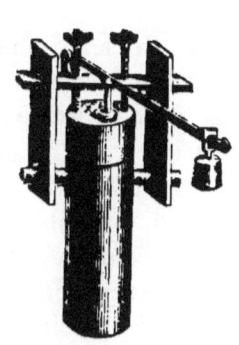
Fig. 60.

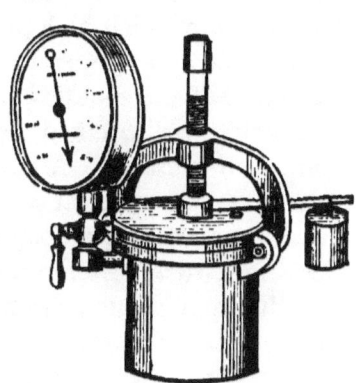
Fig. 61.

One great use to which such an apparatus as the above is applied (and indeed the chief purpose for which Papin designed it) is the extraction of gelatine from bones. These latter consist mainly of an earthy skeleton (of calcium phosphate and carbonate) quite insoluble in water at any temperature, and of a network of animal tissue chiefly composed of gelatine, which, though insoluble in water at ordinary temperatures, and not very soluble even at 100°, is easily acted on by water at higher temperatures. Thus the extraction of it by water, a tedious process when carried on in an open vessel, is very rapid and complete when the bones are put with water into a digester and the temperature raised to 150° or so. A simple form of digester is employed for this purpose in kitchens, for making soups and jellies, the basis of which is usually some form of gelatine.

Many chemical actions which do not occur, or occur very slowly and imperfectly under ordinary conditions of pressure are conveniently carried on by heating the necessary substances in a strong closed vessel, which on the small scale is usually a sealed glass tube, but on the large scale is a wrought-iron cylinder called an 'autoclave.' Several of the processes employed in making dyes from the products of coal-tar are thus carried out; and 'soluble glass' (sodium silicate), a solution of which is extensively used as a varnish for fresco paintings, and for protecting stone from the action of weather, is made by dissolving crushed flint pebbles in a solution of caustic soda placed in a large boiler, into which high-pressure steam is admitted so as to keep the whole at a temperature of $150°$.

The action of the spouting hot-springs called 'Geysers,' which occur in volcanic districts, depends on the variation of the boiling-point with pressure. A geyser is a spring which discharges torrents of boiling water at somewhat irregular intervals through a vertical tube extending deep into the earth. At the bottom of the tube is a rocky reservoir, and both this and the tube are filled with water heated strongly from volcanic sources. The pressure of the column of water in the tube, however, added to that of the atmosphere, prevents the water in the reservoir from boiling at the usual temperature: and thus its temperature rises until the corresponding tension of vapour becomes so great as to overcome the whole pressure. Then a large volume of vapour is suddenly formed in the reservoir and upheaves the column of water in the tube, which overflows at the top. The pressure below is thus relieved, and more vapour is quickly formed which drives up most of the remaining water in a column sometimes 90 or 100 metres in height. The rest of the water loses so much heat in the formation of this vapour that it sinks below the boiling-point, and the action subsides until the reservoir is again filled with water percolating through the rocks, which is gradually heated until it boils over as before.

GEYSERS.

Expt. 61. The action of a geyser may be easily illustrated by the apparatus shown in fig. 62. Fit a tube about 1 cm., or rather more, in internal diameter, and 60 or 70 cm. in length, to a globular flask, such as that used in previous experiments. Into the same cork fit a much smaller tube reaching nearly to the bottom of the flask, bent as shown in the figure, and connected by an india-rubber tube (furnished with a screw-clamp or 'pinch-cock') with a bottle containing water, supported at about the same level as the upper end of the wide tube. This latter should project 6 or 8 cm. through the bottom of a rather large tin basin intended

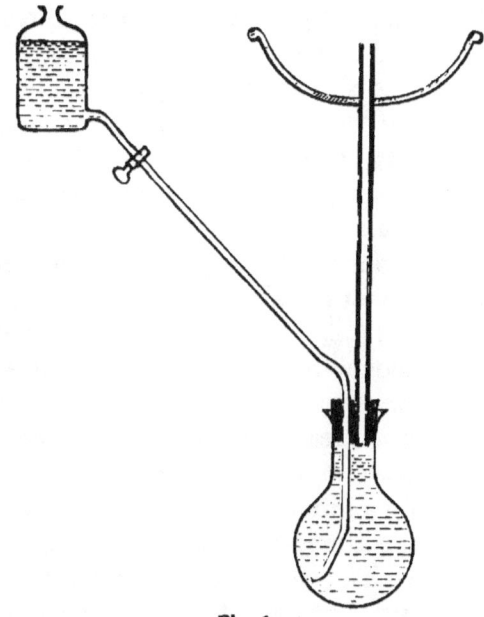

Fig. 62.

to catch the ejected water. Fill the whole apparatus with water nearly up to the level of the top of the wide tube, regulating the screw pinch-cock so that a slow stream of water may flow into the flask whenever the level of the water in the main tube is lowered. If heat is now applied to the flask, a point will be reached at which a sudden burst of vapour will, as above explained, drive out all the water in the tube followed by some of the contents of the flask, in the form of a fountain. Then the action will subside for a minute or two, and the flask will be filled up again with water through the supply-tube: and when this has become sufficiently hot another outburst will take place as before.

Measurement of the Height of Mountains, etc., by Observations of the Boiling-point of Water.—The stratum of air surrounding the earth's surface may be looked upon as a vast fluid sea kept upon the earth by gravitation, and the pressure of this upon objects immersed in it will vary at different levels; being greater at low levels, since here they have above them a longer column of air-particles which by their weight give rise to the pressure (just as the lowest in a pile of books is more compressed than a book near the top, having to sustain the weight of more books above it). Hence it is clear that by determining the pressure of the atmosphere existing at two different places we can ascertain the difference in level (if any) between the two places.[1] The most direct way of doing this is, of course, the employment of a barometer; and although in exact calculations many things, such as temperature, humidity of the air, latitude of the place, etc., must be taken account of, it may be stated that, for heights not exceeding 1500 metres above the sea-level, a difference in pressure represented by 1 cm. of mercury column corresponds to a difference in level of 110 metres (a difference of 1 inch to 900 ft.). But a barometer (except in the form of an aneroid) is a cumbrous apparatus to carry up a mountain, and it is easy to see that if we know the exact temperatures at which water boils under different given pressures, the observation of the boiling-point by a good thermometer at different places will give as accurate an indication of the difference in level between them as the direct readings of a barometer. Thus it is found that, for temperatures between 95° and 100°, a difference of 1° in the boiling-point corresponds approximately to a difference of 26 mm. in pressure, and to a

---

[1] The observations must be made as nearly simultaneously as possible, the conditions of weather must be settled and uniform, and the places must not lie far apart horizontally; for many causes, *e.g.* storms, may produce variations in pressure independently of difference in level.

difference of 286 metres in level. This method has the advantage of readiness and simplicity, pure water (or snow) being obtainable almost everywhere; and very portable forms of apparatus similar to that represented in fig. 43, p. 96, are constructed for such observations, provided with extremely delicate thermometers, graduated to tenths of a degree. These instruments are called Hypsometers (ὕψος, *height;* μετρεῖν, *to measure*).

The approximate boiling-points of water at several elevations are given in the following table, the height of the barometer at the sea-level being assumed to be 760 mm.

Place.	Boiling-point of Water.	Height above Sea-level.
		Metres.
Sea-level	100°	0
Top of Ben Nevis (the highest point in Great Britain)	95.5°	1343
City of Quito (on the Andes)	90°	2900
Top of the Peak of Tenerife	88°	3720
Top of Mont Blanc . .	86°	4810

Even at the same place the boiling-point of a liquid may vary as much as 2° owing to changes of pressure which accompany changes of wind and weather; and hence in determining the boiling-point of a liquid, which is often of great scientific importance, the barometric pressure must always be observed and specified. The same precaution must be taken in graduating or verifying the accuracy of a thermometer: for the level of the top of the mercury column when the instrument is placed in the steam from boiling water (p. 96), can only be marked '100°' correctly when the barometer stands at 760 mm. Thus if the reading of the barometer is 733 mm., the position of the top of the mercury column in the thermometer denotes a temperature of 99°, and must be so marked: if the reading is 788 mm., the corresponding temperature is 101°.

It may be useful to give here a short summary of the facts and laws which have already been observed and explained respecting the formation of vapour.

Laws of Vaporisation.

1. Everything gives off vapour continually into any space beyond its surface, unless the temperature is very far below its boiling-point.

2. All vapours exert a 'tension,' or outward pressure, which varies with the temperature and the nature of the substance.

3. The rapidity with which vaporisation takes place depends on

   (*a*) the tension of the vapour at the existing temperature;
   (*b*) the extent of surface of the substance;
   (*c*) the amount of vapour already existing in the space beyond the surface;
   (*d*) the temperature of the substance;
   (*e*) the pressure upon the point where the vapour is being formed.

4. When a space contains as much of a vapour as can exist there under the given conditions, it is said to be 'saturated' with the vapour.

5. When several substances emit vapour into the same space, each does so independently of the others, so that (*a*) the amount of each vapour which is formed, (*b*) the tension of each vapour is ultimately the same as if nothing else was present in the space (*Dalton's Law*).

6. When the tension of a vapour is equal to the pressure existing at the point where it is being formed, vaporisation takes place very quickly, and the liquid is said to 'boil.' The temperature at which this occurs is called the 'boiling-point.'

7. The boiling-point varies, when all other conditions are unaltered, directly with the pressure upon the substance.

Section IV.—Liquefaction of Vapours and Gases.

It is evident from what has been said already that in order to reduce a vapour or gas (the latter being simply the vapour of a very volatile substance, p. 177) to the state of liquid, we must in some way enable cohesion to exercise its controlling force on the movements of the molecules. We have the choice of two ways of doing this.

1. We may take away heat energy from the vapour (lower its temperature, in fact), and thus reduce the velocity of the molecules until it falls below that which is necessary to counteract their natural attraction. Then the latter asserts its power, and the vapour collapses into a coherent liquid.

2. We may, without altering the temperature, force the vapour to occupy a smaller and smaller space until the utmost tension that it can exert at the existing temperature is overcome, and its molecules are driven within the range of each other's attraction. Then, as before, the vapour usually condenses into a coherent liquid.

Liquefaction of Vapours by reduction of Temperature alone.—Examples of this method, as applied to such vapours as steam, are so common as hardly to need illustration by experiments. If a cold glass tube or beaker is held in the invisible steam rising from boiling water, small drops of liquid water are quickly formed upon it. The steam which comes out of the spout of a kettle or the chimney of a locomotive engine is invisible as long as its temperature remains at or above 100°; but it is almost immediately cooled down by contact with the air so far that its molecules collect together to form a visible cloud of minute particles of liquid water. Similarly the water-vapour which comes out of the mouth at the temperature of the body is, on a frosty day, cooled down into a liquid, visible condition.[1]

The methods of condensing vapours in the process of distillation, which will be described in the next section, also afford examples of the method.

When sulphur dioxide, the substance obtained by burning sulphur in air or oxygen, which is in these latitudes a fairly per-

---

[1] In these and similar cases evaporation commences immediately from the surface of each drop, and goes on more or less quickly according to the amount of vapour already present in the air; and so the white cloud soon disappears.

manent gas, is passed through a bent U-shaped tube as represented in fig. 63 (or, better, through a spiral tube like that shown in fig. 64) surrounded by a freezing mixture of ice and salt, and thus cooled below $-10°$ (its boiling-point under ordinary pressure) it condenses into a colourless liquid. It may be preserved in this condition by closing the ends of the tube before the blowpipe-flame so as to prevent the liquid expanding into vapour when the temperature rises.

Fig. 63.

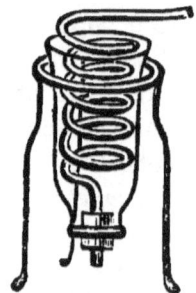
Fig. 64.

Liquefaction of Vapours by Increase of Pressure.—This method is employed practically on a large scale in reducing to the liquid condition not only sulphur dioxide, the gas mentioned just above, but also such very permanent gases as carbon dioxide ('carbonic acid') and nitrogen monoxide ('nitrous oxide' or 'laughing gas'). The gas, dried and purified, is forced by a powerful pump into a strong cylindrical bottle of wrought-iron or steel kept cool in water,[1] until such a crowd of molecules is confined in the space that all the additional gas which is forced in assumes the liquid state. In the case of sulphur dioxide this occurs when three times as much gas has been forced into the bottle as it would contain under ordinary conditions (*i.e.* under a pressure equal to

---

[1] This is requisite because, as already explained on p. 164, vapours in becoming liquids necessarily give out a large quantity of potential heat; and this, unless at once withdrawn, would increase the tension of the rest of the vapour and make liquefaction more difficult.

LIQUEFACTION OF VAPOURS.

that of the atmosphere) and a temperature of $15°$.[1] But carbon dioxide resists liquefaction until the pressure due to the blows of the collected molecules against the vessel is equal to that of nearly 40 metres of mercury column, *i.e.* about 52 times that of the atmosphere (or 52 'atmospheres,' as it is called for brevity) at the ordinary temperature of $15°$. Liquid carbon dioxide thus obtained is used for the production of very low temperatures, as already explained, p. 171, and also in the manufacture of 'soda water,' which is a solution of the gas in water. In the case of nitrogen monoxide a pressure nearly as great, *viz.* 50 'atmospheres,' is required to liquefy it at $15°$. This substance is sent out, stored as a liquid in strong iron bottles, for the use of dentists who employ it for producing insensibility to pain during the extraction of teeth. For this purpose a small quantity of the liquid is let out from the bottle, and when relieved from pressure it quickly becomes gas, which is collected in an indiarubber bag; and rom this it can be breathed in the usual way.

There is a simpler method of collecting the necessary number of molecules into the confined space, *viz.* by generating the gas in the vessel itself and carrying on the action until a sufficient quantity is liberated from the materials. This was, in fact, the original mode in which Faraday, in 1823, reduced many gases, previously thought to be uncondensable, to the liquid condition. His plan was to place the required materials in one branch, *a*, of a very strong glass tube closed at one end and bent into the form shown in fig. 65. The other end of the tube, *b*, was then drawn out and sealed by a blowpipe-flame, and this part of the tube was kept cool by water or a freezing mixture while the gas

---

[1] This is so moderate an increase of pressure that the liquefaction of sulphur dioxide may be shown without employing an expensive pump, by fitting an indiarubber plug, well greased, into a strong glass tube about 30 cm. long and 1.5 cm. in internal diameter, drawn out to a point at one end and sealed: then filling the tube with dry sulphur dioxide (for the method of preparing it a text-book on chemistry must be consulted), and forcing the plug down the tube by a wooden rod inserted into or cemented to it. A small drop of the liquefied gas will form in the narrow, drawn-out end of the tube, and will expand into gas again when the pressure is removed.

was generated in the part a: the liquefied substance soon began to collect in the cooled end of the tube. A similar method has been employed for obtaining liquid carbon dioxide on a large scale, but it is dangerous on account of the difficulty of controlling the action between the substances from which the gas is prepared.

Fig. 65.

Liquefaction of Gases by simultaneously cooling and compressing them.—There is an obvious advantage in combining both the methods above given: and it was by doing so that Faraday succeeded in liquefying nearly all known gases. Oxygen and nitrogen, however, the two gases which compose air, and also hydrogen, a gas present in large proportion in ordinary coal gas, and one or two other less common gases, withstood all attempts to turn them into liquids until the end of the year 1877. At that date M. Pictet, of Geneva,[1] by the employment of a pressure of about 400 atmospheres combined with a reduction of temperature to $-130°$ (produced by the rapid evaporation of liquid carbon dioxide) succeeded in obtaining oxygen in the liquid state. Hydrogen was also liquefied by him in a similar way, but a still higher pressure was required, *viz.* 650 atmospheres, as well as a temperature of $-140°$. Since then means have been found to obtain still lower temperatures, *viz.* by the rapid evaporation of liquid ethylene (the gas to which coal gas owes the brightness of its flame), and M. Olszewski, a Russian chemist,[2] has prepared comparatively large quantities of liquid

---

[1] *Annales de Chimie et de Physique* (5th series), vol. xiii. 145. Roscoe and Schorlemmer's *Treatise on Chemistry*, vol. ii. Part II. p. 516.

[2] *Comptes Rendus*, vol. ci. 238. *Journal of the Chemical Society*, vol. xlviii. 1101. *Nature*, vol. xxxvi. 105.

oxygen and nitrogen and used them as a means of obtaining still lower temperatures: so that now there is not a single gas which has not been reduced to the liquid state. Hydrogen has presented the greatest difficulty of all gases, requiring a temperature of $-213°$ (nearly the lowest point as yet reached) and a pressure of 190 atmospheres for liquefaction.

Critical Temperature of Gases.—It will be interesting to examine what will be the effect of reversing, in part, the above process, and increasing the temperature of a liquid and the pressure upon it simultaneously. We shall then have two influences acting on the liquid, one favourable, the other unfavourable to vaporisation: which of the two will ultimately prevail? Experiments on this point were made by M. Cagniard de la Tour in 1822.[1] He filled a strong glass tube about two-fifths with alcohol, sealed the end of the tube and heated it cautiously. As the temperature rose the vapour which was formed, having only a small space to occupy, exercised a continually increasing pressure on the liquid and prevented it from boiling. As still higher temperatures were reached, the liquid expanded to at least double its original bulk, and at 259° the definite surface between it and the vapour suddenly disappeared, and the whole tube appeared to be filled with vapour only. Similar experiments were made with ether and water, with similar results; ether at 188° and water at about 412° disappearing altogether as a liquid within the tube, but suddenly appearing again with a definite surface as the temperature sank. Since then Dr. Andrews[2] has treated liquid sulphur dioxide and carbon dioxide in the same way, in an apparatus by which he could, while the high temperature was still maintained, vary the pressure within wide limits. He found that at a temperature of 31°, and a pressure of 48 atmospheres, liquid carbon dioxide lost its definite surface, and became undistinguishable as a liquid, and that if the temperature was raised ever so slightly above 31°, no amount of pressure that he could apply would cause the

---

[1] *Annales de Chimie et de Physique* (2d series), vols. xxi. 127, and xxii. 410.
[2] *Philosophical Transactions of the Royal Society*, vol. clix. 565.

liquefaction of the vapour. This temperature, *viz.* 31°,[1] he called the 'critical temperature' of carbon dioxide, since this temperature marks a definite stage or crisis in the behaviour of the substance; for while it can be liquefied by sufficient pressure at any temperature below 31°, at any higher temperature no amount of pressure seems able to produce the effect. The critical temperatures of a few substances are given below.

Substance.	Critical Temperature.	Substance.	Critical Temperature.
Ether	188°	Oxygen	−113°
Sulphur dioxide	155°	Nitrogen	−146°
Carbon dioxide	31°	Hydrogen	−200° (?)

We can now understand why so much difficulty was formerly experienced in liquefying some gases, such as oxygen. Unless the temperature is reduced below the critical point (in the case of oxygen, −113°) it is of no use whatever to compress the gas with a view to reduce it to a liquid. Oxygen and nitrogen were exposed to pressures estimated as at least 2000 atmospheres; but since the temperature was not also reduced sufficiently, no signs of liquefaction occurred.

The explanation of the above facts is not difficult. In order that a substance may appear in the liquid state, the conditions must be such that cohesion is capable of acting between the molecules. But cohesion is a limited force, and we may communicate to the substance so much energy in the form of heat (as indicated by the temperature) as to make it totally impossible for cohesion to act, however near the molecules are brought to one another by pressure on the mass. The substance will then retain the properties of a gas under all pressures.

SECTION V.—**The Density of Vapours and Gases.**

The term 'density,' as already explained (p. 54) has reference to the quantity of matter in a substance; and in order to give

---

[1] More exact determinations give 30·92° as the critical point.

definiteness to our ideas of it, a certain volume of some substance is taken as the standard, and the weight of this is carefully ascertained. Then a comparison between this weight and the weight of the same volume of any other substance under exactly the same conditions will indicate how much more or less matter there is in the latter substance, *i.e.* what its relative density is. Thus 1 c.c. of hydrogen gas, at a temperature of 0° and under a barometric pressure of 760 mm., is found to weigh 0·0000896 grm., and one c.c. of the vapour of water under the same conditions would weigh 0·000806 grm.; and since the latter is nine times as great as the former weight, we conclude that there is nine times as much matter in 1 c.c. of water-vapour as there is in 1 c.c. of hydrogen, and therefore the density of water-vapour is said to be nine times that of hydrogen gas.[1]

The determination of the relative densities of different vapours is of the highest importance, since it affords the most conclusive evidence of the relative weights of the molecules of the substances. For, if we once for all admit the truth of Avogadro's hypothesis that 'equal volumes of different substances in the state of gas contain under the same conditions the same number of molecules,' it is clear that the weights of these equal volumes will bear the same proportion to one another as the weights of the single molecules of the substances (just as the weight of 100 cricket-balls bears to the weight of 100 racquet-balls the same proportion that the weight of one cricket-ball bears to that of one racquet-ball). For example, since the density of water-vapour is found to be nine times that of hydrogen, we are justified in concluding that the weight of a molecule of water is nine times as great as the weight of a molecule of hydrogen.

---

[1] Air was formerly taken as the standard for densities: but it is now for many reasons considered preferable to take hydrogen, which (1) has the lowest density of any known form of matter, (2) is a definite chemical substance, (3) is the accepted standard to which the weights of atoms are referred.

206 DENSITY OF VAPOURS AND GASES.

The following are the principles of the chief methods devised for ascertaining the densities of gases and vapours: for fuller details a larger treatise on physics must be consulted.

1. *Regnault's Method*[1] (applicable to permanent gases only).—A large glass globe is exhausted and its weight ascertained. It is then filled with hydrogen at a known temperature and pressure, and weighed again; the increase in weight shows, of course, the weight of the hydrogen which fills the globe. It is next exhausted and filled with some other gas, such as oxygen, at the same temperature and pressure, and weighed: the increase in weight above that of the empty globe gives the weight of the oxygen which fills the globe. Thus the weights of equal volumes (since the same globe is used) of hydrogen and oxygen are ascertained, and a simple proportion sum gives the relation between them. Thus—

Wt. of hydrogen : Wt. of equal volume of oxygen : : 1 : Density of oxygen.

2. *Dumas' Method.*[2]—(Principle,—to determine the weight of a known volume of the vapour.)—A small globular flask, with long, drawn-out neck, containing a little of the liquid, is heated until the liquid has all boiled away and the vapour which fills the flask has a temperature 30° or 40° above the boiling-point of the substance. Then the end of the neck is sealed by a blowpipe-flame, and when cool the flask is weighed. Deducting from this weight the weight of the empty flask, we have the weight of the vapour contained in the flask, and this compared with the weight of the same volume of hydrogen, at the same temperature and pressure, gives the density of the vapour referred to that of hydrogen.

For example, 200 c.c. of the vapour of alcohol, at 160° and under a barometric pressure of 760 mm., were found to weigh 0·26 grm.; and 200 c.c. of hydrogen at the same temperature

---

[1] *Annales de Chimie et de Physique* (3d series), vol. xiv. 211.
[2] *Ibid.* (2d series), vols. xxxiii. 337, and xxxiv. 326.

DENSITY OF VAPOURS AND GASES. 207

and pressure weigh 0·0113 grm. (nearly). Then, by the proportion,

Weight of 200 c.c. of hydrogen.	:	Weight of 200 c.c. of alcohol.			
0·0113 grm.	:	0·26 grm.	: :	1 : 23	(nearly)

we learn that the density of alcohol vapour is 23 times that of hydrogen.

3. *Meyer's Method.*[1]—(Principle,—to measure the volume of air which escapes from a heated vessel into which a known weight of the liquid or solid substance is introduced.)—This, which is now superseding the older methods, is carried out as follows. A glass tube, fig. 66, about 60 cm. in length, ending below in a cylindrical bulb, closed at the top by a cork, and having a short bent delivery-tube attached at the side near the top, is surrounded by a wider tube containing oil, and heated to a temperature much above the boiling-point of the substance to be examined; the contained air, as it expands by the heat, being allowed to escape freely from the delivery-tube, the end of which dips under water in a small trough. When the air ceases to escape, a weighed quantity of the substance is dropped down the tube, and the cork quickly replaced. The substance soon turns into vapour, and the tension of this vapour is added to that of the enclosed air. The result of this increase of pressure in the vessel is that some air forces its way out through the delivery-tube, and this is collected in a jar full of water, which is inverted over the end of the tube. No vapour escapes, for the distance from the bulb to the delivery-tube is considerable, and the vapour would take a long time to diffuse upwards and reach the latter. The volume of the air which escapes is carefully measured, and it is pretty easy to see that this volume must be

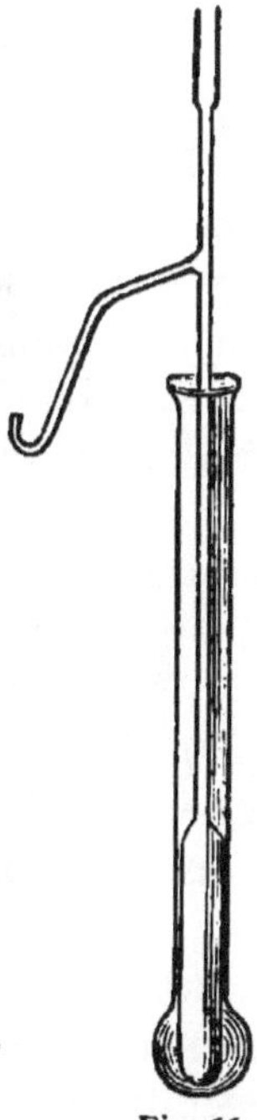

Fig. 66.

---

[1] Roscoe and Schorlemmer's *Treatise on Chemistry*, vol. iii. Part I. p. 100.

exactly equivalent to that of the vapour which has been added to the contents of the vessel, if that vapour could be cooled down to the temperature of the measuring tube: for by Gay Lussac's law (p. 85) the air contracts on cooling to just the same extent as the vapour would do if it could retain the state of gas. The weight of this known volume of vapour must be equal to that of the substance originally taken: and it only remains to compare this weight with that of the same volume of hydrogen, as above explained.

Section VI.—Distillation.

This is the name given to the process of converting a volatile liquid into vapour, condensing the vapour in a separate vessel by cooling it, and collecting the liquid product, or 'distillate.' It is employed chiefly for the purpose of separating the more volatile ingredients of a mixture from those that are less volatile: for if such a mixture is gradually heated, the substance which has the lowest boiling-point will vaporise when the temperature reaches this point, and absorb nearly all the heat supplied, so that, until it has all volatilised, comparatively little vapour of the other substances present in the mixture will be formed.

An apparatus for distillation consists of three parts :—

1. The boiler, or 'still,' in which the mixture is heated.
2. The condenser, or arrangement for cooling the vapour until it becomes liquid, connected with the boiler by a pipe.
3. The receiver, for collecting the condensed liquid.

For distillation on a small scale, the boiler is generally a glass 'retort,' fig. 67, a form of apparatus much used by the ancient alchemists, which is merely a pear-shaped flask with the long neck bent back (Lat. *retortus*) and sloped downwards so that no condensed liquid can run back into the body, where the rest is being heated. It should have a short neck, or 'tubulure,' with stopper, a little below the highest part of the bend, as shown in the figure, through which liquids can be poured, or a thermometer introduced. Such a retort is called a 'tubulated retort.'

DISTILLATION.

For many purposes the long neck itself of the retort, if surrounded with a cloth or some blotting-paper kept wet with cold

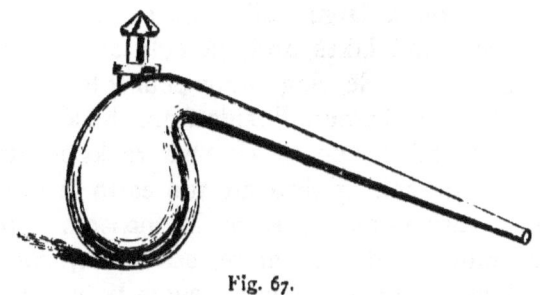

Fig. 67.

water, forms a sufficient condenser: but a preferable apparatus is shown in fig. 68, called a 'Liebig's condenser.' It consists of

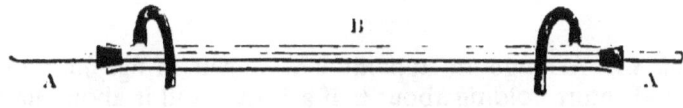

Fig. 68.

a long glass tube, AA, supported in a slanting position, its upper end being connected with the neck of the retort, while its lower end projects into the receiver. This tube is fixed by corks in the middle of a much wider tube B, having side-openings near each end through which a stream of cold water is kept constantly flowing, entering at the lower end from a reservoir at a high level, and escaping from the upper end. The vapour in passing through the inner tube is effectively cooled and condensed into a liquid.

As a receiver a flask or bottle answers well, placed so that the end of the condenser-tube passes down its neck, and immersed in cold water if the distillate is very volatile.

The two following examples will illustrate the applications of distillation :—

DISTILLATION.

1. *Purification of Water.*

Ordinary water, owing to its remarkable powers of dissolving substances, is seldom or never quite pure. It is true that Nature distils it for us on a large scale, the sun's heat raising it in vapour from seas and lakes, and the cold of the upper regions of the atmosphere condensing the vapour into clouds and rain; but the rain dissolves carbon dioxide from the air as it falls, and this increases its solvent power on such rocks as limestone and chalk, so that in sinking through the earth it dissolves these rocks besides various salts present in the soil. Hence spring water, river water, and, still more, sea water, contain a large number, and in some cases a large quantity of substances such as compounds of calcium and magnesium, besides common salt, of which latter there are about 29 grms. in a litre of sea water. All these, however, are less volatile than water itself, and hence the latter can be separated and obtained pure by distillation.

Expt. 62. Arrange an apparatus as shown in fig. 69, taking a tubulated retort holding about half a litre. Fill it about half full (not more, or some of the liquid may in boiling be thrown over into the neck) of common water, the dirtier the better, so as to illustrate the removal of *all* impurities and not only those in solution: and heat it by a lamp. Connect the lower side-tubulure of the condenser by an indiarubber tube, arranged as a syphon, with a pail or jug of cold water supported on a stool. Place another jug under the end of the upper exit-tube, and regulate the flow by a pinch-cock so that a slow stream of cold water passes through the outer tube, or jacket, of the condenser. While the water in the retort is getting hot, some of the same water may be examined for impurities if any simple chemical tests are at hand. For example,—

(*a*) Add to a little of it, in a test tube, (1) a drop or two of pure hydrogen nitrate (nitric acid), (2) a drop of solution of silver nitrate ('lunar caustic'). If a white turbidity or 'precipitate' is formed, sodium chloride (common salt) or some other chloride is present in the water.

(*b*) Add to another portion of the water an equal volume of lime-

DISTILLATION.

water.[1] If a white precipitate is produced, some carbonate (probably calcium carbonate) is present.

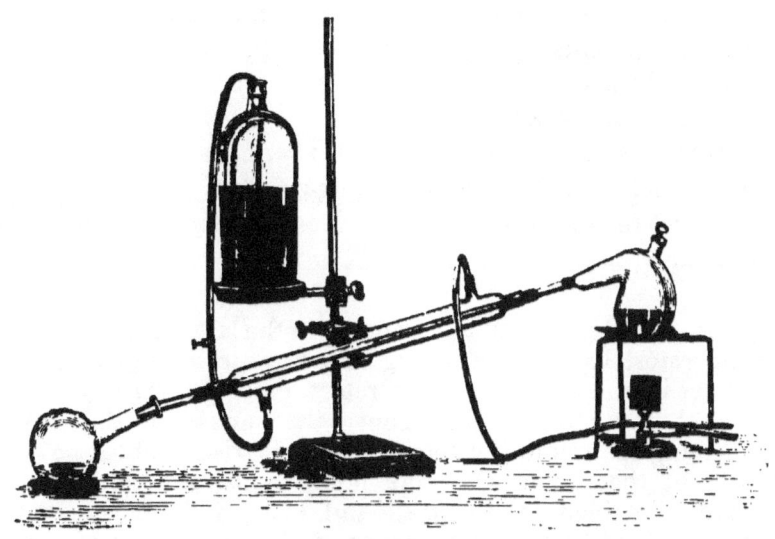

Fig. 69.

(c) About half a litre of the water may be evaporated to dryness in a porcelain dish. The amount of residue left will indicate the quantity of impurities in the water.

When the water in the retort boils, the steam will pass into the cooled tube and be condensed, and pure distilled water will collect in the receiver. When about 30 or 40 c.c. have come over, it may be examined by the above tests[2] to prove its freedom from impurities.

This experiment illustrates the process always used to obtain pure water for scientific and especially chemical purposes;

---

[1] Directions for making this are given in the Appendix.
[2] For other tests, *e.g.* barium chloride for sulphates, ammonium oxalate for calcium salts, and potassium permanganate for organic matter, a chemical text-book must be consulted.

nothing but distilled water should be used for exact experiments, although rain water is often pure enough for many purposes.

It also shows how water wholesome for drinking may be obtained from sea water on board ship: and most vessels carry a distilling apparatus of some kind to be used if the supply of fresh water should fail.

2. *Separation of Alcohol from Water.*

Alcohol, as seen from the table, p. 185, boils at a temperature of 78°, which is 22° below the boiling-point of water; so that if a mixture of the two is heated most of the alcohol distils over first.

Expt. 63. Arrange an apparatus as in the last experiment, and fill the retort half full of strong beer or wine. Support a thermometer in the tubulure of the retort, passing it through a cork so far that its bulb is just above the surface of the liquid. On heating the retort, the temperature will rise to about 90° and remain at that point while the liquid boils, and fairly strong alcohol, identified by its smell and taste, will collect in the receiver. Watch the thermometer, and when the temperature begins to rise above 96°, change the receiver for another. Almost all the alcohol will now have passed over, and only pure water will be collected, the temperature rising to 100°, and remaining there.

This experiment illustrates a process called 'fractional distillation,' very frequently resorted to in order to separate liquids which have different boiling-points from a mixture of them. In distilling such a mixture, the product is collected in one receiver as long as the boiling-point remains constant; but when the temperature rises and becomes constant at another point, the receiver is changed and another of the liquids is collected; and so on for the other ingredients of the mixture, each portion of the distillate being labelled to show the temperature at which it passed over. It must be observed, however, that one distillation alone does not yield the substances in a state of absolute purity. Vapours have, as has been already explained, a decided amount

FRACTIONAL DISTILLATION.

of tension at temperatures much below their boiling-points, and hence some of the substances of higher boiling-point will vaporise and pass over with the more volatile ingredients. Moreover, the formation of vapour of one liquid has a remarkable effect in promoting the evaporation of other liquids mixed with it, one set of molecules seeming to aid others in freeing themselves from the mass of liquid;[1] so that the distillate always contains appreciable quantities of substances of higher boiling-point. Hence each separate portion must be again fractionally distilled by itself, the receiver being changed as soon as the boiling-point shows any tendency to rise; and by a repetition of this process a very nearly pure product may be obtained.

PRACTICAL APPLICATIONS OF FRACTIONAL DISTILLATION.

1. The most important of these is the process by which ordinary alcohol or 'spirits of wine' is manufactured. When grape juice, or infusion of malt, or a solution of sugar is mixed with a little yeast, and kept at a temperature of about 25°, it ferments, and most of the sugar contained in the liquid is converted into alcohol, which remains mixed with the water. This mixture is put into large copper boilers, or 'stills,' usually of the shape shown in fig. 70, and heated by steam passed into an outer metal case, or 'jacket,' sur-

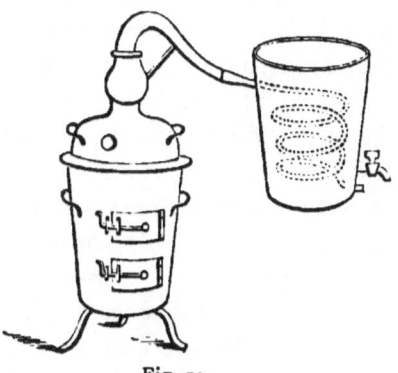

Fig. 70.

---

[1] A good instance of this is afforded by the process of purifying glycerine on the large scale. Glycerine boils at 290°, so high a temperature that some of it is decomposed and lost. But it is found that by passing steam heated to 170° or 180° through the impure glycerine, the substance is readily vaporised and passes over with the steam, yielding a product which can afterwards be freed from water by evaporation in a vacuum pan, like sugar.

rounding the still. The vapour of alcohol which comes over is passed through a long spiral tube, called the 'worm,' coiled in a vessel of cold water. It is thus effectively condensed, and by a repetition of the process, a strong spirit, containing about 80 per cent. of alcohol, can be obtained. The process is called the 'rectification' of the spirit, and to carry it out economically and quickly, much more elaborate condensers than the one described above are now used, by means of which fractionation is effected in one operation only.

2. The separation of coal-tar into its constituents is a very complete example of fractional distillation. The tar, as it comes from the gas-works, contains a very large number of substances of very various degrees of volatility; for example,—benzoline or benzol (boiling-point 80°), paraffin lamp-oil (boiling-point 170°), naphthalene (boiling-point 218°), paraffin lubricating oil (boiling-point 300°), paraffin wax (boiling-point 370°). All these, and many more, are roughly separated by gradually heating the tar in large iron boilers communicating with condensers, the process being dangerous on account of the inflammability of the substances.

3. The preparation of essential oils or 'essences' for perfumes, flavouring food, etc., is another instance. When various flowers, such as roses, lavender, rosemary; or leaves, such as those of peppermint and thyme; or fruits, such as lemons and almonds, are bruised to a pulp, mixed with water, and the whole is distilled, a fragrant oily substance passes over with the steam and condenses in the receiver, together with the water. It is easily separated from the water, as it either floats on, or sinks to the bottom of the latter without dissolving in it.

SUBLIMATION.

This is a term applied to those particular cases of distillation where the process starts and ends with a body in the solid state, and not in the liquid state as usual. An instance of this has already been given in Expt. 9, p. 40, where the effect of heat on iodine was tried. The solid became a liquid, then a vapour, and on cooling passed through the same states in the reverse order, finally appearing in small crystals of 'sublimed' iodine on the neck of the flask. In some few instances the liquid state is never assumed at all by the substance, its boiling-point (as defined on p. 183) being identical with, or even below its melting-point, under ordinary pressure. Sal ammoniac (ammonium chloride) and calomel (mercury protochloride) are examples of such substances, as the following experiment will show.

Expt. 64. Place a small lump of sal ammoniac in a large test-tube, and heat it over a Bunsen burner. The substance will give off vapour without showing any signs of liquefaction, and this vapour will condense in the form of a white solid in the cooler parts of the tube.

SECTION VII.—The Spheroidal Condition of Liquids.

It is a fact, however contrary to expectation it may seem, that a liquid, even when heated nearly to its boiling-point, may be put into a vessel heated very far above that temperature, without flashing at once into vapour, or even boiling violently. The following experiment may be made, to prove this.

Expt. 65. Obtain a flask about 200 c.c. in capacity, of the shape shown in fig. 71, made of copper with the exception of the bottom, which should be of silver.[1] Adapt to the flask a cork (which should fit accurately but not very tightly, and should not enter far into the neck), and insert into the cork a small jet of glass or metal having a very small aperture (such as the smallest needle that is sold will just pass through). Support the flask in a holder over a Bunsen burner, remove the cork, and

---

[1] The silver bottom is not absolutely essential: a flask made entirely of copper will answer, though not quite so well, since the surface soon gets oxidised and roughened at the high temperature to which it is exposed. Silvered copper is sometimes used, but the coating of silver does not last long.

heat the bottom nearly to redness (not higher, at any rate, than a low red heat, or the silver may be melted). Meanwhile, heat some water in a beaker nearly to boiling, and then transfer some of it to the strongly heated flask, putting it in drop by drop from a pipette with a fine jet. A slight hissing will be heard as each drop falls on the heated metal, but there will be no burst of

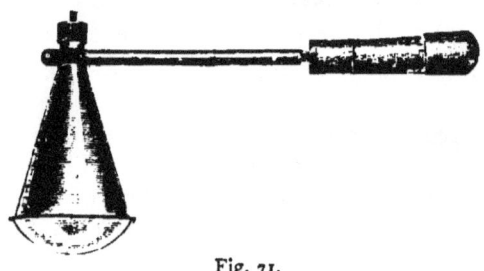

Fig. 71.

steam; and 6 or 8 c.c. of water may be thus gradually collected in the flask without showing much sign of boiling. Now insert the cork, not very tightly, and remove the lamp. Only a slight jet of steam will at first issue from the tube as the flask cools; but at a certain point in the cooling a sudden rush of steam will commence, usually sufficient to blow the cork high into the air.

Thus we learn the curious fact that water boils slightly and slowly in a strongly heated vessel, but much more violently in a cooler one. In order to ascertain the reason of this, it will be best to make the experiment on an open dish instead of in a closed flask, so as to be able to see exactly what is going on.

Expt. 66. Support a shallow silver dish, about 8 or 9 cm. in diameter, over a Bunsen burner, and when it has become nearly red-hot, drop on it slowly some hot water, as in the last experiment. Observe that the water does not give off much steam, but remains on the surface of the metal in the shape of a large, flattened drop, in ceaseless motion, with its edge and surface broken up into beautiful curves and ripples. This will go on, if the heat is maintained, for several minutes without the water becoming much lessened in quantity, proving how slowly

SPHEROIDAL CONDITION OF LIQUIDS. 217

evaporation is taking place; but if the lamp is withdrawn, at a certain point of the cooling the drop will suddenly lose its shape, spread over the dish, and pass off almost entirely into steam and spray.

A similar experiment may be made with alcohol; but care must be taken, owing to the inflammability of the vapour.

Expt. 67. Turn the dish upside down, so that its convex surface is uppermost, heat it as before, and sprinkle on it a little water, or alcohol. The liquid, instead of wetting the surface, will rebound from it and fall off in scattered portions almost like hailstones in appearance.

A similar thing may sometimes be observed when a kettle of water boils over on a hot fire. The overflowing water, as it falls on the hot bars of the grate, dances over them in detached drops without wetting them at all.

These phenomena were first noticed by Leidenfrost, in 1756, and the liquid during the experiment is said to be in the 'spheroidal condition,' from the shape which the mass assumes (Gk. σφαιροειδής, *ball-like*). Several additional facts have been since observed which have helped to suggest and establish the true theory of the action.

1. The surface must be heated to a certain definite extent above the boiling-point of the liquid before the spheroid can be formed on it.

2. The liquid, all the time that it is in the spheroidal condition, is actually several degrees *below* its boiling-point, as will be clear from the following table:—

Substance.	Lowest point at which the spheroid is formed.	Temperature of spheroid.	Boiling-point.
Water	171°	96.5°	100°
Alcohol	134°	76°	78°
Ether	60°	34°	36°

3. The spheroid is not in actual contact with the heated surface.

This is demonstrated by forming the spheroid on a flat, or very slightly convex surface, and holding it in its place by a wire dipped into it. If a light is then placed beyond it and looked at across the spheroid, a clear interval is seen between the liquid (which should be darkened with ink) and the metal below it. Also if the wire and the metal surface are connected respectively with the two poles of a galvanic battery so that the circuit must be completed through the spheroid (which should be formed of some conducting liquid, such as dilute sulphuric acid), no current passes as long as the spheroidal condition is maintained.

The reasons now accepted as accounting for the above curious phenomena are two in number.

1. The intense heat overcomes, more or less, the adhesion between the surface and the liquid (much in the same way as it has already been stated, p. 41, to overcome cohesion).

2. The liquid is supported on a cushion of its own vapour, which is formed, of course, most abundantly on the under side, which is nearest to the source of heat ; the molecules by their energetic movement raising the mass above the surface and keeping it there (somewhat in the same way as in some fountains a cork ball is held up by a jet of water). This vapour finally escapes laterally at the edge of the drop, and in its struggles to do so, causes the peculiar scolloped outline and the ripples already alluded to.

Since, then, it is clear that the substance is not in contact with the surface, but quite independent of it, there is no difficulty in explaining the rounded form assumed by it : for the cohesion between the molecules will naturally cause them to arrange themselves in a spheroidal mass (which would be spherical if it were not for gravitation), just like a drop of water on a greasy plate, or a dewdrop on a smooth leaf.

Under the above circumstances the spheroid can never, of course, get very hot, for it is separated from the heated metal by vapour, which is a very bad conductor of heat. Thus only the heat radiated from the surface can reach it, and a great deal of this is either reflected from or transmitted through the mass without being absorbed. Moreover heat is continually being abstracted from the liquid by the formation of vapour, and hence the temperature may easily fall below the boiling-point, as it has been proved to do.

A similar phenomenon has been observed in the case of some solids, such as sal ammoniac and ammonium carbonate ('smelling salts').

Expt. 68. Put a few small lumps of either of the above salts into a large test-tube or a globular flask; shake the flask, and observe the rattling sound made as the lumps strike against each other or the glass. Heat the tube or flask by waving it through the flame of a Bunsen burner; at a certain point the rattling sound will cease entirely, as the flask is shaken, and the fragments of the salt will glide over the glass and each other easily and noiselessly, like fragments of ice over the surface of a frozen pond: the film of vapour, which is formed at a high temperature, keeping them from actual contact with other things.

The spheroidal condition has been applied to explain several remarkable facts. For instance, it is found possible to pass the hand, if thoroughly moistened with water, through lead heated much above its melting-point, without injury to the flesh (in fact, the hotter the metal is, the safer it is to handle it); for the film of vapour which is immediately formed, prevents actual contact between the metal and the skin.

The ancient 'ordeal by fire' consisted in compelling the accused person to walk blindfold among red-hot ploughshares, or to carry a red-hot piece of iron in his hand for a certain distance: if he escaped without burns he was considered innocent of the crime. It appears that some passed through this ordeal unscathed, probably from some ointment or lotion having been applied, which gave off vapour enough to protect the skin.

Steam boilers have been known to explode shortly after the admission of cold water into them; and it is thought that in such cases the boiler-plates may have become, through neglect, intensely heated; so that the water brought into contact with them would be thrown into the spheroidal state, and then little steam would be generated until the plates had cooled down. At a certain point a sudden burst of steam would take place, in too great a quantity to pass off through the safety valve.

Section VIII.—The Water-Vapour in the Atmosphere. Hygrometry.

The air which surrounds the earth may be considered, for physical purposes, as made up of two parts,—
1. Permanent gases (oxygen and nitrogen, with a little carbon dioxide).
2. Vapour of water.

These exist quite independently of one another. The air does not 'dissolve' or 'absorb' water-vapour, as it is often incorrectly said to do; the formation and condensation of this vapour goes on (so far as quantity and tension are concerned) just as if no other gases or vapours were present, and so we may, in considering it, neglect the presence of the other substances entirely, since they only affect the rate of formation and diffusion of it.

That the vapour of water is really present, though invisible, in all ordinary air, may be readily demonstrated. When air is cooled down more or less, it is invariably found that a deposit of liquid water is formed in it, either in free space, as in the case of clouds and rain, or in the shape of drops of dew upon the surfaces which by their low temperature have caused the cooling of the air near them.

Expt. 69. Place a little water in a flask, drop into it some ice or snow,[1] and leave it for a minute or two supported in a Bunsen holder so that it hangs freely in the air. The cold liquid will

---

[1] If none of this is at hand, mix about 10 grms. of sal ammoniac with an equal weight of saltpetre, powdering both finely, and add the mixture to the water.

cool the glass, and this in its turn will cool the air immediately surrounding it : and a deposit of fine drops of dew will form on the outside of the flask, sharply limited by the level of the liquid within.

An experiment giving a similar result has been already described (p. 171), in which the evaporation of ether was used to reduce the temperature.

The explanation of several frequently occurring phenomena will be suggested by the above experiment. Thus when a decanter of cold spring water, or a bottle of wine from a cellar, is brought into a warm room in summer, it is soon covered with a copious dew. Again, when, after a long frost, a thaw accompanied by a warm wind comes on, the walls of passages and floors, which have become thoroughly cold during the frost, generally stream with moisture condensed from the air. In fives-courts and racquet-courts much inconvenience is caused by this 'sweating' as it is called (an incorrect term, since the moisture comes from without and not within the walls) ; and there is no way of preventing it except the obvious but expensive one of never letting the walls get cold at all.

Lastly, we have evidence on a much larger scale of the presence of water-vapour in the air from the formation of clouds and mist round hill-tops, and in valleys where the ground is colder than the surrounding air.

The source of this vapour is mainly the constant evaporation which is going on from all surfaces of water, and also from damp soil and the leaves of plants. The breathing of animals, the combustion of coal, wood, etc., and that slower combustion of organic matter which is termed putrefaction or decay, also contribute something to the supply.

The amount of vapour in the atmosphere varies very much; more, in fact, than that of any other of the constituents. In 100 grms. of air there is on an average about 1·5 grm. of water-vapour; and if all that exists in this form was condensed, it would be sufficient to cover the earth with a stratum of liquid about 1 decimetre in thickness.

In hot air much more vapour can exist than in cold air; since, as already explained (p. 174), increase of temperature causes an increase of vapour-tension or ability to resist condensation. Over the surface of water and over land swept by winds which have travelled across large areas of sea, such as islands like our own, there is usually much more vapour than in the air over barren continents and deserts. In all cases there is a maximum limit to the quantity which can be present, for if the molecules are crowded beyond a certain point, or possess less than a certain amount of heat energy, condensation sets in, and further increase in quantity is impossible. When this maximum limit is attained, the air (or, more properly, the space) is said to be 'saturated.'

Such a mass of saturated, or nearly saturated air is called 'moist'; not necessarily because there is a large absolute quantity of water-vapour in it, but because there is as much, or nearly as much, as there can be under the existing conditions.

As a rule, however, there is not so much vapour present as there might be; and the tension of it is decidedly greater than that which is necessary to prevent condensation. This is proved by the fact that it is generally requisite to cool down the air several degrees at least before any condensation of moisture takes place. In such a case the air is said to be 'dry'; not necessarily because there is absolutely little water-vapour in it, but because there is much less than might be present under the existing conditions.

Thus the terms 'dryness' and 'moistness' of air are purely relative, indicating the proportion of vapour actually present to that which *could* be present under the circumstances. In winter, the air is often, indeed generally, 'moist,' although there is really only a small weight of vapour present, because the temperature is so low that the tension of even this small quantity is hardly more than sufficient to enable it to exist as vapour. In summer the air is often 'dry,' even though there is a considerable weight of vapour in it, the temperature being high enough to give a tension greater than is necessary to keep even this large quantity in the form of vapour.

HUMIDITY OF THE ATMOSPHERE.

The term 'humidity' is applied to the state of the air with respect to the degree of its saturation with water-vapour; and this humidity is usually expressed by stating how many parts of vapour a given weight of air contains out of 100 parts which it could contain under the existing conditions. Thus perfectly moist or saturated air would be said to have a humidity of 100, implying that it contained 100 per cent., or the maximum amount of vapour. Again, in stating that the humidity of the air on a certain day is 50, we mean that the air contains only 50 per cent. of the vapour that it could contain. Such air would be considered decidedly dry, even though the absolute quantity of vapour in it might be considerable.

The influence of the degree of humidity of the air upon both health and weather is very great. First, as regards health: air should have about two-thirds of the maximum quantity of water-vapour in it (humidity 60-70) in order to be wholesome and pleasant to live in. If it contains less than this, evaporation goes on with undue rapidity from the skin, which becomes parched and shrivelled, while the temperature of it may be so far reduced that a 'chill' is caught. The harsh, unpleasant, parching character of east winds in this country, even when their temperature is high, is only too well known; and the far less endurable and more dangerous simooms of tropical countries are instances of the results of extreme dryness of the air produced by its passage over heated tracts of waterless desert, which raise its temperature without supplying a corresponding amount of vapour. The air in a room heated by a stove generally shows a similar character for similar reasons; and it is a common and scientifically correct practice to put a vessel of water on or close to the stove, the vapour rising from which supplies what is wanting in the air.

If, on the other hand, the humidity of the atmosphere is nearly or quite 100, *i.e.* if the air is very moist, little or no loss of moisture can take place from the skin, and thus the healthy action of the perspiration-glands is interfered with, and a 'cold' is caught just as certainly as under the opposite conditions.

Such moist air feels raw in winter, and sultry, close, and oppressive in summer.

The effect of the dryness and moistness of the air upon the coming weather is equally important, and so obvious that it hardly needs explanation. If the air already contains nearly or quite all the moisture that it can contain, a very slight fall of temperature will cause condensation to begin, and clouds and rain will follow. If, on the other hand, the air is pretty dry, its temperature may be much altered without producing condensation, and rainless, if not actually sunny weather is probable.

We see, then, that in forecasting weather the determination of the humidity of the air is as valuable an assistance as the observation of its pressure by means of the barometer: and the instruments by which this is done will form the next subject for consideration.

HYGROMETERS.

A hygrometer ($\upsilon\gamma\rho\grave{o}\varsigma$, moist, $\mu\epsilon\tau\rho\epsilon\hat{\iota}\nu$, to measure) is an apparatus for finding out the humidity of the air.

The earlier forms of hygrometer, which depended on the shortening in length of a hair or thread (the tightening of window-blind cords in wet weather is a common instance of this), or on the untwisting of a piece of catgut when exposed to moist air, are too rough and inaccurate to be worth a full description. One of these, however, is still occasionally seen, in which an index attached to one end of a short piece of catgut moves over a dial with the words *Rain*, *Fair*, etc., upon it. The other end of the catgut is attached to the frame of the instrument, and in moist air the fibres swell, owing to absorption of water, and the whole gut untwists slightly and moves the index. In dry air the absorbed moisture evaporates from the gut, which again twists up and moves the index in the opposite direction.

The principle of the best modern hygrometers is, to determine how far the air must be cooled down before the tension of the vapour is so far reduced that condensation into liquid water begins. This has been already illustrated in the experiment last made, p. 220; and, in fact, if a thermometer was placed in the water contained in the flask and ice was added little by little so as to lower the temperature gradually, then, by observing the temperature at the moment when a deposit of dew first began on the surface of the flask, a fair estimate of the dryness or moistness of the air might be made.

The point of temperature at which condensation begins is called the '**dew-point.**' It is obviously that temperature at which the tension of the vapour is only just, or barely sufficient to enable it to retain the gaseous state,—is, in fact, its *maximum tension*, as explained on p. 179.

One further observation must be made in order to calculate the humidity, *viz.* the actual temperature of the air at the time; for this enables us to ascertain how much vapour could then possibly exist in the air. It does so because by a reference to a table of tensions, such as that given on p. 226, we learn the tension that the water-vapour would have if there was enough of it present to saturate the space at the existing temperature. Thus, supposing that the temperature of the air on a given day was 20°, then it may be gathered from the table that if the air was saturated, the maximum tension of the vapour in it would be 17·4 mm. If at the same time the dew-point was found to be 10°, we should know that the quantity of vapour actually existing in the air was that which would be sufficient to saturate it at 10°, when it would (see the table) have a tension of 9·1 mm. And since the mass of a particular vapour is (when all other conditions are kept equal) proportional to its tension, we have,

Tension of vapour at 20°.	.	Tension of vapour at 10°.	: :	Maximum humidity.	.	Actual humidity of the air.
mm.		mm.				
17·4	:	9·1	: :	100	:	52·3 (nearly).

Or we may, without considering the tension, look at the matter in the following way. In a litre of air at 20° (when the barometric pressure is 760 mm.) there can exist an amount of vapour weighing 0·0171 grm. The fact that the dew-point is 10° shows that the air really only contains enough vapour to saturate it at

P

10°, and 1 litre of this weighs nearly 0·00895 grm.[1] Then we have the proportion,

Wt. of vapour which each litre of air could contain.	:	Wt. of vapour which each litre does contain.	: :	Maximum humidity.	:	Actual humidity of the air.
grm. 0·0171	:	grm. 0·0895	: :	100	:	52·3

TENSION AND WEIGHT OF 1 LITRE OF WATER-VAPOUR AT DIFFERENT TEMPERATURES.

Temperature.	Tension.	Weight of 1 Litre.
	Mm. of mercury column.	Milligrammes.
0°	4·6	4·8
5°	6·5	6·8
10°	9·1	9·2
15°	12·7	12·7
20°	17·4	17·1
25°	23·5	22·8
30°	31·5	30·1
35°	41·8	39·2

The following are the chief forms of instrument which have been devised for determining the dew-point with accuracy:—

1. *Regnault's Hygrometer.*—The construction of this is shown in fig. 72. It consists of two tubes, A and B, similar in form to test-tubes, about 2 cm. in diameter and 15 cm. in length, supported vertically near each other on a stand. The lower part of each of the tubes is made of polished silver, and the upper ends are closed by corks through which pass the stems of delicate thermometers. A small tube, C, also passes through the cork in the tube A, and reaches nearly to the bottom. At the side of A there is an exit-tube, D, which serves the purpose of an arm for supporting the tube, and can be connected by the flexible pipe E with an arrangement (called an 'aspirator') for withdrawing air from the apparatus.

---

[1] The actual weight of a litre of vapour at 10° is rather greater than this, owing to the increase in density which takes place as the temperature is lowered.

In using this hygrometer, some ether is poured into the tube A, and air is very gently drawn out through D. The external air then enters through C, passing in a stream of bubbles through the ether, the evaporation of which is by this means greatly quickened and the temperature reduced. The silver tube is thus gradually cooled, and at a certain point[1] a film of dew

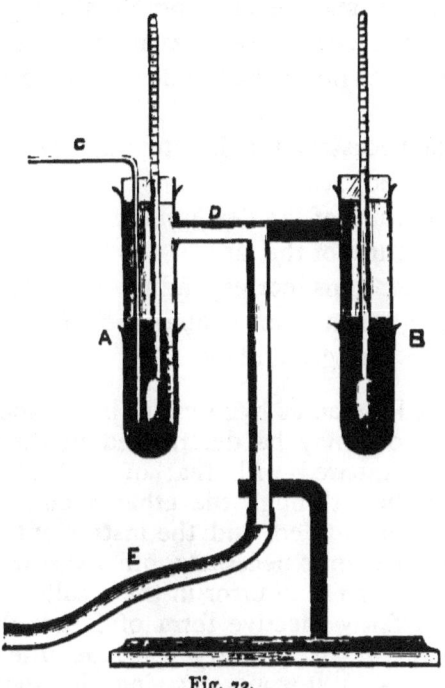

Fig. 72.

clouds its polished surface, showing that the surrounding air is cooled down so far as to be saturated. At this moment the reading of the thermometer in A is taken; and this temperature (since the agitation of the ether by the stream of bubbles

---

[1] It will assist in the detection of this film if, when its appearance is suspected, a small part of the surface is wiped with a camel-hair brush, or a momentary touch of the finger. If any dew exists there, it will be wiped away, and the contrast between the wiped portion and the rest of the surface will be clearly seen.

equalises its temperature and that of the tube throughout) may be assumed as the correct dew-point. The reading of the thermometer in B is also taken, and this gives the temperature of the external air. The current of air is then stopped, and the temperature of the tube A rises again, and the film of dew disappears. As it does so, the temperature indicated by the immersed thermometer may be again noted, although the point of disappearance of the dew is not so easily observed as that of its appearance, and does not mark the dew-point so definitely as the latter.

The two data necessary for determining the humidity of the air are thus obtained, *viz.*—

1. The temperature of the dew-point;
2. The temperature of the air;

and the vapour-tensions corresponding to these temperatures respectively being taken from a table, the humidity is calculated by the proportion sum given above.

Expt. 70. If a Regnault's hygrometer is at hand, the humidity of the air in a room may be determined in the manner above described, and compared with that of the external air. The stream of air-bubbles through the ether should be very gentle, not too quick to be counted, and the instrument should not be approached nearer than is necessary, otherwise the heat radiated from the body will cause an error in the results.

A simple and fairly effective form of Regnault's hygrometer may be made out of an ordinary test-tube fitted with a cork having three holes in it; one carrying the thermometer, the second having a tube resembling C (fig. 72, p. 227) fitted through it, and the third carrying a tube similar to C in shape, but having the *short* branch passed through the cork.[1] Some ether having

---

[1] It is easy to coat the end of the test-tube internally with silver, by filling it about one-fourth full of some chemical silvering solution (preferably Martin's), an account of which will be found in most treatises on chemistry. When a sufficient deposit of silver is obtained, the spent solution should be poured out, and the tube rinsed with water and gently dried in a current of warm air before the ether is put in. Such a coat of silver is not necessary, though it certainly makes the observation of the deposit of dew much easier.

DANIELL'S HYGROMETER.

been poured in, the tube may be held near the upper end in a Bunsen holder, and air very gently blown through the tube dipping into the ether, from a pair of blowpipe-bellows or other blower, a screw pinchcock being placed on the indiarubber connecting-tube to regulate the current. The air and ether vapour escape through the short open tube. No second tube with thermometer is necessary, as the temperature of the air can be readily ascertained by an ordinary thermometer, hung freely exposed at a little distance. Several observations of the appearance and disappearance of the dew-film should be made, and the mean or average[1] of all these may be taken as indicating the true dew-point.

2. *Daniell's Hygrometer.*—In this, as in Regnault's, the evaporation of ether is used as the means of reducing the temperature of the air so as to admit of the dew-point being observed; but the evaporation is promoted by the same means as in Wollaston's cryophorus (p. 179), *viz.* by condensing the vapour as fast as it is formed. The instrument, which was invented and used many years before that of Regnault, is shown in fig. 73. It consists of a Γ-shaped, rather wide tube, with a bulb at each end. The bulb A is half filled with ether, and a small thermometer is fixed within the tube so that its bulb just dips into the ether. Before the apparatus is finally closed up, the ether is boiled for a short time so as to chase out all air, and leave the bulbs and

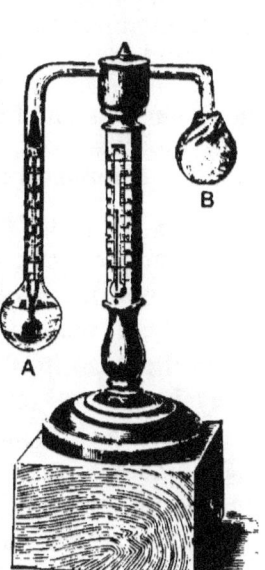

Fig. 73.

---

[1] The mean result of a series of observations is obtained by adding together the results of the separate observations, and dividing the sum by the number of observations. The quotient gives the arithmetical mean. Thus, supposing that in five trials the dew-point was observed to be 9·5°, 9°, 9·5°, 10°, and 9°, respectively, we shall have as the mean result,—

$$\frac{9\cdot 5 + 9 + 9\cdot 5 + 10 + 9}{5} = \frac{47}{5} = 9\cdot 4°.$$

tube filled with nothing but pure ether-vapour. A broad band of bright gold leaf is placed round the equator (so to speak) of the bulb A,[1] and over the bulb B is tied a loose covering of muslin. The whole is supported, as shown, on a stand which also carries a thermometer for showing the temperature of the air.

Expt. 71. In order to observe the dew-point by this instrument, pour a little ether, in successive small portions, upon the muslin which envelops B. The quick evaporation of this cools the bulb, and the ether-vapour in it soon condenses. Thus, as in the cryophorus, the evaporation of the ether in A is hastened, and the temperature of the liquid, the bulb, and the stratum of air round it is reduced until dew appears on the polished gold band. As soon as this is visible, the reading of the enclosed thermometer should be taken; this, as in Regnault's instrument, gives the dew-point. The temperature of the air is ascertained by the thermometer attached to the stand.

Either Regnault's or Daniell's Hygrometer is capable of giving very accurate results, but only after considerable practice in the use of it. The chief sources of error arise,

(1) From the difficulty in determining the exact moment at which the deposit of dew commences;

(2) From the temperature being in a state of continuous change, so as to render it doubtful whether the reading of the thermometer gives the exact temperature of the stratum of air surrounding the bulb.

A preferable form of dew-point hygrometer has been lately devised by Mr. G. Dines, in which the thermometer is in close contact with a very thin plate of black glass forming one side of a flat, shallow cell, through which a slow current of water cooled by ice is made to flow.

3. *Mason's Hygrometer.*—In this instrument advantage is taken of the fact already explained, p. 178, that evaporation takes place more or less quickly, according as there is less or more vapour

---

[1] The bulb is sometimes made of black glass, but a bright metallic surface shows a deposit of dew rather better.

Mason's Hygrometer.

already in the space above the liquid. Hence an observation of the rate of evaporation will obviously afford a means of estimating the humidity of the air.

This hygrometer is shown in fig. 74, and consists simply of two thermometers, as nearly alike in size and shape as possible, supported on the same stand about 8 or 10 cm. apart. One of them has its bulb freely exposed to the air, and is called the 'dry-bulb thermometer.' The other (the right-hand one in the figure), which is called the 'wet-bulb thermometer,' has its bulb covered with fine muslin, to which is attached a piece of loose cotton thread, such as is used for the wicks of oil-lamps, long enough to dip into some water contained in a little cistern placed on the base of the stand. The muslin and cotton are, to begin with, thoroughly wetted with distilled (or clean rain) water, and then the water from the cistern rises up the thread by capillary action and the muslin covering the bulb is thus kept constantly wet.

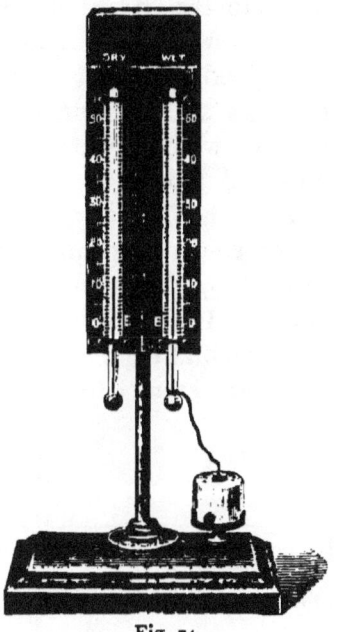

Fig. 74.

From this surface of liquid evaporation is continually going on, and all the potential heat which disappears in the formation of the vapour is taken from the liquid, the muslin, and the thermometer. It is clear, then, that the latter will indicate a lower temperature than the dry-bulb thermometer in all cases except when the air is absolutely saturated with water-vapour, so that no reduction in quantity of the liquid can take place (since as many molecules return to it as rise from it, as already explained, p. 178): for then, and then only, the heat energy given up by the returning molecules just makes up for the heat withdrawn by the outgoing ones, and no difference in temperature occurs. Further, the reduction in temperature of the wet-bulb will be greater, in proportion as more water is being withdrawn from the muslin by evaporation. And, lastly, as mentioned above, the loss of liquid will proceed more quickly, the drier the air is at the time. Thus a mere comparison of the readings of the two thermometers will show whether the air is moist or dry; if the former, the readings will be nearly or quite the same; if the latter, the reading of the 'wet-bulb' will be lower by $2°$, or more, than that of the 'dry-bulb.' There is, indeed, so exact a relation between the difference of the readings and the humidity of the air that tables have been constructed which give the humidity of the air corresponding to any observed depression of the 'wet-bulb' reading below the temperature of the air, through a range of $50°$ of the scale. These tables, constructed in the first instance by calculation, have been verified by direct observations with the dew-point hygrometers already described, and are found to agree closely with the experimental results.

A short extract from one of these tables is given below.

HYGROMETRIC TABLE.

Readings of Dry-bulb thermometer, *i.e.* temperature of the air.	Depression of the Wet-bulb thermometer below the temperature of the air.					
	$0°$	$1°$	$2°$	$3°$	$4°$	$5°$
	Humidity.	Humidity.	Humidity.	Humidity.	Humidity.	Humidity.
$0°$	100	81	63	45	28	10
$5°$	100	84	69	54	39	27
$10°$	100	87	74	62	50	39
$15°$	100	89	78	68	58	49
$20°$	100	91	81	72	64	55

The dew-point, though not indicated directly by Mason's hygrometer, can be readily found by calculation from the results obtained by it: and it may be worth notice that when the temperature of the air is $11·6°$, the dew-point is always as far below the reading of the wet-bulb thermometer as the latter is below the reading of the dry-bulb.

Mason's hygrometer has many advantages. In the first place, it is very simple and inexpensive; any two thermometers properly mounted side by side, with bulbs hanging freely in the air, will answer, and one of the common 'bird-fountains' (constructed to hang at the side of a bird-cage) will give a supply of water sufficient to keep the muslin wet for months.[1] In the next place, it can be used promptly and without trouble; mere inspection of the thermometers, followed by a reference to a set of tables, being sufficient to determine accurately the humidity of the air at the time. No manipulation with ether is required—a rather expensive liquid, always difficult, and sometimes dangerous to deal with.

One more method of determining the amount of water-vapour in the air must be alluded to, as being the most direct and accurate of all, though better adapted for special researches than for ordinary practical use. It consists in passing a measured volume of air (not less than 500 litres, or so) through a U-shaped tube filled with fragments of pumice soaked in strong sulphuric acid (a liquid which has a very strong affinity for water). The tube with its contents is weighed before the air is passed through it, and again afterwards; the increase in weight gives the amount of vapour present in the measured volume of air.

FORMATION OF CLOUDS.

A cloud is a swarm of extremely small particles of liquid water, formed by the condensation of some of the invisible vapour in

---

[1] The water-reservoir should be fixed at the back of the stand, not very near the thermometers (lest it should influence their readings), and the opening into which the thread dips should be about 8 cm. from the bulb, but not directly below it, otherwise the rising vapour may partially saturate the air round the bulb. The muslin and lamp-cotton should, before use, be boiled in water containing a little washing-soda (to get rid of starch and grease), and should be changed every half-year, or oftener if the pores are noticed to be clogged with dust and the muslin is not kept thoroughly wet. Distilled water, or clean filtered rain-water should be used.

the air, and rendered visible (in spite of the perfect transparency of water) by the reflection and scattering of light from the surfaces of the globules. These minute water-drops are constantly on the move, wafted by winds, and falling through the air in consequence of their weight, but falling very slowly because they have to push aside the molecules of the air; and thus they remain suspended for a long time, just as even solid particles of dust remain long floating in the air although their density is, of course, much greater than that of the gases surrounding them.

'Mist' and 'fog' are terms applied to clouds lying close to the surface of the earth.

In order that a cloud may be formed, some reduction in temperature of the air must take place; and the principal and most obvious cause of this is the rise of moist air, either in the natural course of its diffusion or from the formation of upward currents (p. 284), into the higher regions of the atmosphere where the temperature is lower, owing to the increasing distance from the earth's warm surface. Moreover the vapour, as it rises, is exposed to less and less pressure (p. 196) and therefore expands, *i.e.* its molecules move farther apart, and the energy required to enable them to do this is abstracted in the form of heat from the mass of vapour itself, which therefore becomes colder.

Another frequent cause of the formation of clouds, as well as of mist and fog, is the contact of moist air with a colder surface, such as that of a mountain-top, or of a valley or plain which has lost heat by radiation in a way which will be more fully explained in the next chapter.

But there is still another set of conditions under which clouds form, the explanation of which is not quite so obvious. We may not unfrequently observe the sky to become overcast when there is no wind to carry up clouds from other regions, and no evidence of any ascending current or contact with a colder surface to account for it. In such cases the clouds are no doubt occasioned by the meeting and mixing of two horizontal currents of air, one cold and moist, and the other hot and moist. Still it is not at once clear how the action between two such currents can cause any of the vapour in either of them to condense; for

Formation of Clouds.

although the warm air will undoubtedly be cooled by contact with the colder current, and so tend to deposit some of its vapour, yet, at the same time, the cold air must become warmer in consequence of the heat taken from the other current, and therefore more water-vapour can exist in it. It would seem, in fact, that whatever vapour was condensed out of the warm current would be at once taken up by the other current, and that no permanent deposit of cloud could take place. This would certainly be the case if the amount of vapour which can exist in air varied *exactly* with the temperature of the air; but a reference to the table on p. 226, will show that the quantity of moisture required to saturate a given space increases much more rapidly than the temperature. For example, a rise of temperature from 0° to 5° causes an increase from 4·8 to 6·8 milligrammes, *i.e.* of 2 mgrms. in the weight of vapour which can exist in 1 litre; while in passing from 30° to 35° the same rise of 5° causes an increase from 30·1 to 39·2 mgrms., *i.e.* of 9·1 mgrms. in the weight of vapour in a litre. The result is that the warm air-current, in cooling, gives out more moisture than the cold air can take up in becoming warmer by the same number of degrees, and the excess must, of course, be deposited in the liquid form.

Suppose, for example, that a litre of saturated air at 20° is mixed with a litre of saturated air at 0°. The warm air will fall in temperature just as much as the cold air will rise, and the final temperature of the mixture will be 10°. Then from the table we ascertain that—

The weight of 1 litre of vapour at 20° = 17·1 ⎫
 ,,   ,,   ,,   10° = 9·2 ⎬ Difference = 7·9
and further—

The weight of 1 litre of vapour at 0° = 4·8 ⎫
 ,,   ,,   ,,   10° = 9·2 ⎬ Difference = 4·4

Thus the litre of warm air will, in sinking to 10°, give up 7·9 mgrms. of moisture, while the litre of cold air in rising from 0° to 10° can only take up 4·4 mgrms. Hence the remainder, *i.e.* (7·9 − 4·4 =) 3·5 mgrms. of moisture will be deposited as cloud.

APPENDIX TO CHAPTER VII.

PROBLEMS RELATING TO HYGROMETRY.

[The method of finding the humidity of the atmosphere when (1) the dew-point, (2) the tension of water-vapour at the dew-point, (3) the tension of water-vapour at the existing temperature of the air are known, has been given already on p. 225.]

PROBLEM 1.—**Given the volume, temperature, and pressure of a quantity of water-vapour, to find its weight.**

The weight of 1 c.c. of air at 0° and under 760 mm. pressure is known to be 0·001293 grm.

The density of water-vapour, referred to that of air as unity, is approximately 0·623, when the two are under the same conditions of temperature and pressure.[1]

Hence the problem can be solved by

  (i) Finding the weight of 1 c.c. of air under the specified conditions of temperature and pressure,
  (ii) Multiplying the result by the density of water-vapour,
  (iii) Multiplying the product by the given volume of the water-vapour expressed in cubic centimetres.

EXAMPLE.—What is the weight of a litre of water-vapour at 27° and under a pressure of 200 mm.?

(i) According to the method explained on p. 119, the weight of 1 c.c. of air at 27° and 200 mm. is

Wt. of 1 c.c. at 0° and 760 mm. $\times \dfrac{P \times 273}{760 \times (273+t)} = 0 \cdot 001293 \times \dfrac{200 \times 273}{760 \times 300} = 0 \cdot 00030964$ grm.

---

[1] It must be noted that this is only true on the assumption that both gases alter equally in volume when the temperature and pressure are altered (obey, in fact, the laws of Gay Lussac and Mariotte). Water-vapour near its liquefying-point can hardly be expected to do this.

Problems on Hygrometry. 237

(ii) 0·00030964 grm. × 0·623 = 0·00019290572 grm. = wt. of 1 c.c. of water-vapour at 27° and 200 mm.

(iii) There are 1000 c.c. in a litre. Hence
0·00019290572 grm. × 1000 = 0·19290572 grm.

Therefore the required weight of the litre of water-vapour is 0·1929 grm. (nearly).

PROBLEM 2.—**To find the weight of a given volume of air saturated with water-vapour, measured at a specified temperature and pressure.**

This is a calculation which has often to be made in scientific work, since in accurate weighings allowance has to be made for the effect of air in buoying up the weights, etc., used, and thus masking their true mass; and ordinary air always contains more or less water-vapour.

By Dalton's Law (p. 181) the dry air and the water-vapour in it act quite independently of one another, each exerting its own pressure and behaving generally as if the other was not present. Hence we may consider that there are, in the given volume, two distinct atmospheres:—

(1) One of water-vapour, exerting a pressure corresponding to its temperature, which can be found from any table of vapour-tensions (such as that given on p. 226).

(2) One of dry permanent gases (chiefly oxygen and nitrogen), which is exerting the rest of the observed pressure.

And the problem is solved by finding, by the method explained in the preceding problem,

(i) The weight of the given volume of water-vapour, at the specified temperature and the corresponding pressure.

(ii) The weight of the given volume of dry air, at the specified temperature and under a pressure = (the given pressure − the pressure of the water-vapour).

The sum of the two weights thus found will be the required weight of the moist air.

EXAMPLE.—Find the weight of a cubic metre of saturated air, measured at a temperature of 17° and a pressure of 700 mm.

[The tension of water-vapour at 17° = 14·4 mm.]

(i) The weight of 1 c.c. of the water-vapour in the space at 17° and 14·4 mm. will be,

$$0 \cdot 001293 \times 0 \cdot 623 \times \frac{14 \cdot 4 \times 273}{760 \times 290} = 0 \cdot 000014368 \text{ grm. (nearly).}$$

A cubic metre contains 1000 litres, or 1,000,000 c.c.

Hence the weight of water-vapour in the cubic metre = 14·368 grms.

(ii) The weight of 1 c.c. of the dry air in the space, at 17° and under a pressure of (700 − 14·4 =) 685·6 mm., will be

$$0 \cdot 001293 \times \frac{685 \cdot 6 \times 273}{760 \times 290} = 0 \cdot 001098 \text{ grm.}$$

And the weight of a cubic metre of the dry air will be 1,000,000 times this; *i.e.* 1098 grms.

Hence the total weight of the cubic metre of saturated air will be (14·368 + 1098 =) 1112·368 grms.

PROBLEM 3.—**To find the weight of the dry gas in a given volume of a gas saturated with water-vapour.**

This, like the last, is a calculation often required in scientific work; for in many experiments on gases it is important to know what quantity of a gas exists in a measured volume, independently of any water-vapour which may be also present.

The method of proceeding is the same as that explained in the second step of the last problem; the real pressure under which the gas exists being found by subtracting the tension of the water-vapour from the total pressure.

EXAMPLE.—500 c.c. of hydrogen gas, saturated with water-vapour, were measured at a temperature of 8° and under a pressure of 600 mm. What is the weight of pure hydrogen present?

[The density of hydrogen, referred to that of air, is 0·0693. The tension of water-vapour at 8° is 8 mm. (nearly).]

The pressure on the gas is (600 − 8 =) 592 mm.

Then 1 c.c. of hydrogen at 8° and 592 mm. will weigh

$$0 \cdot 001293 \times 0 \cdot 0693 \times \frac{592 \times 273}{760 \times 281} = 0 \cdot 00006781 \text{ grm.}$$

And 500 c.c. will weigh (500 × 0·00006781 =) 0·033905 grm.

In the above examples the air or gas has always been considered as *saturated* with water-vapour, in which case the pressure will be its *maximum* tension.

Problems on Hygrometry.

If the space is not saturated, a determination of the humidity (p. 225) will give the percentage of water-vapour present, and we may consider it as exerting this same percentage of the pressure which it could exert at the given temperature, *i.e.* of its maximum tension.

Thus, if the humidity of the air is 70 and the temperature 20°, then the vapour is to be considered as exerting $\frac{70}{100}$ or $\frac{7}{10}$ of its maximum tension at 20°, *i.e.* $\frac{7}{10}$ of 17.4 mm.; and the reduced pressure obtained in this way must be taken in working out problems.

CHAPTER VIII.

THE CONVEYANCE OR TRANSMISSION OF HEAT.

Heat has hitherto been considered as passing from one body to another, or from one part of space to another, without reference to the exact mode in which this transference takes place. This latter point must now be examined more minutely: and it will be advantageous to begin by making the following experiment.

Expt. 72. Heat an iron ball, about 5 cm. in diameter, to redness or nearly so in a common fire, or by hanging it from the ring of a retort-stand by a stout iron wire in the flame of a Bunsen burner: then place it upon an iron stand like that shown in fig. 75. It will rapidly lose heat, which is conveyed away from it in several distinct ways.

In the first place, apply the hand (or a thermometer) to the iron stand, a little below the part on which the ball rests. Notice that the iron is warm and gets more so every moment, showing that heat is being transmitted from the ball where it touches the iron stand along the latter: each molecule of the solid iron taking up heat continually and passing it on to the next molecule, and this again to the adjacent ones, and so on until heat from the ball has travelled from the top to the bottom of the stand. This mode of transference of heat is called '**conduction**': and it may be roughly likened to the way in which water is often carried from a pond to a fire-engine by a number of men standing in a row and handing on buckets of water from one to the next in the row, without moving from their

Fig. 75.

places: or to a train of cog-wheels, fixed in the frame of a clock, transmitting the motion of the weight to the hands.

Next, hold the hand (or a thermometer) directly above the heated ball. Heat will be at once felt, and will be easily referred to its real cause, *viz.* the current of air which becomes heated by contact with the ball and hence expands and rises (see p. 88), carrying heat away with it. This mode of conveyance of heat in which the molecules, as they receive the heat, immediately move away, and their places are taken by other molecules which carry away heat in like manner, is called '**convection.**' It may be roughly compared to the conveyance of water to a fire-engine by a procession of men, each filling a bucket at a pond and then running off to the fire-engine and returning to the pond for another supply.

Lastly, hold the hand, or a thermometer (an air thermometer, p. 107, or Leslie's differential thermometer, p. 109, is very suitable for this purpose) near the side of the heated ball; anywhere, in fact, except directly above it or in contact with the stand. Heat will be felt in these positions also, although it obviously cannot be conveyed either by conduction or by convection. So there must be some third mode in which heat is passing away from the ball, totally distinct from both of those above mentioned. It is, in fact, darting off on all sides in straight lines, with enormous velocity, by a process called '**radiation,**' which probably consists in a vibratory or swinging motion of the molecules of the intervening air or other suitable medium, forming a succession of waves like those which spread in circles round the place where a stone has been dropped into still water.

Thus we are able to recognise at any rate three distinct modes in which heat energy may be transmitted from one place to another, *viz.*—

1. **Conduction**: in which heat is communicated from one molecule to the adjacent ones by contact, without any of them moving appreciably from their places.

2. **Convection**: in which heat is conveyed (Lat. *convectio*, conveyance) by the actual passage of molecules, after being heated by contact with a body, from one place to another.

3. **Radiation**: in which heat is transmitted along a row of molecules in the form of waves; each molecule swinging to and

fro across the direction of transmission, but not otherwise moving from its place.

Section I.—Conduction of Heat.

This, as will be evident from the definition given just above, is the chief way in which heat is conveyed through solid substances. In a solid the molecules are held together so firmly and compactly by cohesion that they cannot move from their places; so that all they can do is to take up heat motion and pass it on to others which lie close to them.

The first point to be noticed in reference to conduction is that by it heat is transmitted steadily and regularly, but comparatively slowly, through bodies, whatever may be their form; for instance, along an iron bar, or the handle of a tea-kettle, or a spoon, whether it is curved or straight. The heat will follow most readily the direction in which there are most molecules in actual contact, or close proximity.

In the next place, it is a matter of common observation that some things conduct heat much better than others. The handle of a copper kettle containing boiling water soon gets too hot to be touched, and requires to be held with a piece of flannel or wood interposed between the metal and the hand, the heat being found not to pass so readily through these materials. The handle of an earthenware tea-cup never becomes too hot to hold, while the handle of a tea-spoon (especially if of real silver), when placed in the tea, rapidly becomes hot.

The differences in conductivity of substances may (when the bodies do not differ much in mass or specific heat) be approximately estimated by observing the rate at which heat passes through them. The quicker it does so, the greater (as a rule) is the conductivity of the substance.

Expt. 73. Take a strip of glass and another of iron of the same size and shape, about 15 cm. long, 2 cm. broad, and 1·5 mm. thick, and secure them in a wooden frame, as shown in fig. 76, their inner ends just touching one another. Attach to the under side of each, 6 cm. from the point of junction, a small wooden ball about the size of a marble (or a marble itself will

Fig. 76.

do), with soft wax;[1] this will serve to indicate, by dropping off, when sufficient heat to melt the wax has reached that part of the strip to which it is attached. Heat the adjacent ends of the two strips by a Bunsen burner with a rather small flame, adjusting the position so that each strip gets an equal share of the heat. In a very short time the wooden ball attached to the iron will fall off owing to the melting of the wax, while the ball attached to the glass will not show any signs of being loosened. Moreover, the glass, on being touched by the finger at a point still nearer the source of heat, will not be found to have become perceptibly warm, even if the application of heat to the end be continued for a couple of minutes longer.

This result shows that heat travels much quicker along the iron than along the glass; in fact, that iron is a much better conductor of heat than glass.

Expt. 74. We may next compare the conductivity of iron with that of copper in a similar way: substituting a strip of copper for

---

[1] A mixture of equal parts of beeswax and lard melted together.

the glass. Or longer bars of iron and copper may be used, riveted or brazed together at the ends, as shown in fig. 77. When the

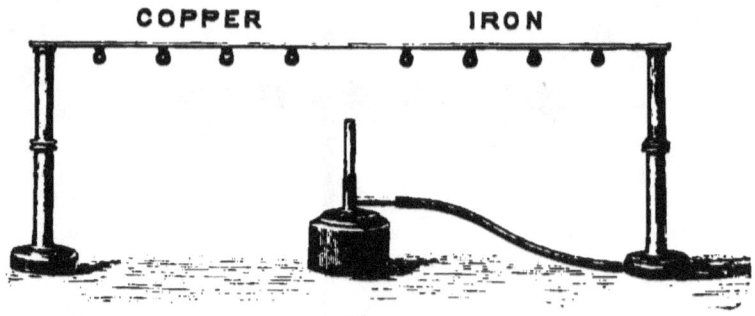

Fig. 77.

point of junction is heated, it will be found, (1) that the balls drop off one by one from *both* bars, showing that both metals are fair conductors of heat; (2) that the balls will drop off more quickly from the copper than from the iron, proving that metals differ in conductivity and that copper is a better conductor than iron.

The method illustrated in the two last experiments, of which the principle is to determine the time required for so much heat to travel along a substance as will raise a distant part of it to a given high temperature, is liable to give erroneous results for more than one reason.

In the first place, no account is taken of the loss of heat by convection and radiation which goes on from the surface of the substance. Bodies differ very much in radiating power, as will be presently shown, and if the surface of one of the substances under trial radiates away heat rapidly, it will appear a worse conductor than it really is, on account of the great leakage, as it were, of heat before it reaches the point where the temperature is determined.

In the next place, no account is taken of any difference in specific heat of the substances. To take the case of glass and iron, more heat is required to raise a given mass of glass 1° in temperature than is required for an equal mass of iron (see the table of specific heats, p. 135); hence the glass, in the experiment lately made, appears even a worse conductor than it really is. Similarly, in the other experiment, in which copper and iron were compared as to conductivity, the copper, having a slightly lower specific heat than iron, appears a rather better conductor than it really is, because less heat is required to raise its temperature to the melting-point of wax.

In some cases it is possible to obtain an altogether erroneous result, and to make a substance appear to be superior in conductivity to another one, while it is in reality much inferior to it. This happens when bodies which differ so much in specific heat as iron and lead are tested for conductivity by the above method.

Expt. 75. Take two short round bars, as shown in fig. 78, one

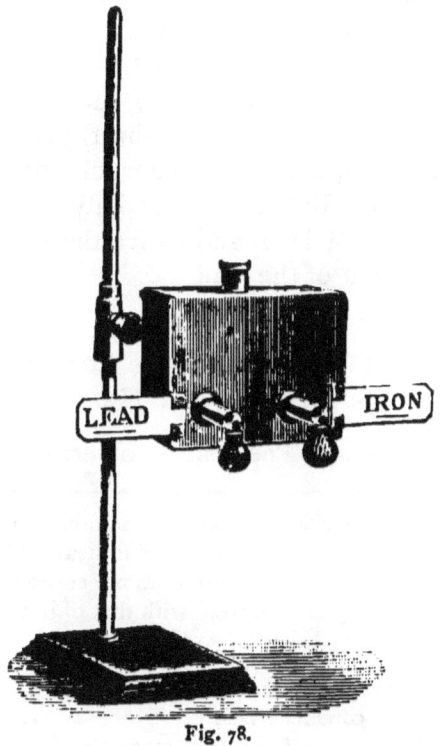

Fig. 78.

of lead, the other of iron, 6 cm. in length and 1 cm. in diameter; attach their ends by solder to the side of a small copper cistern, so that the bars project at right angles to the side. To the outer end of each affix a wooden ball by means of soft wax, and fill the cistern with boiling water. The ball attached to the lead will fall off first, although it has been clearly ascertained that lead is a decidedly worse conductor of heat than iron.

The reason of this apparently anomalous result is that the specific heat of iron is nearly four times that of lead, and hence so much more heat must enter the iron in order to raise its outer end to the melting-point of wax, that a longer time elapses before this temperature is reached than in the case of lead, which, like a smaller vessel, is sooner filled up to a given level. Thus, on reference to the table of specific heats on p. 135, we find the ratio of the specific heats of the two metals to be,

Iron.	Lead.		Iron.	Lead.
0·114	0·031	or	3·7	1 (nearly).

If the conductivity of iron was 3·7 times that of lead, this superiority would (when the bars are short) just compensate for its higher specific heat, and the balls would drop off both bars at the same moment. But the conductivity of iron (see p. 249) is only 1·4 times that of lead, and hence the outer end does not get hot so soon as that of the lead.[1]

Methods of Determining the Conductivity of Substances for Heat.

1. *Despretz's Method.*—The sources of error mentioned above

---

[1] Bismuth is in many books mentioned as suitable for the above experiment; but a reference to the tables of specific heat and conductivity will show that the low specific heat of bismuth does not entirely compensate for its greatly inferior conductivity as compared with that of iron, so that the wax is melted on the iron first.

	Iron.		Bismuth.
Thus : Ratio of specific heats,	3·79	:	1.
,, conductivities,	6	:	1.

Strictly speaking, the masses of the bars must also be taken into account, since the specific heats given are those of equal masses of the metals, not of equal bulks.

are avoided in this method, which has also been employed by Professor Forbes and later experimenters. A long bar of the substance, as shown at C D in fig. 79, is heated at one end by a steady lamp-flame, and after some time has elapsed the temperatures of different points in its length are determined by means of thermometers placed in holes drilled into it. As the heat travels along the bar, the thermometers will, of course, rise : but after some time it is found that the different parts of the bar assume a constant temperature, unchanged as long as the application of heat to one end is steadily maintained. The temperature of the part nearest to the source of heat is, of course, highest ; while if the bar is fairly long (about 1·5 metre or more) the temperature of the farther end does not much differ from that of the surrounding

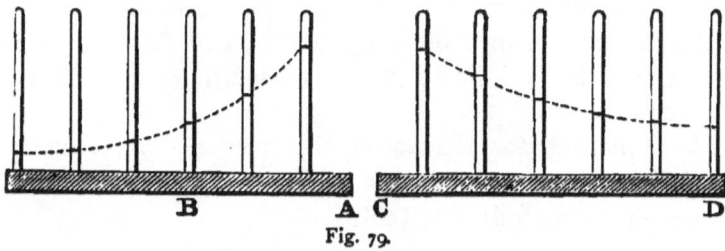

Fig. 79.

air. The reason of this will be seen, if we compare the flow of heat along the bar to that of water passing at high pressure into a pipe which has a long narrow slit in it from end to end. The water when it enters will at once begin to leak away through this slit, and thus the pressure will become less and less along the tube until a point is reached where the escape by leakage is just as great as the supply of water which arrives at that point, and there a gauge will show no appreciable pressure outwards at all. Similarly in the case of heat, no sooner does it begin its course along the bar from the heated end than it escapes from the surface by convection and radiation, so that the temperature shown by the thermometers gradually falls off, and at a certain distance along the bar the temperature will not be affected at all by the influx of heat at the end. It is obvious that this distance will be greatest in those substances which conduct heat best, since more heat will be able to travel along them to make up for the loss: and, to put it generally, the point of the bar where a given high temperature, *e.g.* of 30°, is maintained constant will be farther from the source of heat in proportion as the conductivity of the substance is greater. In order, then, to compare the conductivity of two substances the distances from the source of heat to the points where the bars show a temperature of, say, 30° are measured: the squares of these distances express the relative conductivities of the substances.[1]

Thus, supposing that in a bar of lead, C D, fig. 79, the point D where the attached thermometer acquires and maintains a temperature of 30° is 4 decims. from the end C, while in a bar of bismuth, A B, the point B, which has a temperature of 30°, is only 2 decims. from the end A, then the conductivity of lead will be to that of bismuth as $4^2 : 2^2$, or as 16 : 4, or as 4 : 1.

By such methods as the above, the relative conductivities of a number of solid substances have been ascertained, with the results given in the following table :—

---

[1] That the conductivities must be proportional, not to the simple distances thus measured but to the squares of the distances, will be evident from the following considerations. The amount of heat which passes into the ends A C of the bars (fig. 79) will vary with the conductivity of the substances. In the case of the bar A B this amount of heat is sufficient to reach the point B and maintain it at 30°; while in the case of the bar C D it is sufficient to reach the point D and maintain it at 30°. If the distance C D is twice the distance A B, the heat must, in order to produce the observed effect at D, (1) have travelled twice as far, so that twice the quantity must have entered the bar, (2) have been sufficient to make up for twice the loss by radiation and convection. So that altogether $(2 \times 2 =)$ 4 times as much heat must have been able to pass into the bar at C as has passed into the other bar at A. Thus the material of the bar C D must have four times the conductivity of the material of the bar A B. Similarly if the distance C D had been 3 times A B, the conductivity of the bar C D would have been $(3^2 =)$ 9 times that of A B: and so on.

CONDUCTIVITY FOR HEAT.

TABLE OF RELATIVE CONDUCTIVITIES FOR HEAT.

(Silver is taken as the standard of comparison = 100.)

Substance.	Conductivity.	Substance.	Conductivity.
Silver	100	Ice	0·2
Copper	74	Marble	0·15
Iron	12	Glass	0·05
Lead	8·5	Wood	0·01
Platinum	8·4		
Bismuth	2		

It will be noticed that all the best conductors are metals, though they differ greatly in conductivity among themselves. It may also be worth remarking that the order of the conductivity of substances for electricity has been found to be precisely the same as that of their conductivity for heat.

ABSOLUTE CONDUCTIVITY OF SOLIDS.

In determining this we have to consider the actual number of calories which pass through the substance in a given time, and the number called the 'absolute conductivity' of a substance is that which expresses the number of calories of heat which pass through a block of the substance 1 cm. thick and 1 square cm. in area in one second, when the surface at which the heat emerges is maintained 1° below the temperature of the surface at which it enters.

Determinations of the absolute conductivity of lead and one or two other metals have been made by Peclet, and the general nature of the apparatus he used is shown in fig. 80. A is a metallic vessel containing cold water and surrounded by a larger one of similar shape, the interval between them being filled with cotton-wool to

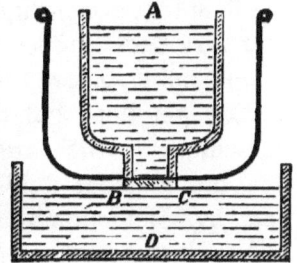

Fig. 80.

prevent loss of heat. At the bottom of this vessel is fixed a plate, B C, of the substance of which the conductivity is to be determined, and the whole is supported in a large basin, D, into which hot water is poured until its surface just touches the lower surface of the plate B C. This water, as well as the cold water in A, is stirred up and kept in motion by revolving paddles (not shown in the figure), so that fresh portions of water are continually brought into contact with the plate B C. The temperature of the water in each vessel having been ascertained (in most of the experiments the water in D was at about 24°, and that in A about 9° lower), heat was allowed to pass through the plate for a certain time, and then the rise in temperature of the water in A was ascertained. From this it could be calculated how many calories had passed through the plate from D to A in the given time. The area and thickness of the plate were measured, and thus all the necessary data were obtained for determining the absolute conductivity of the substance. In the case of lead Peclet's experiments gave 0·039 as the absolute conductivity, implying that 0·039 calorie per second passes through a plate of lead 1 cm. thick and 1 square cm. in area, when the water in A is 1° below that in D. This being known, the absolute conductivities of other substances can be calculated from their relative conductivities given in the table on p. 249.

Conductivity of Liquids and Gases.

All fluids, except liquefied metals, such as mercury, conduct heat very badly indeed. It is true that when a fluid is heated at any point below the top of it, the parts above that point quickly get hot: but this is fully accounted for by the convection-currents which are at once set up. It is, indeed, very doubtful whether fluids can be rightly considered to convey heat by conduction at all: their molecules are so free to move, and so incessantly on the move (as shown by the rapidity with which diffusion takes place in them) that the passage of heat

CONDUCTIVITY FOR HEAT.

through them can hardly take place in the manner specified in the definition of conduction already given on p. 241, *i.e.* without any motion of the molecules from place to place.

The real test of conductivity would be to see whether heat passes through the substance under conditions where convection is impossible and diffusion is greatly hindered (it cannot be prevented entirely). The following experiments will show that under such circumstances the passage of heat through fluids is extremely slow: mercury, however, and also liquid alloys of metals, show fairly high conducting powers, as well as the other characteristic properties of a metal.

Expt. 76. Support two tubes, closed at one end, about 4 cm. in length and 1 cm. in diameter, in wire holders attached to a retort-stand. Bend a piece of stout copper wire, about 3 mm. or more, in thickness, into the shape shown at A B in fig. 81, so that one end of it dips into each tube. Attach a wooden ball by soft wax to the bottom of each tube, and fill one of the tubes with mercury, the other with water. Heat the coil in the middle of the wire by a Bunsen burner. The heat will be conducted along the wire quickly and equally to the liquid in each tube, and in a minute or so the ball attached to the tube containing mercury will drop off, proving that heat has passed readily through the metal.

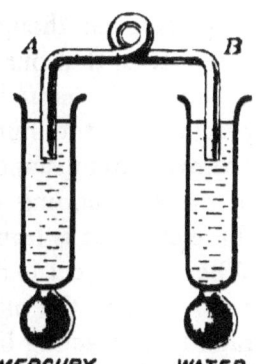

Fig. 81.

The upper part of the water in the other tube will boil, but the ball below will not drop off, and the lower part of the water will remain cold to the touch even if the heat is continued for five or ten minutes. This shows that water, even allowing for diffusion, is an extremely bad conductor of heat.

That gases, such as those which compose common air, are bad conductors may be shown in the following way:—

Expt. 77. Place a sheet of bright, polished tin plate, about 20 cm. square, upon the largest ring of the retort-stand: put at each corner a small piece of wood, about 1 cm. in thickness, and lay upon these a sheet of tin similar to the first. We have thus a layer of air 1 cm. thick between two conducting surfaces. Attach to the under surface of the lower plate a wooden ball by soft wax, and place upon the upper tin plate an iron ball heated nearly to redness, just above the part of the lower plate where the wooden ball is fixed. The latter will not drop off, at any rate for several minutes; proving that even a thin layer of air is an extremely bad conductor of heat.[1]

Results and Applications of the Conductivity of Substances.

I. One result of the great difference in conductivity between wood and metals has been noticed at the commencement of the book (p. 4), *viz.* that metals feel colder to the touch than wood or glass, even though both are really at the same temperature. The fact is, that our sense of touch simply informs us of the *rate* at which the skin is gaining or losing heat. Now, the temperature of the human body is about $37°$, which is, under ordinary circumstances, decidedly higher than that of the air and of the surrounding objects, which are seldom above $25°$; and therefore heat is transferred from the skin to the objects touched. When the object is made of metal, this heat is quickly conducted away through the mass, and the surface goes on for some time rapidly abstracting heat from the skin, and therefore 'feels cold.' But in the case of a bad conductor, such as wood, the heat is not thus conveyed away from the surface, and the latter almost immediately becomes as hot as the skin, from which it then ceases to take away heat, and therefore no sensation of cold is felt.

For the same reason, floors of stone or marble appear much colder than wooden ones, or carpeted floors, and are much more likely to cause a chill or 'cold' to be caught.

Expt. 78. Obtain a smoothly turned cylindrical rod, about 20

---

[1] Some heat will eventually reach the lower plate, but this is conveyed by radiation, not conduction.

CONDUCTIVITY FOR HEAT. 253

or 25 cm. in length and 2 cm. in diameter, one half of the length of which is made of wood, the other half of brass (or, preferably, copper). Wrap round it a piece of thin writing-paper, retaining the paper closely in contact with the rod by pinching the projecting part of it between the fingers, or by spring clips; then hold it at a little distance above (not in) the flame of an Argand burner, moving it about so that heat is evenly applied to the rod at the junction between the metal and the wood, and for a distance of 4 or 5 cm. on either side of it. When the paper shows signs of blackening or charring, remove it, and observe that the charring effect of the heat is entirely confined to that part of the paper which was in contact with the wood; the metal having conveyed away the heat so quickly from the paper which was in contact with it that the temperature never rose high enough to decompose the material and separate carbon.

For a similar reason, a piece of hot metal is much more painful to handle than an equally hot piece of wood or glass, since the surface of the latter soon becomes cooled to the temperature of the hand, and heat can only pass slowly from the interior of the mass to make up for the loss.

II. It is obvious that copper will, in spite of its expensiveness, be a better material for boilers, kettles, and saucepans than iron or tinned iron, since heat can pass through it much more readily to the liquid within. It has lately been proposed to have short rods of copper projecting from the bottom of a vessel to be heated; these penetrate into the flame and convey to the liquid much heat which would otherwise never reach it at all.

The fact that the 'singing' of a liquid (p. 186) begins at a lower temperature in a copper vessel than in an iron one, depends on the greater quickness with which the heat passes through the copper and causes bubbles of vapour to form in contact with its inner surface. So also the reason why singing never occurs in a glass vessel is that heat passes through it so slowly that bubbles cannot form and collapse in such quick succession as to cause a musical note.

III. The stony incrustation, or 'fur,' which is formed on the inner surface of boilers and kettles in which hard water is boiled, is, from its bad conductivity, a serious nuisance, and even a source of danger. Much of the heat of the fire never reaches the liquid in the vessel at all, and is absolutely wasted.[1] Moreover in a steam boiler the heat of the furnace, not being transferred to the water as soon as it has passed through the metal plates, remains in the latter, which may thus become nearly or quite red-hot, and give way under the steam pressure. In any case such excessive heating of the plates causes much damage by rusting or oxidation.

IV. The handles of kettles are often covered with wood or glass, to protect the hand from the heat conducted freely through the copper or iron itself: and with a similar intention small blocks of ivory or wood are inserted between the body of a teapot and its handle.

V. It is found that flame is immediately extinguished by contact with a coil of wire or an open network of wire such as wire-gauze. This curious effect may be shown in many ways.

Expt. 79. Form a piece of moderately thin copper wire into a spiral about 1 cm. in diameter and 2 cm. in length, by coiling it round a tube or rod, leaving 8 or 10 cm. straight at one end, to serve as a handle. The coils of the spiral should be about 1 mm. apart, so that air may be able to pass freely through it. Light a candle and lower the coil of wire, held vertical, over the flame until it nearly touches the wick. The flame will appear to shrink from contact with the metal and will be extinguished altogether, although air can obviously be supplied freely, so that the spiral cannot act like an ordinary extinguisher.

Expt. 80. Re-light the candle and depress upon the flame a piece of wire-gauze about 14 or 15 cm. square, and having about 2 meshes to the millimetre, bringing it finally close down to the wick. The flame will be stopped by the gauze as effectually as by a solid plate of metal.

---

[1] A few years ago one of the steamers plying between India and England accumulated so much scale in her boilers (a crust nearly 3 cm. thick) that sufficient steam could hardly be made for the engines. The whole of the stock of coal was used up, and even some of the wood-work of the ship had to be burnt in order to bring her into an English harbour.

ACTION OF WIRE-GAUZE ON FLAME.

The experiment may be repeated with the flame of a large Bunsen burner (the air-holes being stopped so that the flame is luminous and more easily seen).

It may be shown that un-ignited gas is passing through the interstices of the gauze by holding a lighted match just above it while it is depressed on the flame, when the gas will catch fire and burn almost as if no obstacle was interposed.

Expt. 81. Put out the flame by turning off the gas-supply or by pinching the indiarubber connecting-tube, and lay the wire-gauze on the mouth of the burner. Then allow the gas to flow again, and hold a lighted match just above the gauze, so as to light the gas which passes through it. Now raise the sheet of gauze gradually: the burning gas will not pass downwards, and the flame may be lifted, as it were, from the orifice of the burner, leaving a clear, flameless interval of 5 cm. or more between them.

Expt. 82. Place a piece of tow moistened with alcohol upon the sheet of gauze, lay it on the ring of a retort-stand and hold a lighted candle or gas burner underneath. It will be found impossible to set the alcohol on fire.

These, and other experiments which may be readily devised, give abundant proof of the fact that wire-gauze destroys flame; and we have next to consider the reason why it does so.

In the first place, the phenomenon which we call 'flame' is that which occurs when gases or vapours are undergoing combustion, *i.e.* are combining with each other so rapidly as to produce intense heat, which raises the temperature of the gases to such an extent that they emit light. Flame, in fact, is simply glowing gas; its brightness, in cases of ordinary combustion, being increased by the presence of particles of solid carbon, as already explained, p. 32. Now, a mixture of coal-gas or (in the case of a candle) wax-vapour with the oxygen of the air requires to be heated pretty strongly before they will combine and produce flame;[1] at any temperature below a full red heat no

---

[1] This may be easily and effectively shown as follows: Lay a piece of fine wire-gauze upon the chimney of an Argand burner, and place upright upon it a small cylinder of platinum foil (made by folding a bit of the foil about

chemical action occurs, and the gases remain simply mixed. Anything, then, which keeps the mixture cool prevents or extinguishes flame. The iron wire of which the gauze is composed, being a good conductor, and moreover exposing a very large surface, carries off heat from the burning gases to the cooler parts of the gauze, and also gives off the heat by radiation so quickly that the gases are cooled below the temperature at which their chemical combination can take place, and the flame ceases to exist in the meshes of the gauze.

A very important application of this property of wire-gauze is made in the construction of a lamp which can be used with safety by men who have to work in coal-mines, where there is always more or less of the dangerously inflammable gas called 'fire-damp'[1] mixed with the air. When more than 10 per cent. of this gas is present in the mixture, the introduction of a lighted candle or match causes it to inflame suddenly with a violent explosion. Many terrible accidents have occurred from this cause; indeed, some coal-mines had to be abandoned from the impossibility of using a light in them, until Sir Humphry Davy, in A.D. 1816, from an examination of the property of wire-gauze illustrated above, invented the lamp which is still known as the Davy Safety Lamp. One of these is shown in fig. 82, and consists of an ordinary oil lamp surrounded by a cylindrical cover of fine wire-gauze, closed at the top with a double thick-

---

5×3 cm. loosely round a pencil). Turn on the gas and light it above the gauze; and when the platinum cylinder is red-hot, turn off the gas. As soon as the flame goes out, but before the platinum is cool, allow the gas to flow again. The platinum foil will again begin to glow, in consequence of an action which need not be explained here: the point to be noticed is, that no flame is produced although the inflammable mixture of gas and air is in contact with the red-hot platinum, and must therefore be at about the same temperature. If the platinum foil, held in a pair of crucible tongs, is heated to bright redness in a lamp and then replaced on the gauze, inflammation will occur at once.

[1] This is a compound of carbon with hydrogen, called by chemists 'methane.' More than one-third of the volume of ordinary coal-gas consists of it.

THE DAVY SAFETY-LAMP.

ness of the material, and fitting accurately round the lamp below, so that no gas can reach the flame without passing through the meshes of the gauze, the interstices of which must not be more than half a millimetre square. The action of the lamp is plain. Sufficient air for the combustion of the oil can pass freely through the gauze, but when it is brought lighted into a dangerously explosive mixture of fire-damp and air, the mixture simply burns quietly within the gauze cover, none of the flame getting through to ignite the mass of gases outside. Indeed, if the proportion of fire-damp to air exceeds a certain amount, the flame of the burning oil is extinguished, and thus the miner has a useful indication of the condition of the atmosphere in which he is working.

Fig. 82.

Expt. 83. If a Davy lamp is at hand, its action may be illustrated by lighting it, screwing on the cover, and directing a stream of gas from a burner at the flame. The gas will burn within the gauze, but it will be found impossible to communicate the flame to the burner.

A more striking illustration of the 'safety' of the lamp is the following. Hang a large bell-glass, about 30 cm. in diameter, and the same in height, mouth downwards, from a retort-stand. Pass up into it some coal-gas from a tube, taking care that no light of any kind is near. The gas will soon, from its slight density, collect in the jar, mixing freely with the air: and when its smell is perceptible at the mouth of the jar, pass up slowly into the mixture of gas and air a lighted safety lamp. The inflammable mixture will take fire within the gauze, but no explosion will occur, and when the lamp is raised higher in the jar, the flame will quietly go out.

Wire-gauze is, from this property, employed for several other purposes in scientific work. Thus a good form of burner, where a large flame is required, is made by covering the chimney of an Argand burner with a disc of fine wire-gauze. The ascending

R

mixture of gas and air can be lighted above the gauze, and burns with a large smokeless flame. The lamp may in fact be regarded as a number of Bunsen burners, each represented by a mesh of the gauze. Fletcher's burner, in which a similar use is made of wire-gauze, is shown in fig. 9, p. 34.

In heating glass vessels over a lamp, it is always advisable to place the vessel on a piece of wire-gauze, laid on the ring of a retort-stand or iron tripod. The heat is thus distributed over a large area of the glass, greatly diminishing the chance of fracture.

VI. That glass cracks when either quickly heated or quickly cooled is an unfortunately too familiar experience. The reason of its doing so is connected with the extremely bad conductivity of the material for heat. When a piece of glass is suddenly exposed to a high temperature, the outside layer expands before the heat can be transmitted by conduction to the interior portions. These latter therefore remain unexpanded, and a state of strain is set up between the inner and the outer parts of the glass, which may easily become sufficiently great to tear the mass to pieces.

So again, when a piece of strongly heated glass is cooled quickly, the outer portions contract at once, while the interior, not having been able to part with its heat, remains expanded. The outer layer, becoming too small to fit, as it were, over the inner portions, splits into pieces and determines the fracture of the whole.

Expt. 84. Take a strip of thick window glass, and plunge it into a blowpipe-flame. The glass will fly to pieces at the heated end, for the reason above given.

Again, heat another strip of glass slowly and carefully over a Bunsen burner, holding it at first in the hot air 10 or 12 cm. above the actual flame, turning it constantly round, and gradually lowering it into the flame. Under these circumstances the heat has time to travel leisurely through the mass, so that no part is ever at a very different temperature than the rest, and no great strain and consequent fracture will take place. When the glass is nearly red-hot, plunge it quickly into a jug of cold water: the

Effect of sudden Heat on Glass.

heated portion will break up into a multitude of small irregular fragments, like the pieces of a dissected puzzle.

A useful application of the above property of glass is often made in laboratories and glass-works. If heat is applied locally by means of a pointed rod just in front of the end of a crack which has been commenced in a piece of glass, the strain thus set up in the heated part will cause the crack to extend in that direction in preference to any other; and if the heated rod is moved along the surface of the glass, the crack will obediently follow its lead. Thus the glass may be cut into any form desired, almost as accurately as with a diamond.

Expt. 85. Take a flask or test-tube, of which the neck has been chipped or broken, and press upon the glass, close to the broken part, the end of a piece of thick iron wire[1] previously heated to redness. If a crack is not at once started, withdraw the rod and quickly touch the heated glass with a drop of water. This will pretty certainly begin a small crack, which may then be led by again applying the red-hot rod, first a little way downwards and then round the neck, so as to cut off evenly a ring of glass, including the damaged portion. As a specimen, a cylindrical lamp-glass may be taken, and a crack led spirally round and round the glass from one end to the other; or a useful evaporating dish may be made by cutting off the bottom of a damaged flask or retort.

The properties of glass illustrated above have been applied to the production of a kind of ornament. If a glass jug or vase is, immediately after being made, and while still extremely hot, dipped for a moment only into cold water, cracks are of course formed in it in every direction. Owing, however, to the irregular interlacing of the fragments, the mass does not usually fall to pieces at once, and if it is then carefully held in a furnace, the surface of the glass melts and forms a continuous layer over the

---

[1] A very good substitute for this is a bit of glass rod or tube drawn out to a blunt point in the blowpipe-flame. Or a minute gas-flame may be used, such as is obtained from the jet of a mouth-blowpipe.

cracks, which are still visible underneath. Thus the vessel regains a good deal of its former strength, and can be used for ordinary purposes, while the network of cracks is considered to make it ornamental.

It will be evident from what has been said that the particles of a bad conductor which has been strongly heated must, unless it has been allowed to cool very slowly, be always in a state of strain, owing to the surface cooling and contracting before the rest of the material, and thus compressing the inner portions somewhat in the manner of the tyre of a wheel (p. 56). This strain may not be sufficient to cause fracture, but it must exist and give rise to weakness and liability to fracture by any sudden blow, or by contact with a hard substance which may scratch the surface. The existence of such a strain in quickly cooled glass is well shown by what are called 'Bologna phials.' These are merely small glass flasks, very thick at the bottom, which after being blown into shape are allowed to cool at once in the air. If a small sharp splinter of flint is dropped into one of these flasks, the microscopic scratch it produces is sufficient to determine the cracking to pieces of the whole vessel.

Another good illustration of the same fact is afforded by the toys called 'Rupert's drops.' These are made by slowly pouring melted glass from a ladle into cold water. Some of the drops fly to pieces at once, as might be expected, but a good many remain uninjured in the form of small pear-shaped lumps with long, thin, straggling tails. There will evidently be a state of intense strain throughout the mass of one of these drops, no single particle being on easy terms with its neighbours. Its rounded shape, however, gives it considerable strength, and it can be roughly handled, and will even stand a blow without damage; but if the smallest portion is broken off the tail, the sudden concussion causes the whole of the drop to fly to pieces, becoming, in fact, a coarse powder.

Expt. 86. Hold the body of one of these 'Rupert's drops' tightly and securely in one hand with the tail projecting, and break a small piece off the latter. The hand will feel, as it were, a blow, and on opening it a crumbling mass of small bits of glass will be found in place of the drop.

The experiment may also be made by holding the 'drop' inside a large thick glass bottle and breaking a piece off the end of the tail as before. If a smaller flat-shaped bottle, about 200 or 300 c.c. in capacity, is filled with water and the 'drop' inserted and exploded as above, the bottle will be broken by the violent concussion transmitted through the water; the latter, owing to its inertia and slight elasticity, conveying a sudden impulse almost without loss of intensity (a property utilised in the 'water-shells' used in war).

The preceding experiments abundantly prove the absolute necessity of allowing glass articles, after they have been blown or moulded into shape from the melted material, to cool with extreme slowness, in order that no part may during the cooling be at a very different temperature to the adjacent parts; otherwise a state of strain must infallibly be produced, and the article may break with a slight scratch or blow, or even fly to pieces without any apparent cause, on a change of weather or when brought into a hot room. This process of slow cooling is called the 'annealing' of glass, and it is carried out in practice as follows : The glass article, as soon as ever it has cooled to the point at which it ceases to be soft and liable to be bent out of shape, is put into a brick chamber called an 'annealing oven,' which has been previously raised to a red heat, only slightly below the temperature at which glass becomes soft. When the oven is filled, the door is closed, the furnace fire is allowed to die out slowly, and everything is left to itself for about a week, by which time the contents will have cooled down to the ordinary temperature.

It has lately been found that by a modified application of the method of making 'Rupert's drops' additional strength can actually be imparted to glass, just as in making a carriage wheel additional firmness is imparted by shrinking on the tyre. This was discovered by De la Bastie in 1872, whose so-called 'toughened glass' was for some time thought highly of. The glass articles, immediately after being made, instead of being carried off to the annealing oven, were plunged quickly into a cistern of moderately heated oil or grease. Under these circumstances the surface of the glass is at once cooled and contracted, but the viscosity and bad conductivity of the oil allow the interior to cool slowly and regularly; thus preventing so violent a strain on the molecules as to cause actual fracture. In this way the glass is 'case-hardened,' as it were, with an outer skin which binds together and strengthens the rest of the mass. Glass thus treated will withstand pressure and blows which would certainly break up ordinary annealed glass, but if the surface is scratched or chipped the strained state of the material shows itself, and the article flies into a multitude of fragments like a 'Rupert's drop.'

VII. There are many cases in which the bad conductivity of gases is applied in conjunction with that of solids to prevent the free transmission of heat to or from a substance. Thus 'ice-houses'—receptacles in which a store of ice can be kept without melting during the heat of summer—are always constructed with double walls, having an interval of about 20 cm. between them. The air which fills this space, being a much worse conductor than even the brick or stone of the walls, effectively prevents any heat from reaching the ice. The passage leading into such a building has a door at each end, one of which is closed before the other is opened, so that no current of warm air can enter the place where the ice is stored. Similar 'double doors' and 'double windows' are not uncommon in ordinary dwelling-houses, and are very effective in keeping the rooms cool in summer and warm in winter in spite of fluctuations in the temperature of the outside air.

The same 'double wall' principle explains the comfort of thick, loose, woollen clothing, and of furs, feathers, etc., which owe their warmth (or, more strictly, their power of preventing the escape of heat from the body), not so much to the bad conductivity of the material as to the even worse conductivity of the layers of air which exist between the fibres. For a similar

Applications of Conductivity for Heat. 263

reason steam boilers and steam pipes are usually covered with a thick layer of felt or asbestos to prevent loss of heat and consequent waste of fuel. On a smaller scale kettle-holders of flannel or cloth are an application of the same principle.

Expt. 87. Cover the palm of the hand with a thick, *loosely* spread layer of asbestos or finely powdered whiting (chalk) or plaster of Paris, about 2 cm. thick, and lay on it a small block of iron (a pulley sheave 4 or 5 cm. in diameter answers well) heated to full redness. This may be carried in the hand, for a couple of minutes at least, without an inconvenient amount of heat reaching the skin. The block must not, however, be so heavy as to press the material closely together, or heat will soon be felt; showing how much of the bad conductivity of the protecting layer is really due to the air-spaces between the solid particles of asbestos or chalk, and not to the latter themselves.

Fire-proof safes protect their contents from damage by heat partly at any rate [1] owing to the double casing of which they are constructed, and to the loose, porous material which is placed in the interval between the inner and outer cases.

In laboratories, when glass flasks or other vessels have to be heated gently and steadily, they are placed on a 'sand-bath,' which is merely an iron tray containing a layer of fine sand, and heated below by a furnace or gas-burner. The heat only travels slowly through the sand and the air surrounding the particles to the glass vessel resting on it, and thus there is little or no risk of breakage.

Water-pipes and pumps are often wrapped in straw during a frost to prevent the water in them from freezing.

Blocks of ice, when they have to be carried about in summer, or when an ice-house or ice-safe is not available, are closely wrapped in thick flannel, which keeps the heat of the air from reaching and melting them. A large, smooth block of ice may, however, be left in warm air for a considerable time without

---

[1] The other reason for their effectiveness has been already explained on p. 169.

much loss by melting, on account of the bad conductivity of the material itself. This also explains in part the slowness with which the layer of ice on ponds increases in thickness during even a severe winter.

Section II.—The Convection of Heat.

The general meaning of this mode of transmission of heat, implying, as it does, actual mechanical movement of the molecules themselves from one place to another, has been already explained and illustrated on p. 241. We have now to consider more minutely the conditions under which convection of heat can take place.

In the first place, it can obviously only occur in fluids, *i.e.* in liquids and gases, since it is only in them that the molecules are sufficiently independent of cohesion to move from one point to another in the mass.

In the next place, it can only occur when heat is applied to a particular part of the fluid, *viz.* to those portions which lie below the surface. In an experiment made already, p. 251, it has been shown that water, when heated at the top, does not get hot throughout; yet it is a matter of common experience that all the water in a vessel quickly gets hot when heat is applied to the bottom. The reason is that in the latter case conditions are favourable for setting up convection-currents by which heat is rapidly diffused through the whole mass.

The production and course of these convection-currents may be shown by the following experiments :—

1. *Convection-currents in Liquids.*

Expt. 88. Fill a large globular flask, at least 15 cm. in diameter, with clean water up to the bottom of the neck, and support it on wire-gauze placed on the largest ring of the retort-stand. Cover two or three small crystals of the dye known as 'magenta' (or some other equally soluble and intense colouring-matter) with glycerine or thick solution of gum, and drop them one by one into the flask of water. They will fall through the water without colouring it perceptibly, owing to the protective viscous coating, and will lie at the bottom slowly dissolving and forming a deep-coloured stratum there. Heat the flask by a gas-burner (with a rather small flame, in order that the heat may be localised). A convection-current will be almost immediately set up at the heated spot, rising in streams of coloured liquid to the surface, and then turning outwards and descending at the sides of the flask until it returns to the place it came from. The general course of these currents is shown in fig. 83, by which it will be seen that a regular circulation is set up in the mass of liquid; each portion, as it becomes hot, rising straight upwards and parting with heat to the surrounding water; then, on reaching the surface, it is pushed towards the sides by the portions which succeed it, and falls to the bottom as it cools, pressing upwards, in its turn, other portions of heated water.

Fig. 83.

In a short time the whole of the contents of the flask will have become pretty uniformly coloured, and this may be taken as a proof that heat has been distributed by currents through every part of it.

Similar convection-currents can be produced in a downward direction by cooling a portion of water at the surface.

Expt. 89. Fill a large glass jar or beaker about three-fourths full of cold water, and pour on it carefully some warm water (at about 40°) coloured with indigo or ink, precisely as directed in Expt. 21, p. 76, in sufficient quantity to form a layer about 2 cm. in thickness. Place very gently in the water a flat lump or two of ice: this will cool the surface layer, and very shortly streams of coloured water will be seen descending through the mass of liquid.

2. *Convection-currents in Gases.*

These are much more rapid than those which occur in liquids, not only on account of the greater freedom of movement possessed by the molecules of a gas, but also because the coefficient of expansion of a gas is much greater than that of any ordinary liquid, so that the changes in volume which accompany the heating and cooling of a gas are very considerable; and these, as will be shown very shortly, determine the force of the currents.

In order to render visible the movement in a colourless gas, such as air, some light solid particles must be introduced, which (like straws in the wind) may show the direction in which the streams of gas are moving.

Expt. 90. Take a large glass bell-jar (such as one of those used for protecting plants, or an aquarium-glass) and place it inverted near a plate. Put on the plate a small, carefully dried piece of phosphorus, set it on fire by a lighted match, and place the bell-jar over it. The phosphorus in burning forms a white solid (phosphorus pentoxide), the particles of which are wafted quickly upwards, showing that the air, heated by the combustion of the phosphorus, rises until it comes into contact with the cold surface of the glass: then, as it cools, it falls down by the sides of the jar, and ultimately again reaches the source of heat. This circulation of air occurs so quickly that the whole jar will be in a few seconds filled with white clouds, proving that heat has been conveyed to every part of the air.

Similar convection-currents must obviously exist in the air over every heated substance, as may be very easily shown.

Expt. 91. Place a small block of iron (or the iron ball used in Expt. 72, p. 240), heated nearly to redness, upon a brick or a thick layer of sand, and hold over it a strip of brown paper which has been lighted, and then gently blown out so as to leave it still smouldering. The white smoke, which rises with the convection-currents caused by the heat of the paper, will be hurried much more swiftly upwards by the larger and more powerful currents rising from the heated metal.

The same effect may be shown more strikingly by holding over the hot metal one of the small sets of mica vanes sold for fixing over gas-lamps, balanced on a needle so that it can rotate freely.[1] The invisible currents will soon set it in rapid motion.

---

[1] Such a set of vanes can be made in a few minutes out of thick paper or thin cardboard. Cut a circle of the material, 10 cm., or more, in diameter, and make eight or ten radial cuts in it, not reaching quite to the centre. Set

Causes of Convection-Currents. 267

It may also be held over a lighted candle, or a gas-flame, or a vessel of hot water, with the same results.

We must now inquire more closely into the cause of these movements in the fluid.

The actual force which produces the motion is that of gravitation, which is always acting upon every molecule of a fluid, dragging it downwards towards the earth's centre. Under ordinary conditions this downward pressure of the molecule is met by equal and opposite pressures of the molecules below it (according to the Third Law of Motion and Pascal's Law of Fluid Pressure), so that each molecule remains in its place[1] and the whole mass of the fluid is in equilibrium. This, however, is only the case while the temperature is uniform throughout the mass. If heat is communicated to a small portion of the fluid at some distance below the surface, this portion of the mass increases in volume without alteration of absolute weight. Thus its surface is increased, and more molecules can get at it to exercise pressure upon it, and this increased pressure more than balances the downward force of gravitation; hence the portion of fluid is forced upwards and rises to the surface, the result being practically the same as if, instead of an increase in volume, there had been a decrease in weight of the heated portion of the fluid.

Similarly, when a portion of the fluid at or near the upper surface is cooled below the temperature of the rest, it contracts in volume and diminishes in surface: hence fewer molecules can get at it below, and gravitation causes it to fall through the warmer fluid.

---

each of the sectors thus made somewhat obliquely to the plane of the circle, and make a slight dent or hollow (not a hole) in the exact centre with a blunt-pointed pencil or bit of wood. Balance the whole on a pin stuck into a cork (the dent will serve to keep it steady on the pin), and hold it at some little distance above the heated metal block or a lighted candle.

[1] Except, of course, as regards the movements due to diffusion, which need not be considered here.

The following experiment may serve to demonstrate that mere alteration in volume without change of absolute weight is sufficient to cause upward or downward movement in a mass immersed in a liquid.

Expt. 92. Take a small indiarubber bag (one of those now used instead of bladders in footballs answers well[1]); put some shot into it until it sinks readily in water, and connect to it a piece of indiarubber tubing about a metre in length. Place it in a large deep jar filled with water (such as an aquarium-glass), and as it lies at the bottom, blow air gently into it from the mouth or a pair of bellows, so as to increase its size. At a certain point of expansion it will rise through the water and float at the top, although its actual weight has not been lessened (it is, in fact, slightly heavier than before, by the weight of the air which has been blown into it). The increase in volume implies a larger surface; and the pressure of the liquid, being proportional to the area of the surface, has become sufficient to overcome the force of gravitation and raise it through the water. If a little air is now let out, the bag shrinks in size and falls to the bottom again.

The elastic bag, of variable size, in the above experiment may fairly represent a portion of any fluid which differs in temperature from the rest; and it is easy to see that the forces acting on such a portion will push it upwards or pull it downwards in the fluid according as its volume is greater or less than that of an equal weight of the surrounding fluid. Heat, then, by producing such changes in volume, is competent to cause movement in portions of a fluid; in fact, to give rise to currents which will continue until either by diffusion among the molecules all parts of the fluid are brought back to the same density, or by communication of heat the temperature is equalised throughout the mass.

---

[1] A bladder, or an air-tight sponge-bag, with its mouth tied tightly round a short bit of glass tubing, will answer.

Results and Applications of the Convection of Heat in Liquids.

I. The facility with which masses of liquid can be heated throughout by setting up convection-currents in them has been sufficiently alluded to already. If the liquid has a high co-efficient of expansion, the convection-currents will be more rapid and powerful, and the process of heating much hastened. If the liquid is viscous, the convection-currents will be sluggish and the rise of temperature slow. Thus soup, chocolate, etc., require to be stirred continually in order to distribute the heat through the mass and prevent overheating of one part, which would lead to charring or 'burning' of the substance. It is equally obvious that such viscous liquids will cool slowly, and scalds produced by them will be rather severe, owing to the heated matter remaining long in contact with the skin.

II. Liquids can be made to travel very great distances by a judicious production of convection-currents in them.

This is the principle of all the forms of hot-water apparatus for warming houses, etc., in which the heat produced by burning coal in one place is conveyed throughout the building, however large. Such an apparatus consists of three essential parts, *viz.*: —

(i) A boiler with furnace attached; in which water is heated, and thus a disturbance of equilibrium is produced which starts it upon its travels.

(ii) Distributors of heat; usually a cluster of pipes placed in every room and passage to be warmed.

(iii) Pipes connecting the boiler with the distributors.

The arrangement and the working of a hot-water warming apparatus may be shown by a small model like that represented in fig. 84, in which the boiler is the glass bulb A, connected with a vertical tube, B, from which (near the top) a horizontal branch-pipe, C, leads to the set of three vertical pipes, D. These latter represent the much more numerous assemblage of pipes which usually compose the 'heating coil' in a room. From the lower ends of these another horizontal pipe, E, serves to convey the cooled water back to the boiler.

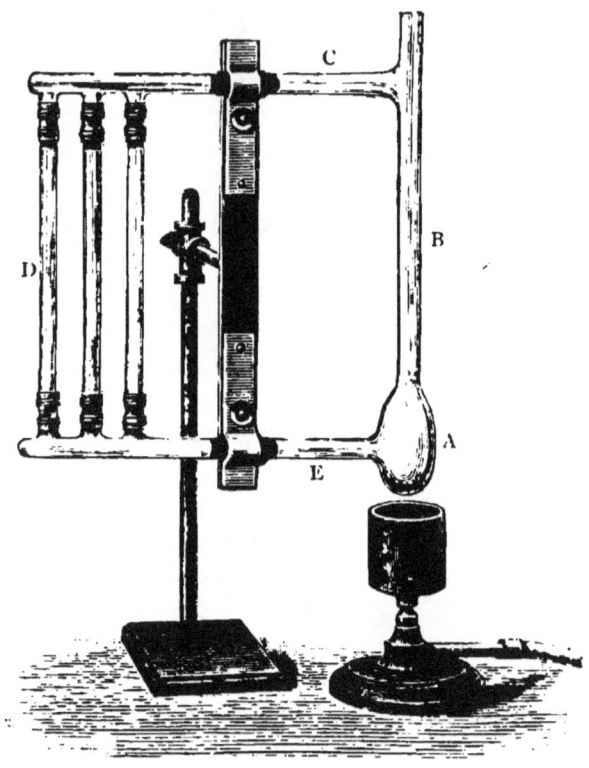

Fig. 84.

Expt. 93. Arrange the apparatus as shown in the figure, and fill it with water up to about 2 cm. of the mouth, getting rid of any air-bubbles by inclining the tubes in one direction or another. Drop down the vertical tube B two or three bits of magenta covered with glycerine or gum (see Expt. 88, p. 264), which will fall to the bottom of the bulb without colouring the water in their passage. Now place the lighted lamp below the bulb: a stream of heated and coloured liquid will soon rise up the vertical tube, pass along C, and enter the distributor D. Here it will rapidly part with its heat to the surrounding air, causing convection-currents in the latter, and will also give off heat by radiation. Thus it will become denser and descend, making its way along E and forcing more currents of hot water up through B. In a short time the currents will become imperceptible owing to the uniform colouring of the water, but they will still go on as long as the water in the boiler and the vertical tube B is kept hotter than that in the rest of the apparatus.

The points to be attended to in the construction of an effective and scientifically arranged warming apparatus will be pretty plain from what has been said upon its principle.

(*a*) The boiler must be placed as low as possible; lower, at any rate, than any part of the piping connected with it. The course of the heated water is always an upward one, and that of the colder water is always a downward one; any position of the boiler, therefore, which reverses the natural direction of the currents lessens their force.

(*b*) The pipe conveying the heated water from the boiler should proceed from the top of the latter in as straight a vertical line as possible up to the highest part of the system, where it usually enters a small cistern from which pipes are led to the various heating coils.

(*c*) The connecting-pipes should be fairly large, and as free from bends as possible, at any rate from abrupt angles: the object being to avoid any baffling of the rapid flow of water through the apparatus.

(*d*) The coils or clusters of piping which serve as distributors of the heat should be as freely exposed as possible, so that a constant and rapid renewal of the air at their surfaces may go on. A great deal of heat is lost by enclosing them in a case, as is often done for the sake of appearances: indeed, none of the heat which would otherwise be diffused through the room by radiation is utilised at all.

(*e*) Allowance must be made for the changes in size which must necessarily occur as the pipes change in temperature. A free space must everywhere be left between the pipes and the walls: and where long, straight lengths of piping have to be used, one part must be made to slide into the other, like the tubes of a telescope; the joint being made watertight by a 'stuffing-box' (a contrivance explained in Chapter x.).

III. We may next give our attention to convection-currents on a much larger scale, which are going on in lakes and oceans. They constitute, as it were, a warming and cooling apparatus arranged by Nature for equalising temperature on different parts of the earth.

In the first place, the cooling of large masses of water in winter is caused in the manner illustrated by Expt. 89, p. 265, already made. Heat is radiated freely from the surface, and thus downward convection-currents are set up, which (as explained on p. 78) reduce the temperature of the whole of the water to $4°$, at any rate.

Again, there is undoubtedly a constant movement going on throughout the whole mass of the water in an ocean, caused primarily by the cooling of the surface-water in high latitudes, which loses heat partly by radiation but mainly by contact with the masses of ice in the Polar regions. Since the temperature of the water in the Tropics is kept up by the sun's heat (in spite of loss by radiation), there is a disturbance of equilibrium in the whole mass of water, and the cold Polar water sinks, and finding its way slowly along the bottom displaces the warmer water in the Tropics, which latter necessarily moves in the direction of the Poles to take the place of the water which has sunk there.

That there is really such a general, though slow, circulation of ocean water going on in the manner above indicated, is sufficiently proved by observations of the temperature of the water at different depths, a few of which are given in the following table:—

TEMPERATURES AT VARIOUS DEPTHS IN THE NORTH ATLANTIC OCEAN, Lat. $36°$ N., Long. $48°$ W.[1]

At the surface	$24°$ } Gulf Stream.	At 1000 fathoms	. .	$4°$
,, 100 fathoms[2]	$18°$	,, 1800 ,,	. .	$3°$
,, 250 ,,	$17°$	,, 2700 ,,	. .	$1·5°$
,, 500 ,,	$11°$			

[1] From soundings made on board H.M.S. *Challenger* in 1873.
[2] A fathom = $1·83$ metres.

Convection-Currents in the Sea.

These results show clearly that at depths below 900 metres (about 500 fathoms) there is, in spite of what we know respecting the high temperature of the interior of the earth (p. 14), a stratum of water much colder than the average temperature of the surface in the latitude referred to. This cold water could only have come there from the Arctic regions; and although withdrawal of the Tropical water by evaporation, and also the mechanical impetus given by winds sweeping over the ocean surface contribute to cause an influx of colder water towards the Equator and a corresponding flow of surface-water towards the Poles, yet inequalities of temperature are competent to account for a great deal of the observed movement.[1]

Besides the above general movements taking place throughout ocean-water, there are several more or less extensive surface currents which are, in part at any rate, caused by inequality of temperature. One of the best known of these is called the 'Gulf Stream,' which is of especial interest from its effect in modifying the climate of the British Isles. Its direction is so well defined, and its boundaries so sharply marked, that it may be said, in the earlier part of its course, to resemble a river of warm water floating upon the colder water of the Atlantic. Although the currents which maintain its volume are traceable as far south as Cape San Roque in South America, its course as the

---

[1] The effect of surface inequalities of temperature in causing currents throughout the mass is well shown by an experiment due to the late Dr. W. B. Carpenter, F.R.S. A rectangular trough, with glass sides, about 1 metre, or more, in length, 40 cm. deep, and 12 or 14 cm. broad, is filled with water. At one end a basket of wire-gauze is placed, just dipping below the surface and containing ice. At the other end a thick strip of copper about 8 cm. broad is supported, just touching the surface of the water. Its outer end is bent up and heated by a gas-flame. Thus we have the surface-water cooled at one end of the trough and heated at the other end; and if a little solution of indigo or ink is dropped on the ice, and a little warm solution of magenta or red ink is carefully added to the water close to the heated copper, the two currents—a colder one sinking and passing along the bottom, and a warmer one passing along the surface from the copper to the ice—are distinctly shown.

Gulf Stream begins in the Gulf of Mexico, which is, as it were, the boiler of the heating apparatus; the rays of the tropical sun heating the water to as high a point as 28° or 30°, and thus enabling it, as it passes out through the Florida Channel, to float upon the colder and denser water of the open sea. The current then proceeds for some distance along the North American coast with a velocity of about 6 kilometres per hour (the average pace of a good walker), a breadth of about 60 kilometres, and a depth of less than 200 metres. At the latitude of New York, it sets out on its course across the Atlantic, still for a great distance preserving its distinctive characters of high temperature (24°), well marked boundaries, and deep blue colour, strikingly contrasted with the greenish tint of the adjacent water. A vessel may be placed so exactly on the edge of the Gulf Stream that thermometers let down into the water from the bows and stern respectively will show a difference of 6° or 8° at least. Owing to its lower density, the middle of it is appreciably higher than the level of the rest of the ocean, as proved by the fact that weeds and driftwood collect at its edges, having slipped down, as it were, from the ridge of its centre-line. As it gets farther eastward, it spreads out and becomes less sharply defined, but the main portion of it appears to keep together in a north-easterly course, washing the shores of Ireland, Scotland, and Norway. At any rate, either the current itself, or the winds which have swept over it and acquired warmth and moisture from it, exercise a remarkable influence on the climate of those coasts, raising the average temperature very considerably above that which corresponds to the latitude. Thus the Shetland Isles, though 9° of latitude farther north, are in winter as warm as Ilfracombe in Somersetshire; and the harbour of Hammerfest, in Norway (lat. 70° 40′ N.), is free from ice throughout the year; while the mouth of the Baltic, 12° farther south, is completely frozen up during four of the winter months.

Results and Applications of the Convection of Heat in Gases.

I. The ascent of heated air and smoke in chimneys is due to the disturbance of the equilibrium between the contents of the chimney and the external air, caused by the heat communicated to them by the fire in the grate at the bottom. The mode in which a convection-current is thus set up in the flue will be so plain from what has been said already respecting the currents in liquids, that it only remains to show how the force of the upward draught may be calculated: The problem may be most simply treated as analogous to that relating to the ascensive power of an air-balloon (p. 88): *i.e.* we may consider the force of the draught as depending upon the difference between the weight of the column of hot air in the chimney and that of a similar column of the cold, denser air outside. These two columns of equal height may, on hydrostatic principles, be regarded as having the same relation to one another as weights placed in the two scales of a balance.

Suppose that a flue is 1 decimetre square, and 10 metres, or 100 decimetres, in height: it will hold 100 cubic decimetres, *i.e.* 100 litres of air. If the air in the chimney is under the same conditions of temperature and pressure as the outside air, a similar column of each will have the same weight, and each of them will just balance the other, so that there will be no movement of air upwards or downwards in the flue. But suppose that the air in the flue is raised to an average temperature of 100°, while the air outside is at 0° (the barometer standing at 760 mm.). Since a litre of air at 0°, and under a pressure of 760 mm., weighs 1·3 grm. (nearly), the 100 litres of cold air will weigh (100 × 1·3=) 130 grms., while the 100 litres of hot air in the flue will only weigh (100 × 0·95[1]=) 95 grms. Thus the cold

---

[1] For the method of calculating the weight of a litre of a gas at any given temperature, see p. 118.

air will overbalance the hot air by ($130 - 95 =$) 35 grms., and will press the latter upwards with a force represented by 35 grms. weight.

This force of draught not only insures a full and free supply of fresh air to keep up the combustion of the fuel in the fireplace or furnace (and consequently the intensity of the heat), but also can be employed to work 'smoke-jacks'[1] in the chimney, and revolving cowls on the top of it.

We may here shortly consider the reasons why, under some conditions, smoke unfortunately does not go up a chimney as it should do, but rushes out into the room. The chimney is then said to 'smoke,' and the occurrence obviously implies a stoppage, or even a reversal, of the direction of the draught. Several causes may contribute to this.

(*a*) The flue may be too short to enable an effective draught to be set up, or too small to carry off all the smoke, or crooked, rough, or clogged with soot to such an extent as to baffle the ascending current.

(*b*) The opening of the fireplace may be so large that a great deal of cold air enters above the fire without having a chance of being warmed by the latter; and this, mixing with the ascending smoke, lowers its temperature, and checks the force of the current.

The remedy for this is obvious. The opening into the chimney should be contracted to such an extent that little or no air can enter it except what has passed close to the burning fuel, and been thus heated strongly. This is always done in the so-called 'register-grates,' which are seldom liable to cause annoyance by smoking: and many cases of smoky chimneys can be cured by fitting an iron plate vertically in the chimney-opening, reaching down like a curtain nearly to the front grate-bars. This is often done temporarily to increase the draught and quicken the combustion of the fire, when burning low.

(*c*) There may be an insufficient supply of air to the room.

---

[1] These are sets of vanes placed on a vertical axis in the chimney, which are turned round by the upward current of smoke (like the vanes in Expt. 91), and are used to turn spits for roasting meat in front of the fire.

Draught in Chimneys. 277

In order that air may go out through the chimney, it must first be allowed to come into the room; and if the windows and doors fit tightly, and no inlet-pipe or ventilator exists, no amount of fire in the grate can rouse a strong draught in the chimney.

(*d*) A feeble draught is easily overpowered by strong, gusty winds or eddies, which prevent the rising smoke from issuing out of the top of the chimney; and many forms of chimney-pots or cowls have been devised with the object either of giving an upward slant to the wind when it strikes the mouth of the chimney, or of keeping the opening always directed away from the wind, as in cowls with side openings, balanced on a vertical spindle, and moved by a vane fixed on the top.

In all furnaces, other than blast furnaces, intensity of heat is gained by keeping the fire-doors closed, so that all the air which forces its way in has to pass actually through the fuel, aiding the combustion by its oxygen, and increasing the upward current in the chimney by the high temperature to which it is raised. 'Wind-furnaces,' as they are called, are used even for melting steel, for which a temperature of at least 1800° is required.

II. The action of the glass chimneys of Argand gas-lamps and paraffin-lamps depends on the strong draught maintained in them by the combustion of the gas or oil, owing to which far more air, and therefore oxygen, comes in contact with the flame than would otherwise do so. Thus (1) more of the combustible material can be burnt; (2) the combustion is more complete, so that little or no soot escapes from the chimney; (3) the carbon separated in the flame (as explained on p. 32) is more strongly heated so as to give a brilliant white light.

Expt. 94. Take off the chimney of an Argand or paraffin-lamp and kindle the gas or oil. Observe the dull yellow, smoky flame which is produced, and then replace the chimney. This will immediately cause an increase in the brightness, clearness, and steadiness of the flame.

III. The ventilation of buildings.

Ventilation (Lat. *ventus*, wind, *ventilare*, to move through the air) is the term applied to the process of continually renewing the air in confined spaces: and it is easy to see how this is a necessary consequence of a draught maintained in a chimney of which the lower end opens into a room or passage, since the real impelling cause of the draught is the rush of cold, fresh air towards the chimney.

Owing to the withdrawal of oxygen from the air by the breathing of living beings as well as the burning of lamps and candles, a constant change of the air in a room is absolutely necessary for health and comfort. A room crowded with people and lighted by gas or candles soon becomes 'close' and oppressive unless not only fresh air is freely admitted, but also the spent air is freely withdrawn, so as to maintain the due proportion of oxygen in the space, as well as keep down the temperature several degrees at least below that of the body: and it is found that about 500 litres of fresh air per minute should be supplied for each person or lamp existing in a room.

Many simple experiments may be made to show that ventilation is necessary for health and life: the burning of a taper being taken as the index of the condition of the air, for it may be safely assumed that where a candle will not burn, no animal can live.

Expt. 95. Take a large paraffin-lamp glass (preferably one of those not widely expanded in the middle): fit a cork to its upper end, and pass through the cork a wire bent like that shown in fig. 86, p. 280, having a short piece of wax-taper stuck upon its upturned end. The length of the wire should be adjusted so that when the cork is in its place the taper may be entirely within the lamp glass. Support the glass in a clip attached to a retort-stand, light the taper and lower it into the glass, pressing the cork closely into its place. The flame will shortly go out, although air can enter very freely at the open mouth of the lamp-glass.

The reason is easily seen. The residue of the air, *viz.* nitrogen, and the products of combustion, *viz.* carbon dioxide and steam, collect (owing to their high temperature) in the upper part of the glass, but cannot escape since the cork stops the opening. Hence air cannot enter at the bottom (except slowly, by diffusion), although there is plenty of room for it to do so.

Expt. 96. Repeat the last experiment, but before the flame has quite expired raise the cork a little so as to allow the gases to escape. The flame will at once become brighter again, and continue to burn well unless the cork is again pressed into its place.

Expt. 97. Arrange an apparatus as shown in fig. 85: a tall cylindrical jar or wide-mouthed bottle, about 25 or 30 cm. in height, fitted with a cork through which pass two tubes about 2 cm. in internal diameter and 20 cm. in length; the upper and lower ends of these should be at the same level. Between these a wire is passed through the cork with a piece of wax taper attached to its lower end, which should be so bent that, by turning it round in the cork, the taper can either be brought directly under either of the tubes or be placed so as to stand entirely clear of both of them.

Having put the taper in this latter position (*viz.* clear of the tubes), light it and lower it into the jar, pressing the cork tightly into its place. The flame will soon become faint and go out, though it is plain that air could enter freely by either of the tubes and escape by the other.

Expt. 98. Relight the taper and replace it in the jar (from which the products of combustion must be previously expelled); then, before the flame is quite extinguished, turn the wire so as to bring the taper under one of the tubes. An upward draught will at once be set up in this tube, while cold, fresh air will descend in the other tube, and the taper will burn brightly until consumed. The direction of the currents may be shown by holding a piece of smoking brown paper close to the mouth of each tube in succession.

Fig. 85.

This experiment proves that the mere existence of a chimney in a room is of no use whatever for ventilation. Some motive must be given to the air to pass up it or down it, otherwise its contents will remain stagnant, as also the air in the room. It also shows how there may arise a down-draught in a chimney opening into a room which communicates with another room or

VENTILATION OF ROOMS.

gallery in which there is a fire. Even if there are fires in both rooms, one chimney may smoke if the fire under it is burning less vigorously than the other.

Expt. 99. A model illustrating more exactly the conditions of an ordinary room is shown in fig. 86. It consists of a flask, A, about 1 litre in capacity, having two side-openings (the wider the better). One of these necks is left open, and to the other is fitted a glass cylinder or lamp-chimney, B, having a side-opening blown in it. In the lower end of the latter is placed a piece of wax taper attached to a cork (this cork should have several deep notches cut in it to allow a little air to enter, otherwise the supply is hardly enough to keep the taper burning). In the large flask is placed a lighted taper on a bent wire attached to the cork. This taper will not burn for more than a few seconds unless the taper in the side glass (which represents an ordinary fireplace and chimney) is lighted, so as to create a draught of air through the flask.

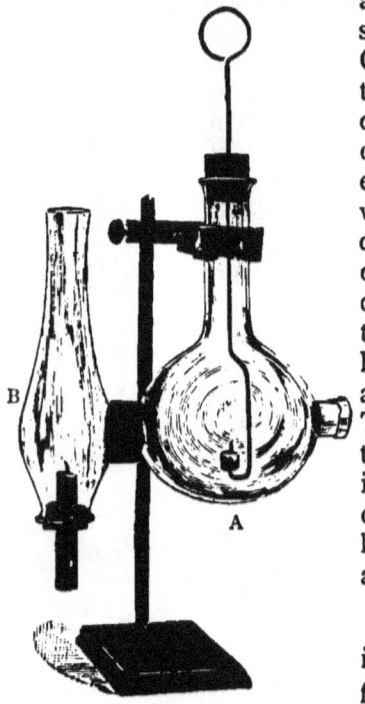

Fig. 86.

From all the above experiments it is clear that the essential requisites for the efficient ventilation of a room are—

1. There must be openings, not only for the egress but also for the ingress of air: and these should be situated at different levels.

2. The inlet should be as low as possible: but, to prevent a chilly draught of air sweeping along the floor, the mouths of the openings may be turned up, as is done in Tobin's ventilators. This will distribute the fresh air more gently and uniformly through the room.

3. The outlet should be as high as possible. The vitiated air,

VENTILATION OF ROOMS. 281

from its high temperature, collects (as illustrated in Expt. 95) near the top of the room: as may be proved by holding a thermometer, or simply the hand, near the ceiling of a badly ventilated room. The heat and oppressiveness of the air in that part, as also in the gallery of a crowded theatre, will be very perceptible. Even if there is no fire, or stove, or lighted gas to warm the air, yet, from the comparatively high temperature of the human body, the mere presence of several persons in the room will give rise to convection-currents of warm air. Every person is, in fact, a warming apparatus in himself.

Properly arranged ventilators, communicating directly with the outer air, are of course the best means of fulfilling the above requirements. But sash windows, made as they should be, to open both at top and bottom, or tall casement windows, such as are common in warm countries, answer very fairly. These latter, as also open doors, afford an entrance for fresh air below, while the vitiated air escapes through the upper part.

Expt. 100. Open the door of a room slightly, so as to leave a clear aperture 10 or 12 cm. wide, and hold a lighted candle just within the opening and pretty close to the floor. The flame will (under ordinary conditions) be blown inwards to the room, showing the existence of a current of fresh air in that direction. Next, raise the candle slowly, still keeping it within the doorway: about half way up the flame will burn undeviated, while near the top of the door it will (unless the room is lofty, and the other arrangements for ventilation very perfect) be blown outwards by the current of warm air making its escape from the room.

In mines a much more perfect and elaborate system of ventilation is necessary, since the lives of the workmen depend entirely upon the supply of air which reaches them through the deep shafts which afford the only communication with the surface. Moreover in coal-mines, not only have the products of respiration and the spent air to be removed, but also a large quantity (in many cases) of the inflammable gas known as 'fire-damp' (see p. 256), which issues from cracks in the seams of coal, and which, besides being irrespirable, would cause accidents if it was allowed to accumulate and mix with the air in the proximity of open lamps. Hence a very large volume of air must be continually forced through all the passages of a mine.

The arrangements for effecting this are, in principle, illustrated by Expt. 97 (p. 279), already made. Every mine has at least two shafts,[1] corresponding to the two tubes used in the above experiment. At, or near, the bottom of one of them, called the 'upcast' shaft, a fire is kept constantly burning; and, owing to the upward draught thus caused, a body of cold fresh air descends the other shaft, which is called the 'downcast' shaft,—forces its way through the galleries of the mine, carrying with it all vitiated air and fire-damp, and escapes through the up-cast shaft, as through a chimney.

A plan of the workings of a mine is given in fig. 87; from

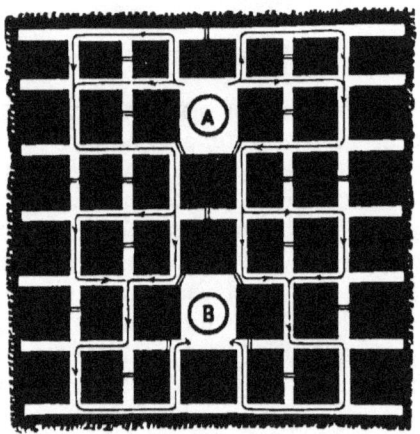

(A) Downcast Shaft.   Fig. 87.   (B) Upcast Shaft.

which it will be seen that the current of air is made to circulate, as shown by the arrows, through all the labyrinth of passages before it reaches the upcast shaft; wooden partitions and tightly

---

[1] Sometimes only a single shaft is used, divided longitudinally into two vertical passages by a wooden partition: but this is dangerous, since if it becomes blocked by any accident, there are no means of getting the men out.

fitting doors being placed in the galleries to direct the course of the air and prevent its taking the shortest road.

In some mines so much fire-damp gets mingled with the air that, if the mixture passed within reach of the furnace flame, an explosion might occur. In such cases the draught is not maintained by a furnace, but by a revolving fan fixed at the top of the upcast shaft and worked by steam or water power, which exhausts the air from the shaft.

Winds.

As in the case of liquids, so in that of gases we have excellent examples of natural convection-currents on an enormous scale in the winds,—Nature's *venti*-lating apparatus, in fact, in the strictest sense of the word. The whole atmosphere is in a state of restless movement, caused almost entirely by the inequalities of the temperature of different parts of the earth's surface. The rays of the sun, passing through the air itself without materially affecting it, when they fall upon the earth raise the temperature of different regions to very different degrees, according to the direction in which they strike the surface and the material of which it is composed. The portions of air in contact with the surface take up its temperature, whatever that may be; and thus we have all the conditions necessary for setting up convection-currents of all degrees of intensity, from a gentle breeze to a tremendous hurricane.

We may consider first some examples of local winds thus produced, such as those called 'land-breezes' and 'sea-breezes.' It is an observed fact that at places along the sea-coast, especially in warm latitudes, there is (except when storms interfere with it) a gentle wind blowing steadily from the sea to the land, commencing in the morning, increasing in force up to 2 o'clock p.m., and dying away at sunset. This 'sea-breeze' is produced in the following way. During the day-time the rays of the sun heat the surface of the land more than that of the sea, partly because rocks and soil have a lower specific heat than water; partly because the heat received by the surface remains there, since it cannot be conveyed away by conduction or convection; partly because land absorbs radiated heat rather more readily than water (a subject which will be more fully treated of in the next Section). The air lying over the land becomes consequently heated, and is forced up by the colder air which flows in from the regions over the sea. It is this latter stream of air which forms the sea-breeze.

During the night the conditions of temperature of land and sea are reversed. No heat is supplied by the sun to either, and the land is lowered in temperature more quickly than the water, for reasons easily deduced from what has been said above. Thus the sea is now relatively the hotter region, and therefore an air-current is set up from the land to the sea, which is well known to coasting vessels as the 'land-breeze.'

The Simooms, or hot, suffocating dust-storms of Africa and Western Asia, originate in a similar way. The sandy soil of the Sahara desert becomes intensely heated by the sun, and the hot air over a large region of it is forced upwards by the pressure of the surrounding colder air. Thus strong currents set in from different sides towards the region of greatest heat; and these, impinging on one another, produce eddies and whirlwinds, precisely like those noticed at the meeting-point of two rapid rivers.

A similar, but less destructive, wind is the Mistral of the south of France, which is the cold sheet of air lying on the slopes of the Pyrenees slipping down, as it were, to the heated plains of Provence. The explanation of all such winds is very simple if the one fact is borne in mind,—that air always tends to move along the surface of the earth *from* a colder *towards* a hotter region.

Such local winds, however, sink into insignificance before the great system of atmospheric convection-currents called the 'Trade-winds.' These correspond in a general way to the oceanic currents already explained (p. 272); but, living as we do at the bottom of the sea of air, we have a more exact knowledge regarding the direction and force of the air-currents than a few soundings enable us to obtain with respect to the movements of the water.

It is a fact, first noticed by Columbus in A.D. 1492, that in tropical latitudes there is a steady wind perpetually blowing over the ocean from an easterly quarter, never sinking to a breeze and never increasing to a hurricane. North of the Equator this wind blows from points N. of east, chiefly from N.E.; while south of the Equator, the direction of it, is more or less nearly S.E. At the Equator, and within 2° or 3° on each side of it, there is no decided wind at all; it is a region of almost complete, perpetual calm, shifting, however, slightly in position as the apparent track of the sun shifts to the N. or S. of the Equator in the course of the year. The northern and southern limits of the zone within which the Trade-winds blow shifts to a corresponding extent; it may, however, be said to comprise a region of (roughly) 30° of latitude on each side of the Equator. The fact that this zone alters its position on the earth's surface with the seasons, being always found on that part where the sun is most nearly overhead, clearly indicates that the sun is the cause of the trade-winds; but there are one or two points about them which must have a fuller explanation. Why, for instance, should there be a dead calm in the region directly under the sun? Why are the Trade-winds experienced only over the open sea, and not, or at any rate only very imperfectly, over the land? Why, lastly, do they not blow *directly* from the colder to the hotter regions, *i.e.* from each Pole to the Equator?

The heat acquired by the part of the earth's surface which lies exposed to the vertical rays of the sun is imparted to the air above it, and, as usual, causes a disturbance of equilibrium in the latter; air from colder N. and S. latitudes gravitating towards the heated portions, which latter are forced almost vertically upwards between the opposing currents.

Thus, considering the motion of the air alone, it is exactly what we should expect, to find a wind from the northward in regions N. of the Equator and a wind from the southward in regions

S. of it, separated by a space where the only movement of the air is directly away from the earth's surface and not along it, and where, therefore, no wind in the ordinary sense of the word is felt. Thus the existence of the equatorial belt of calms is easily and fully accounted for; and the only reason why there is no great regularity and permanency in the winds experienced over large continents is that high mountain-ranges and other irregularities of the surface of the land, and also local air-currents such as those already alluded to, alter and baffle the main circulation of the atmosphere to an extent which cannot occur over tracts of open sea.

It remains to explain why the Trade-winds always blow from directions considerably *East* of North in the northern hemisphere, and *East* of South in the southern one. The cause will be understood if it is borne in mind that the direction of the wind as observed by those on the earth's surface is not necessarily that of the mass of air itself; for instance, a traveller in a train or a rower in a boat will, even on a perfectly calm day, feel all the effects of a current of air proceeding in a direction opposite to that in which he is going, simply in consequence of his own movement against the really stagnant air. So with respect to the Trade-winds. We live on a globe which is rotating as if supported on a spindle or axis, the extremities of which would be at the N. and S. Poles. The consequence is that, while at the actual Poles no progressive movement would be felt, at the Equator things are carried through a space equal to the circumference of the earth (about 39,000 kilometres) in the time of one rotation, *i.e.* in 24 hours, and must, therefore, be moving at the rate of 1600 kilometres (about 1000 miles) per hour. In intermediate latitudes this movement is obviously slower in proportion as the circle of latitude is smaller: thus in lat. 45° the velocity of a point on the earth's surface is 1131 kilometres (about 700 miles) per hour, and at London it is about 1000 kilometres (622 miles) per hour. The reason why we are not conscious of this enormous speed is that everything on or near the earth's surface is, owing to friction and gravitation, compelled to move at the same rate as the surface.

Now, when a mass of air at some distance N. or S. of the Equator is loosened, as it were, from the earth's surface by the causes above mentioned, and transferred to a region nearer the Equator which is moving more quickly than itself, this air, from its inertia, does not at once take up the quicker motion of the surface beneath it; hence objects on the latter, as they are carried *towards* the east, run up against the air-current, which therefore appears to them to be coming *from* the east. The whole case may be put shortly thus. If the earth was motionless and the air alone moved, the direction of the wind would be from the northward (in the northern hemisphere). If the air was stationary and the earth alone moved, the direction of the wind would be from the eastward. But both are in motion; therefore the wind (according to the well-known dynamical Law of the Composition of Forces) appears to blow from a point between N. and E.; more nearly N. where the air is in relatively rapid motion, and more nearly E. in parts near the Equator where the motion of the earth's surface is rapid, and the air is checked as it approaches the region of calms and comes up to the corresponding current from the S.

Expt. 101. The effect of combining the two motions may be easily shown if a blackened globe is at hand on which chalk-marks can be made, and which is mounted so as to rotate within a brass ring, as they usually are. A piece of chalk should be held lightly against the side of this ring as a guide, and a line drawn from about lat. 40° to the Equator, while the globe is turned steadily round. Several curves may be thus drawn to illustrate the effect of varying the relative velocity of the chalk and the surface of the globe underneath it.

It will be easily understood that the S.E. trade-wind in latitudes south of the Equator is produced in a precisely similar way.

We have next to inquire what becomes of the mass of heated air which rises upward from the earth's surface in the region of calms near the Equator. It is clear that the air must become cooler as it rises, not only because it gets farther from the earth's heated surface, but also because the pressure on it is lessened (p. 196), and the mass expands—*i.e.* the molecules move outwards; and heat, as already explained (p. 164), disappears in supplying the energy necessary for this movement. Hence a large quantity of water-vapour is condensed, forming the heavy tropical rains; and the cool air must evidently flow in the only direction open to it, *viz.* towards the north and south poles, to take the place of the air which has travelled towards the Equator. The natural inference will be drawn that there exists above each of the trade-winds a current in the opposite direction—*i.e.* from the S.W.[1] in the northern hemisphere, and from the N.W. in the southern hemisphere, gradually approaching the earth's surface, and ultimately reaching it in latitudes higher than 30°. There are several conclusive proofs of the existence of these return-currents, or 'counter trade-winds,' as they are called, without the necessity of taking upward soundings, so to speak, with a balloon.

1. In the year 1812 a shower of volcanic ashes fell on the island of Barbadoes, which were clearly shown to have come from a volcano in Morne Garou, an island lying at least 160 kilometres to the westward. Now, as Barbadoes lies within the region of the N.E. trade-wind, the ashes could not possibly have been brought by the surface-current; they must have been shot upwards through the trade-wind into a westerly current, which was thus proved to exist above it.

2. In 1835 there was a violent eruption of Conseguina, a volcano on the shore of the Pacific Ocean, not far north of the Equator. Showers of ashes were borne by air-currents, not only to Jamaica,

---

[1] The reason of the westerly direction of these return currents is analogous to that of the apparent easterly direction of the trade-winds themselves. The air at the Equator shares the rapid motion (from W. to E.) of the earth's surface from which it has risen; and when it comes to latitudes where the surface is moving more slowly, it outstrips the motion of the surface, flowing from W. to E. more quickly than the latter. This movement, combined with the movement towards the Poles, gives to the return current a direction more or less from the south-west in the northern hemisphere, and more or less from the north-west in the southern one.

which lies about 1300 kilometres to the N.E., but also upon the ship *Conway*, which was at the time in the Pacific about 1100 kilometres to the S.E.-ward. The only explanation is that the volcanic ashes were projected upwards into a region where the vertical air-current divides, as above mentioned, into two currents, one proceeding from the S.W. over the N.E. trade, and the other passing from the N.W. over the S.E. trade.

3. The Peak of Tenerife rises more than 3700 metres into the air in lat. 28° N., which is near the northern limit of the trade-winds. At the base of the mountain the N.E. trade blows steadily during the summer months, but a traveller ascending the Peak comes, at an altitude of about 2700 metres, to a region of nearly complete calm, and then at greater heights he experiences a S.W. wind, which increases in strength up to the summit of the Peak. This latter wind is obviously the counter-trade, which has fallen thus far towards the surface of the earth, and in the winter months actually reaches the base of the mountain; the trade-wind ceasing to be felt there, since at that season it commences farther southward.

The practical importance of these immense atmospheric convection-currents is very great, though lessened since the introduction of steamers. The sailing course from the Azores to the West Indies is all 'down-hill,' as it were, the N.E. trade speeding ships reliably and swiftly to their destination. Ships bound for Australia find it best to keep in the N.E. trade right across the Atlantic until near the coast of South America; then to get across the belt of calms as best they can, and take a course nearly due south until they reach lat. 30° S.—*i.e.* the limit of the S.E. trade, beyond which they find winds (though not so constant) from the N.W. which take them to Melbourne. The course of ships in the Pacific is shaped with reference to the trade-winds on similar principles, and the longest way is often found the shortest when sailors can avail themselves of such steady, reliable winds.

The counter-trades, which descend to the sea-level in latitudes a little higher than 30°, constitute more or less directly the S.W.

T

winds so prevalent in the British Islands. It is true that most of the storms which arrive from the Atlantic are of the nature of cyclones, the wind circulating round a centre in the opposite direction to that in which the hands of a watch move; but one great cause of this circular movement is that the return trade, coming as it does from quickly-moving parts of the earth, impinges obliquely upon more slowly-moving portions of air, and produces in the whole mass not only the spiral movement of a whirlwind, but also a progressive motion of the centre towards the eastward.

Section III.—Radiation of Heat.

It will have been observed in an experiment which has been made already (p. 240), how remarkably and completely the conveyance of heat by radiation differs from conduction and convection. In fact, all investigations of the subject lead to the conclusion that heat does not, in the process of radiation, pass *as heat* from one point of space to another, but as some form of energy of motion which is translated, as it were, into heat, or light, or electricity, or chemical affinity, according to the nature of the substance which receives it.

The molecules of a hot substance appear to have the power of exciting in the molecules of surrounding media, not only the movement of revolution which probably causes the phenomena of heat, but also an extremely rapid motion to and fro (like that of a pendulum), which is propagated through the mass far more quickly than the former; just as wave-motion in the sea far outstrips any motion due to tides and currents, while it can co-exist with them almost without affecting their course or velocity. The character of this vibratory movement may be roughly illustrated as follows:—

Expt. 102. Lay a piece of stout, flexible cord (or, better, of indiarubber tubing filled with sand), at least 3 or 4 metres long, upon the floor or a long table: take hold of one end and move it quickly up and down once. Each portion of the cord will successively take up the ascending and descending movement of the portion held in the hand; and the form of a wave will travel along the cord to the other end. By properly timing the movement of the hand, a succession of such waves may be kept up.

A rather better illustration of wave-motion is obtained by hanging a row of large bullets (or of the hollow brass balls used as feet for boxes, etc., and procurable from most ironmongers) about 3 or 4 cm. apart, from a horizontal rod by strings 60 or 70 cm. long. The balls should be lightly connected together by a horizontal strip of the thinnest 'elastic' (indiarubber thread), tied to each one in the row just where the suspending string is attached to it. If the end ball is pulled aside and then let go, it will begin swinging, and impart its motion to the next one, and this to the next, and so on: a series of regular waves passing from end to end of the row and back again.

These experiments illustrate roughly the kind of molecular movement which goes on in what is understood by the term 'radiation'; and the energy which is so conveyed, and which is convertible into heat, light, etc., according to the character of the substance which accepts it, will be in future referred to, for convenience and brevity, as 'radiant energy.'

But it must be noted that the curving cord and swinging balls do not give a complete idea of all the stresses and tendencies towards movement which are going on among the molecules. They only show the wave-form as transmitted along a single row of molecules, whereas there is no doubt that each molecule, as soon as it has been set in motion by energy derived from any source, becomes (like a stone dropped into a pool of water) a centre of disturbance from which vibrations tend to propagate themselves in *every* direction, up and down, backwards and sideways, as well as forwards, or directly away from the source. It can, however, be easily proved from the mathematical theory of wave motion in an elastic body, that all the lateral, etc., waves interfere with and destroy each other, so that the only practically effective transmission of energy takes place in a direction straight away from the source, where molecules lie which are hitherto undisturbed.[1]

Thus we can gather, from observing the motion of the cord, or balls, what is the real and effective direction in which the radiant energy travels; and it is easy to see that with reference to this direction the motion of the molecules is transverse, *i.e.* from side to side of the line which we call the direction of the ray. They probably swing in ellipses which may approach indefinitely near to a straight line, or, on the other hand (as some of the phenomena of light prove), may become actual circles.

Before examining the particular phenomena which accompany the conversion of radiant energy into heat, it will be advantageous to consider briefly the general characteristics of radiation.

1. It goes on in spaces which contain comparatively little ordinary matter, and it does so, in many cases, without perceptible alteration of character.

This fact is illustrated by the 'solar radiation thermometer' used in meteorological work, the bulb and stem of which are enclosed in a glass vessel, from which the air is exhausted as far as a good pump can do it. Only about $\frac{1}{1000}$ of the usual quantity of air remains in the vessel, and passage of heat by conduction or convection is impossible; yet the thermometer is affected by the heat of the sun's rays as readily and fully as if the vessel was full of air. The bulbs of the incandescent electric lamps

---

[1] The hypothesis of an 'ether,' or subtle, perfectly elastic, imponderable (or, at least, indefinitely light) fluid for conveying radiant energy, is not insisted on here. Such a fluid appears to be the last survival of several similar conceptions, *e.g.* the heat-fluid, or Caloric, and the electric fluid or fluids, which, while they have served a good purpose not only in giving clearness and definiteness to popular descriptions of phenomena, but also in facilitating refined mathematical calculations, are gradually being dropped as the fact becomes recognised that the molecules of ordinary matter are quite capable of taking up all the varieties of motion which the different manifestations of energy require, and of carrying on these motions independently of one another. Electricity is found to travel, under proper conditions, with a velocity equal to that of radiant heat or light, yet no one now thinks it neces-

now so extensively used are exhausted until they contain only about $\frac{1}{1,000,000}$ of the usual quantity of air, but this diminution in the number of molecules has no appreciable effect on the quantity of heat and light from the glowing filament which reaches the space outside the lamp, or on the ease of its transmission. The nearest approach to a real *vacuum*, or space void of matter, which has yet been made was in one of Mr. Crookes's investigations,[1] when the air was withdrawn from a glass globe until it only contained $\frac{1}{100,000,000}$ of the usual quantity; yet no diminution of the transparency of the interior (which would imply diminished power of transmitting radiant heat and light) was observed. Since, however, even at this high degree of exhaustion there would be at least a billion of molecules remaining in each cubic centimetre of the globe, it is not difficult to conceive that this number is far more than sufficient to transmit radiant energy.

In the regions of space between the planets, the sun, and the stars there undoubtedly exists some substance possessing mass and inertia (as proved by the observed retardation in the motion of at least one comet); and although there is certainly not enough of this to bring to us the sun's heat by conduction or convection, it conveys radiant energy without the slightest difficulty.

---

sary to assume the existence of a special fluid by the motion of which it is thus conveyed.

It is difficult, moreover, to understand how a fluid which has so little in common with ordinary matter (not even being affected perceptibly by gravitation) can influence and be influenced by ordinary matter so decidedly as the phenomena of radiation require it to be. Not only must the massive molecules of a heated piece of platinum be supposed to communicate motion to the ether which surrounds them, but also this imponderable, moving fluid must be considered capable of quickly overcoming the inertia of such molecules and putting them in rapid motion, as when a piece of platinum is melted by the concentrated rays of the sun.

For a full discussion of the subject of an ether, Grove's *Correlation of Physical Forces*, p. 108 *et seqq.* (ed. 1874), should be consulted.

[1] *Proc. Royal Soc.*, vol. xxxi. p. 450.

2. Radiation proceeds with enormous velocity.

The rate at which radiant energy passes through air is 300,000 kilometres per second. Starting from the sun, it only requires eight minutes to reach the earth, and it would travel about eight times round the earth in one second. This velocity is about a million times greater than that of sound, which latter itself travels about twelve times the rate of an express train.

In substances denser than air the velocity of radiant energy is somewhat less. Thus in water it travels 225,000 kilometres in a second; in glass, about 200,000 kilometres in a second. For the methods by which these high velocities have been determined a treatise on Light must be referred to. In the meantime it may be remarked that such enormous velocities would be of themselves sufficient to lead to the rejection of the theory that heat is a material fluid.

3. Radiation proceeds in straight lines only.

A metal bar gets heated throughout by conduction equally well whether it is twisted or straight; and convection currents can, as proved already, be easily made to travel in curves; but radiant energy, from the reason already indicated (p. 291), can only be transmitted effectively in directions straight outwards from the source, all waves, except those which spread in such directions, being destroyed by mutual interference.

One simple fact is sufficient to prove this, *viz.* the fact that the rays of the sun can be cut off sharply and entirely by interposing a screen made of any substance which is incapable of transmitting them. They do not curve round the edges of this obstacle,[1] and the heat-shadow is as definite as the light-shadow. Hence a rock or a tree affords a cool shelter from the heat of a tropical sun, and the direct radiation from a fire is cut off by a screen which may even, as will be explained hereafter, be made of such a substance as glass.

---

[1] The phenomena of diffraction, which have hitherto been investigated rather with reference to light than to heat, come more appropriately within the scope of a treatise on optics.

Characteristics of Radiant Energy. 295

4. The amount of radiant energy received by a surface of given size diminishes very rapidly as the distance from the source increases.

This is simply a consequence of its travelling in straight lines; and while the fact is one of everyday experience (if we find a fire too hot, we get farther from it), the exact law of the variation with distance may be demonstrated in the following way :—

Expt. 103. Place a screen, A, with a small hole in it, in front of a burning candle, as shown in fig. 88. Support a plate of

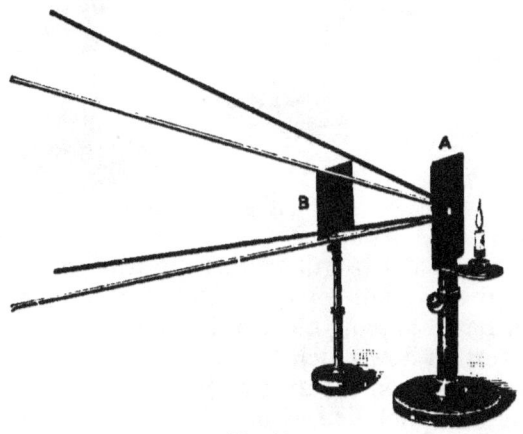

Fig. 88.

blackened metal or cardboard, B, 10 cm. square, opposite the hole at a distance of 2 decimetres from the screen on the other side. This plate will receive a definite amount of the radiant energy which proceeds from the flame through the hole in the screen, and we have to determine how this amount varies with the distance of the plate from the hole, from which latter we may assume the radiation to start.

Support four light wooden rods in sockets attached to the screen close to the hole, in such positions that they just touch the corners of the plate, as shown in the figure. These will serve to mark the boundaries of the definite portion of radiant energy which, diverging from the hole, falls on the plate; and which will, when the latter is removed, proceed onwards in the form of a cone having a cross section similar in shape to the plate.

Now place the receiving plate, B, 4 decimetres from the screen, in the position shown in fig. 89, one of its corners touching one

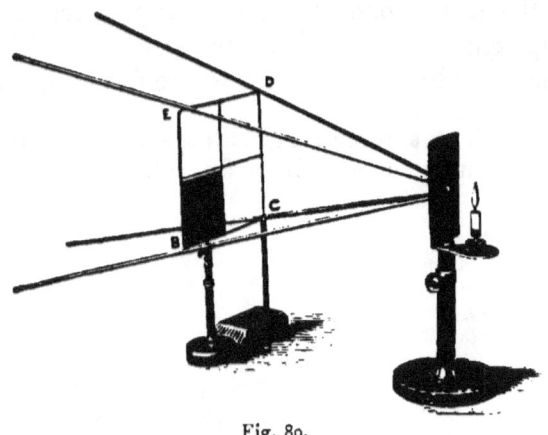

Fig. 89.

of the wooden rods. It will in this position obviously receive much less of the radiant energy than before; a large portion of the divergent beam passing clear of its edges.

We can ascertain how much less it now receives, if we can determine what proportion the area of the plate bears to the whole area of the surface included within the rods (the cross section of the cone, in fact) at the same distance from the source, *viz.* 4 decimetres. To do this, tie threads BCDE from rod to rod in the same plane as the plate; and also stretch threads along the two interior edges of the plate, tying them to the former threads. Or (as is more convenient for demonstration), place a wire frame of the proper size in the position shown in the figure, close behind the plate. We have here an area divided into four spaces, one of which is equal to the plate in every respect, while the three others will be found on measurement to be squares equal in size to the plate and to each other. Thus the whole area BCDE is proved to be four times the size of the plate B; and since the latter only occupies one-fourth of the whole, it obviously can only receive one-fourth of the radiant energy which it did in its original position, the remaining three-fourths passing on without touching it.

Characteristics of Radiant Energy. 297

Next, remove the plate to a distance of 6 decimetres from the hole in the screen, *i.e.* 3 times its original distance, placing it in the position shown in fig. 90. By tying threads on a principle

Fig. 90.

similar to that which was just now explained, or by placing a wire frame constructed as shown in the figure, it will easily be demonstrated that the plate now receives only one-ninth of the whole radiation included in the divergent cone, the remaining eight-ninths passing clear of its edges.

In a similar way it may be shown that at 8 decimetres from the hole in the screen, *i.e.* at four times the original distance, the plate only receives one-sixteenth of the original amount of radiant energy.

These results may be put in the following form, taking the original distance as 1 unit, and supposing that the plate B at this distance receives 1 unit of radiant energy from the source.

Distance from source	1	2	3	4 units.
Amount of radiant energy received	1	$\frac{1}{4}$	$\frac{1}{9}$	$\frac{1}{16}$ unit.

It is easy to see that there is a definite and fairly simple relation between the numbers in the two rows. The square of 2 is 4, and the inverse or reciprocal of 4 is $\frac{1}{4}$; so that the amount of radiant energy received at a distance of 2 units is expressed by the inverse of the square of 2. Similarly, at a distance of 3 units the amount received is $\frac{1}{9}$, which is the inverse of the square of 3 ; and so on. We thus arrive at the general law that

> THE AMOUNT OF RADIANT ENERGY RECEIVED BY A GIVEN SURFACE VARIES IN THE SAME PROPORTION AS THE INVERSE OF THE SQUARE OF THE DISTANCE FROM THE SOURCE.

Or, as it is usually expressed—

> THE AMOUNT OF RADIANT ENERGY RECEIVED BY A GIVEN SURFACE VARIES INVERSELY WITH THE SQUARE OF THE DISTANCE FROM THE SOURCE.

As an example of the application of this important law, we may take the determination of the amount of radiant energy from the sun which reaches the planet Mars, as compared with that which falls upon the Earth; a point which has a close relation to the question whether Mars can be inhabited by living creatures like ourselves.

The distance of the Earth from the Sun is 150,000,000 kilometres; while the distance of Mars from the Sun is 230,000,000 kilometres (for the sake of simplicity approximate numbers are taken). Supposing that Mars was equal in size to the Earth (it is in reality rather smaller), the radiant energy received by the two planets will, according to the above law, be proportional to the inverses or reciprocals of the squares of these distances; that is—

$$\frac{1}{150{,}000{,}000^2} : \frac{1}{230{,}000{,}000^2} :: \text{Radiation received by Earth} : \text{Radiation received by Mars};$$

$$\text{or} :: 1 : 0.425.$$

So that Mars receives less than half as much of the Sun's radiant energy as the Earth.

5. Radiant energy is reflected from a polished surface in a definite direction.

If the direction of the ray approaching the surface (the 'incident ray' as it is termed) makes a right angle with the surface, the radiant energy is reflected straight back along its former path.

If the ray approaches the surface in any other direction, such

Reflexion of Radiant Energy.

as AB, fig. 91,—then if a line CB is drawn perpendicular to the surface through the point where the ray meets it, the ray will be reflected in the direction BD which makes the same angle with the perpendicular that the incident ray does: *i.e.* the angle CBD is equal to the angle CBA. Moreover, the reflected ray lies in the same plane as that which includes the incident ray AB and the perpendicular CB.

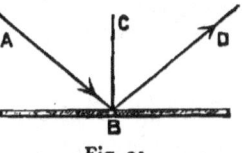

Fig. 91.

The above statement may be illustrated in the following way:—

Expt. 104. Set up vertically a double screen, about 30 × 20 cm., made of two sheets of bright tin plate separated by a small interval.[1] On one side of this, place the differential thermometer, with one of its bulbs about 10 cm. from the centre of the screen as indicated at A, fig. 92 (which is merely a rough ground-plan of the arrangement). On the other side of the screen, and at the same distance from it, place a red-hot iron ball, B, on a stand at the same level as the thermometer bulb. No heat from this will reach the thermometer; since the bright tin surface of the screen absorbs but little radiant energy (as will be hereafter shown), and the layer of air between the plates is an extremely bad conductor (p. 251); moreover, as already stated (p. 294), radiant energy will not bend round a corner.

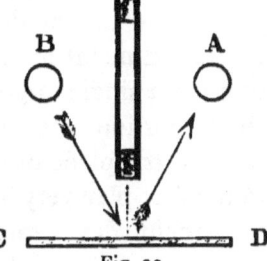

Fig. 92.

Now place a sheet of tin plate (which should be flat, and have a bright untarnished surface) vertically in the position indicated by CD, about 8 or 10 cm. from the edge of the screen and at right angles to it: so that a horizontal line drawn from B to the place where the plane of the screen would, if produced, meet CD, makes the same angle with this plane as is made by a line drawn from A to the same point. The movement of the column of liquid in the thermometer will show that heat is now reaching the

---

[1] This can easily be made by placing narrow slips of wood, about 1 cm. in thickness, between the two sheets of tin, close to opposite edges, and nailing the plates to them by short tacks. About 4 cm. of the lower edge of each plate should be bent outwards so as to form a base on which the screen will steadily stand upright.

bulb A, having been reflected from the surface CD. If the latter is moved so that the rays from B fall on it at any other angle, no heat will be found to reach the bulb unless the latter is also moved into such a position as the laws above stated require.

Several practical applications have been made of the facility with which radiant energy can be thus turned in any desired direction. Apart from optical applications (such as looking-glasses, etc., which hardly come within the scope of this book), a small surface may be most intensely heated by concentrating upon it a large number of rays from the sun or an electric lamp by means of a series of reflectors arranged side by side in such positions that they all throw the reflected rays upon the same spot. This may be done either by fixing a number of flat pieces of polished metal or silvered glass in a curved frame, or by constructing a single large mirror with a concave surface, since each minute portion of such a mirror is equivalent to a very small flat reflector, the plane of which forms a tangent to the curved surface. On the first plan very large burning mirrors have been constructed, and Archimedes is said (though the statement is only a late tradition, not mentioned by Livy) to have used such a mirror to set fire to the Roman ships which were besieging Syracuse in the 2d Punic War, B.C. 212. It is more certain that the ever-burning vestal fire was thus kindled, and that in A.D. 1687 Tschirnhausen of Leipzig made a concave mirror of polished metal 1·5 metre in diameter, by which metals, and even fragments of slate and brick, were melted by the sun's rays. Quite recently Major Ericsson has erected at New York a reflector of long, narrow, flat strips of silvered glass arranged in the form of a portion of a cylinder exposing an area of about 10 square metres, by which the sun's rays are converged upon a cylindrical boiler placed in front. The amount of energy thus communicated to the water in the boiler is said to be sufficient to do about 4,000,000 metre-grammes of work per minute at noonday in summer.

It is easy, from what has been said, to see the advantage of fitting open fire-grates with conical hoods of polished metal, and also of making the backs of fireplaces of white glazed tiles. These receive a great deal of the radiated heat, and throw it forward into the room.

6. Radiant energy, when it enters a medium which can transmit it, and which differs in density from the one in which it is travelling, is changed in velocity; and if it falls obliquely on the surface of the medium, it is changed in direction also.

Since the molecules of substances differ in mass as well as in complexity of chemical composition, it is natural to infer that wave-motion will be taken up and transmitted by them with very different velocities. When a ray has passed from air into a plate of glass, it proceeds through the glass with only two-thirds of the velocity which it had in air.[1] When it has emerged from the other side of the plate into air again, it travels with the same velocity that it had before it entered the glass.

If the course of the ray is at right angles to the surface of the new medium, it goes straight on in the same direction within the latter; but if it meets the surface at any other angle, it proceeds through the medium in a direction differing from its former one. This change of direction of radiant energy is called 'Refraction'; and although it has until comparatively recent times been studied principally with reference to light, its chief facts and laws must be noticed here, as affording the best means hitherto attainable of distinguishing between the components of the very complex system of waves originated by substances at a high temperature.

Expt. 105. Arrange an apparatus as indicated in the diagram, fig. 93. A is an Argand gas-burner or paraffin-lamp with glass chimney, enclosed in a lantern, which may easily be extemporised by bending a sheet of tin (or even thick brown paper) into a cylinder about 15 cm. in diameter and 40 cm. in height; a hole

---

[1] Of course the number of waves per second is not changed, just as the number of swings made by a pendulum in a second is independent of its mass; but the distance from crest to crest of the waves is changed, so that the ray gets through a shorter distance in the time.

about 5 cm. in diameter should be cut in the side, to allow light to pass out. B is a tube about 5 or 6 cm. in diameter and 30 cm. long, made of tin or cardboard, blackened inside, each end of which is closed by a disc having a horizontal slit, about 3 mm. wide and 5 cm. in length, cut in it. The object of this is to

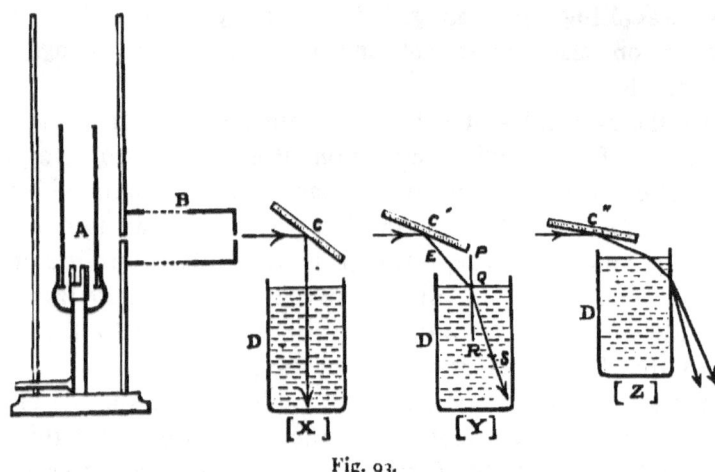

Fig. 93.

obtain a narrow, approximately parallel beam of light by cutting off all the radiation from the lamp except that which, travelling in a straight line, passes through both the slits. C is a piece of silvered glass plate, about 10 × 12 cm. (which may be conveniently held in a Bunsen holder), for the purpose of reflecting the beam emerging from the second slit in any required direction. D is a glass trough with flat sides, not less than about 10 cm. broad and 20 cm. high (one of the ordinary square or rectangular glass 'shades' used for covering ornaments, clocks, etc., will do very well if inverted).[1] It should be filled nearly to the

---

[1] It should be selected with sides as *flat* as possible ; not wavy or distorted, as some of them are, from the mode of manufacture. These shades answer well as troughs for many optical experiments, possessing the great advantage of having no cemented joints, so that they can be used for any liquid without fear of leakage. The rounded bottom may be supported in a shallow box for the sake of steadiness.

A trough may be constructed by cutting away nearly all the sides of a tin

top with clear water, and a piece of white cardboard should be fitted within it, near to, and almost parallel with the hinder side, for the purpose of showing the path of the rays through the water.[1] When this is adjusted so that it makes a very slight angle with the plane of the beam reflected from the mirror, the light glancing along it illuminates the white surface, and shows a bright, well-defined track.

Having lighted the lamp and arranged the flame so as to throw a bright, narrow beam of light on the mirror, the following experiments may be tried.

1. Adjust the mirror so as to throw the beam perpendicularly downwards on the water, as shown in [X]. Observe that no change in direction occurs when the beam enters the water; the track within the liquid being simply a prolongation of the incident beam. The wave-motion is transmitted more slowly through the water, as mentioned already, p. 294; but the velocity is, in any case, so enormous that the change in it cannot be detected by ordinary observation.

2. Alter the position of the mirror so as to throw the beam of light obliquely upon the surface of the water as shown in [Y]. The beam will now be observed to be bent or 'refracted' at the point where it enters the water in such a direction as to be more nearly vertical than the incident ray. If the mirror is moved so that the beam falls more obliquely upon the surface the deviation within the water will be still more decided.

This illustrates the general law that, when radiant energy falls obliquely upon the surface of a medium denser[2] than the one it is travelling in, it is (if transmitted at all) deviated in such a way as to approach (or, rather, make a smaller angle with) a line drawn perpendicular to the surface through the point of incidence. Thus, referring to the figure, let PQR be such a line;

---

biscuit-box, leaving a portion about 1 cm. wide at each of the angles, and cementing glass plates inside the frame thus made, with pitch or shellac.

[1] This seems preferable to the usual plan of rendering the water turbid by the addition of milk or a few drops of spirit-varnish, since the solid particles of the latter obstruct and scatter so much of the light as to greatly weaken the brilliancy of the emergent beam.

[2] It is a general, but not a universal, fact that refractive power increases with the density of the medium. One obvious exception is afforded by benzol, which though less dense than water, has a much higher refractive power.

then the angle of refraction SQR is smaller than the angle of incidence EQP.[1]

3. Raise the trough a little, and alter the position of the mirror until the beam falls very obliquely upon the water and comes out through the farther side of the trough into the air again, as shown in [Z]. Place a strip of cardboard nearly in the same plane as that in the trough, so that the course of the beam after emergence may be rendered evident. Observe that this beam is deviated still farther from its original direction, being refracted *away* from the line drawn at right angles to the surface at the point of emergence. Notice also that the beam is no longer parallel, but decidedly divergent, and that its upper and lower edges show colours, red above and blue below. Hold a piece of cardboard at right angles to the beam, about 18 or 20 cm. from the trough, and observe the coloured band, or 'spectrum,' as it is termed, into which the original white beam is resolved by refraction.

The facts which have been learnt from the foregoing experiments are briefly these :—

1. Radiant energy, when it enters obliquely a medium denser (see note 2, p. 303) than the one it is travelling in, is deviated towards a perpendicular line drawn to the surface at the point of incidence.

2. When it enters obliquely a medium less dense than the one it is travelling in, it is deviated away from a perpendicular drawn to the surface at the point of incidence.

3. The radiation which is emitted from an intensely hot substance (*e.g.* the carbon present in an ordinary gas-flame), is not all deviated to the same extent.

This last point is a very important one; it shows that radiant energy may be made up of several components unlike each other in properties, and that refraction may afford a means of separating these components from each other.

It may be worth while to try the effect of passing the radiation

---

[1] For an account of the simple and beautiful law of refraction, that the sines of these angles have a constant ratio to each other in any two given media, a book on optics must be consulted.

DISPERSION OF RADIANT ENERGY.

from the lamp through a substance which has a stronger refractive power than water, such as glass.

Expt. 106. Remove the trough of water and the reflector, and support horizontally in the direct path of the beam issuing from the slit a wedge-shaped block of glass, with its edge upwards, as shown in fig. 94. A piece of glass (or other transparent body)

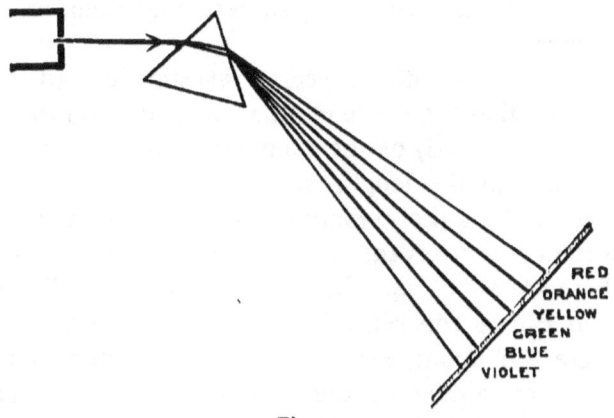

Fig. 94.

of this shape, *i.e.* having two polished surfaces inclined to one another, is called a 'prism.'[1] Hold a piece of white cardboard at right angles to the deviated beam at a distance of 30 cm. or more from the prism. Observe that the original white, narrow, parallel beam is spread out, after passage through the prism, into a broad diverging series of rays forming a coloured spectrum on the screen: those rays which are least deviated from their original direction causing in the eye the sensation we call a red colour: those which are most deviated causing the sensation of violet; while between these extremes appear orange, yellow, green, and blue, in succession, commencing from the red.

The same general result is obtained when the sun's rays are passed through the slit (which may then be made much narrower);

---

[1] It should be about 4 or 5 cm. long, and each of the sides should be not less than 2 cm. in breadth: they should be inclined to each other at an angle, preferably, of 60°. The prism need not, for the present purpose, be of very good quality, and a chandelier pendant may often be found which will answer fairly well.

and, in fact, whenever any very strongly heated solid or liquid is used as the source of radiation. We learn from it that the energy radiated from such sources contains components which affect the sense of sight very differently, and which, though all originally emitted in the same direction, can be spread out (like a hand of cards at whist), by the unequal deviation produced in them by a prism, and examined separately by receiving them on an appropriate surface.

It has been abundantly proved by investigations which cannot be given here, that the cause of this unequal deviation (or 'dispersion,' as it is called) of the components of the radiated beam is a difference in the length of the waves[1] by which they are transmitted. Thus the component which, when it falls on the retina of the eye, causes the sensation of red, consists of waves of which there are about 1400 in the length of one millimetre, while that which causes the sensation of violet consists of waves of which there are about 2400 in 1 mm.; the other waves which produce the remainder of the colours of the spectrum being intermediate in length between these extreme rays. There appears to be no other difference whatever between the various rays; the simple difference in wave-length between $\frac{1}{1400}$ and $\frac{1}{2400}$ of a millimetre causing all the variation in colour and other properties which we observe in the radiated energy, much in the same way as the difference in length of a sound-wave causes the difference in pitch of a musical note.

We may next consider the result of receiving these separated wave-series upon other surfaces than the eye.

1. If a spectrum is formed by using a prism of rock-salt[2] instead of glass, and a thermometer (or other delicate heat-measurer, such as a thermopile, p. 90), covered with lamp-black (a substance which absorbs all rays equally well, and converts the whole of the radiant energy into better-known forms), is held

---

[1] The length of a wave is the distance between one crest and the next crest in the series. Thus the length of each wave in the following set, A⌒B⌒, is the distance AB.

[2] This, as will be mentioned later, is found to transmit *all* the components of a beam of radiant energy far more perfectly than glass.

DISPERSION OF RADIANT ENERGY.

in each part of it successively, commencing at the violet end (the shortest waves), it is found that very little heating effect is produced in the violet and the blue rays, and only slightly more in the green: that the heat regularly increases as the surface is carried on through the yellow, orange, and red rays in succession; but that the maximum effect on the thermometer is only obtained when the surface is placed nearly at the boundary of the extreme red, where the eye almost fails to detect the presence of any radiation at all. If the instrument is moved still farther on through the utterly dark region beyond the visible spectrum, the heat-effect still continues to be considerable for some distance, but eventually falls off in intensity; and it is only when the thermometer has reached a point at least twice as far from the red region as the red is from the blue on the other side, that we fail to obtain evidence of any radiant energy whatever.[1]

Fig. 95 represents the long band into which the radiation proceeding from a single narrow slit is extended by refraction through

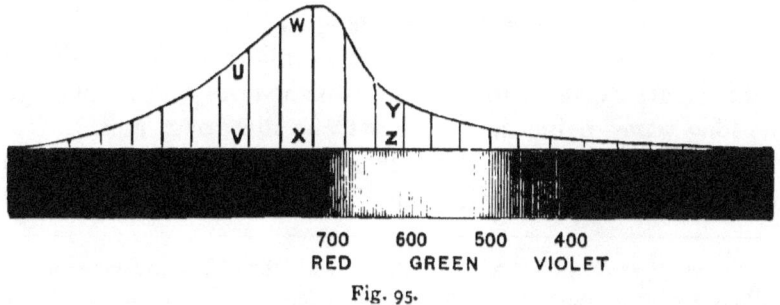

Fig. 95.

a rock-salt prism; the part which affects the eye being distinguished by being more or less lightly shaded. The magnitude of the effect in heating a surface of lamp-black is expressed by drawing lines such as UV, WX, YZ, at right angles to the band; the lengths of these lines (which might be looked upon as the heights of the mercury column in a thermometer placed at these

---

[1] See Professor S. P. Langley's papers in the *Philosophical Magazine* for March 1883 and December 1888.

points) denoting the amount of radiant energy convertible into heat at each point. Then a curve drawn through the tops of these lines will give a clear idea of the amount of energy associated with waves of any given length. The figures placed just below the spectrum, which mean millionths of a millimetre,[1] denote the lengths of the waves at various points.

2. If a slip of paper coated with white silver chloride is exposed to the spectrum, it is blackened in certain parts owing to the chemical decomposition of the silver chloride into its constituents, silver and chlorine. This action takes place most strongly, not in the part of the spectrum where we find the maximum heat-effect, but in and beyond the blue and violet rays. In the rays of longer wave-length than the green, *i.e.* of 500 millionths (about $\frac{1}{2000}$) of a millimetre, the chemical action is hardly perceptible.[2]

The subjects of the last few paragraphs may be summed up briefly as follows :—

1. Radiant energy may be transmitted by molecular motion in the form of waves of very various lengths.

2. These waves, when they fall obliquely on a different medium, are deviated to an extent which varies with their length, the long waves being deviated least, the short ones most.

3. These waves can be converted into other forms of motion by receiving them on various media, and the particular form into

---

[1] Or *micro-millimetres*, a term which, with the above meaning of 'one millionth of a millimetre,' is now generally recognised as a convenient unit for small magnitudes.

[2] It must be noted, however, that this is true only as regards the particular compound, silver chloride. Other compounds of silver, *e.g.* silver bromide, are acted on to a certain extent by the yellow rays, and by associating with them certain colouring-matters which absorb orange and red rays we can readily effect their decomposition even by those rays. Moreover, plants decompose the carbon dioxide of the air chiefly when exposed to these same yellow and red rays. Hence we must conclude that the chemical action of radiant energy depends entirely upon the nature of the particular substance exposed to it, and is not shown by the shorter waves only, as was formerly thought to be the case.

which they are converted depends entirely upon the nature of the substance on which they fall.

4. When received on the retina of the eye, some of them, *viz.* the waves of medium length, cause the sensation of light and colour.

5. When received on a surface of lamp-black, they are all (so far as we know at present) converted into heat.

6. When received on many substances which are susceptible of chemical change, the energy which they convey produces chemical action, *viz.* either decomposition of a compound or formation of a compound. In most cases the shorter waves undergo this conversion most readily, but this is not a universal rule.

It should be noticed, finally, that since lamp-black absorbs (so far as we can tell) *all* waves, whether long or short, equally well, converting them into heat alone, the curve in fig. 95 may be taken to express the sum-total of the energy which is being transmitted by radiation, and its true distribution among the waves of various lengths.

This deviation or refraction of rays accounts for the effect which a convex lens has in converging and concentrating them upon a particular spot called the 'focus.' Such a lens acts as if it was composed of a number of prisms with their refracting edges all turned towards its circumference, so that parallel rays which enter it are deviated inwards and meet at the focus. When all the sun's rays which fall on the lens are thus gathered together into a space of perhaps a square millimetre, and received on a piece of paper or cloth (preferably brown or black) enough heat is developed to set the material on fire. Such 'burning-glasses' have been made of large size; one is recorded to have been 80 cm. in diameter, made in A.D. 1800 by an optician in London, and various metals, and even slate and quartz, were melted by means of it. The only practical use, however, to which they have been put is in obtaining a record of the duration of sunshine, for which purpose a curved strip of paper is held in a frame at the focus of a glass sphere, so that the sun, when it shines, burns a track as its image moves along the paper. The length of this track shows the number of hours that the sun has shone during the day.

Emission of Radiant Energy.

It appears certain that as long as a body has any heat in it at all, it is continually parting with that heat by radiation when surrounded by a medium through which radiation can pass. But there are many causes which modify the character of the waves which set out from its surface, and also their intensity, *i.e.* the extent to which the molecules swing from side to side, or the 'amplitude,' as it is termed, of their vibration.

Let us consider what occurs when a very cold mass (solid or liquid) is gradually raised in temperature. At first the only waves which it excites in the surrounding medium are long ones, which are incapable of causing any sensation of light in the retina of the eye, though they are recognisable by their conversion into heat when they fall upon the skin or a thermometer. When tested by passing them through a prism, they are found to belong entirely to the least deviated part of the spectrum. This is the character of the radiation emitted by such sources as a vessel containing boiling water, or lead heated not much above its melting-point: in fact, by all substances the temperature of which does not much exceed 400°.

Radiant energy of this kind is distinguished by the convenient, though not very scientific, term '**Dark Radiation.**'

When the temperature is raised further, not only do these long waves increase in intensity, the molecules swinging from side to side through greater distances, but new and shorter waves are excited in addition, which, falling on the eye, give the sensation of red light. Most solid and liquid bodies become, as we say, 'red-hot' when their temperature rises to 450° (approximately), and they are then emitting all the rays included in the less refrangible half of the spectrum shown in fig. 95.

The effect of heating a substance still more strongly is of an analogous kind, shorter and shorter waves being added to those already existing, while the latter (though not altered in length) become more intense. Eventually the body emits all the rays included in the visible part of the spectrum, glowing with a bright light, and becoming 'white-hot' at a temperature of 1600°, or rather higher. Such is the radiation emitted by melted iron or platinum, or the carbon filament in incandescent lamps: and the radiant energy which is thus recognisable by the eye is distinguished by the term '**Luminous Radiation.**'

Any further increase in temperature only adds the shortest waves, which are not convertible into light, hardly so into heat, but are chiefly recognisable by their chemical action on some silver compounds.

Solids and liquids, then, if sufficiently heated, excite radiation waves of all lengths between about 300 millionths and 1200 millionths of a millimetre, giving what is called a 'continuous spectrum' when their radiation is passed through a prism. But with gases the case is different. In them the molecules are so free to move that they are able to show their own peculiar properties and excite radiation-waves of special lengths characteristic of the particular substance which is being heated. Thus the spectrum of a heated gas is not, unless the gas is very much condensed indeed (and its molecules thus so crowded as to hamper each other's movements), a continuous one, but consists of a few rays only, occupying definite places in the spectrum. The examination of these, which is called 'spectrum analysis,' affords an easy and extremely delicate means of recognising the chemical nature of a substance; but as it has been hitherto chiefly confined to the visible rays it belongs rather to the subject of Optics, and for further details a treatise on that subject must be referred to.

The next point we have to consider is, how far the nature and condition of the particular substance which is being heated has any influence on the amount and intensity of the energy which it radiates.

It must be borne in mind, to begin with, that radiation in the surrounding medium is excited almost entirely at the *surface* of the heated body; since it is only there, or within an extremely small distance from it, that the molecules are sufficiently near those of the surrounding medium to influence them and throw them into wave motion. This is even true of transparent substances, *i.e.* of those which are capable of passing on radiant energy unchanged through their mass, since it has been conclusively shown that rays excited by the molecules in the interior of a given substance are of such kinds only as are capable of being assimilated by the surrounding molecules of the same substance: so that it is only at the bounding surface between two media that radiation takes a fresh start, so to speak. Hence a thin film of varnish, or even a single layer of gold leaf (only $\frac{1}{10,000}$ mm. in thickness) is as effective in determining the character of the radiation from it as a very much thicker mass of the same substance. In experiments on radiation, then, we need only take such films, without considering what is the substance behind them, so long as the latter will convey heat sufficiently well to the true radiating surface.

The following experiments may now be made to find out the comparative radiating power of various substances; and we may first of all examine cases in which the surface is only heated to such a point as 100°, at which temperature it gives out radiant energy consisting solely of the longer waves, *i.e.* those of 'dark radiation.' The most convenient source will be a cubical metal canister, such as is shown in fig. 96, each side of which is about 12 cm. square, and can be coated with the various substances of which the radiating power is to be determined. This is called a 'Leslie's cube,' since it was first devised and used by Sir J. Leslie in his researches on radiation in the year 1804.

Fig. 96.

Expt. 107. Comparison of the emissive powers of lampblack, white paper, and polished tin for dark radiation.

Take a Leslie's cube made of sheet tin, two adjacent sides of which are coated with lamp-black,[1] while one of the other sides

---

[1] A good dead black varnish for such a purpose is made by putting 2 grms. of lamp-black (or 'vegetable black') into a mortar or earthenware cup, weighing into the vessel 8 grms. of gold size, grinding the whole together

EMISSION OF RADIANT ENERGY.

is covered with white paper,[1] and the fourth side retains its original surface of bright tin. Support it on a ring of a retort-stand at such a height as to be just between the up-turned bulbs of a differential thermometer, as shown in fig. 97. By this arrangement we have the means of comparing the radiative power of the opposite surfaces of the cube, by observing the effect produced in heating the respective bulbs of the thermometer.

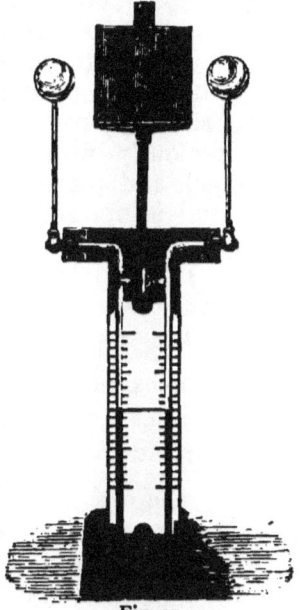

Fig. 97.

(*a*) White paper and lamp-black.

Having made all necessary adjustments of the support and the thermometer, remove the cube and fill it with boiling water: then replace it on the stand in such a position that the white paper surface is opposite one bulb of the thermometer and one of the blackened sides is opposite the other bulb, and at exactly the same distance from it. Scarcely any movement of the column of liquid in the thermometer will be perceptible: proving that white paper and lamp-black at a temperature of 100° emit very nearly the same amount of radiant energy.

(*b*) Polished tin and lamp-black.

Turn the cube one-quarter round, so as to bring the polished side opposite one bulb of the thermometer and one of the blackened sides opposite the other bulb. Observe that the

---

into a smooth paste, and adding 10 grms. of turpentine. The whole must be stirred thoroughly together just before use, and laid on evenly with a soft brush. It adheres better if the surface is slightly warmed previously, and heated more strongly after the coating is applied.

[1] Ordinary drawing-paper may be used. It should be attached to the tin (previously roughened with emery paper) by freshly made paste, and should be pressed closely into contact with the metal, by laying another sheet of paper upon it, and passing a burnisher (such as the smooth handle of a paper-knife) over it in every direction.

column of liquid is at once depressed on the side of the bulb which is receiving radiation from the blackened face of the cube: proving that lamp-black at 100° emits much more radiant energy than polished tin at the same temperature.

Expt. 108. Comparison of polished tin with glass.
Remove the Leslie's cube from its position between the bulbs and allow the latter to regain an equality of temperature. Take a large beaker, one (vertical) half of which is covered with bright tinfoil, smoothly pasted on; fill it with boiling water (rinsing it out previously with some moderately hot water, to prevent breakage), and place it between the thermometer bulbs so that the tinfoil is opposite one of them and the plain glass surface opposite the other.[1] The liquid column will be depressed on the side of the bulb which is opposite the plain glass: proving, of course, that glass at 100° is a more effective radiator than bright tin.

Differences of emissive power are also well shown by the different rates at which equally hot surfaces cool, when exposed to the same conditions.

Expt. 109. Obtain three ordinary cylindrical tin canisters, with lids, about 7 or 8 cm. in diameter and 20 cm. in height. Paint the sides of one of them with the dead-black varnish used for the Leslie's cube (p. 312): surround the second with a closely fitting coat of rather thin flannel; and leave the third with its bright tin surface untouched. Place the three on a table sufficiently far apart to be unaffected, or nearly so, by each other's radiation, and fill them with the same quantity of hot water (at about 70°) taken from the same kettle or jug, to ensure uniformity in the temperature. Observe and record the temperature of the water in each jar with a thermometer, then place the covers on them and leave them undisturbed for half an hour. At the end of this time

---

[1] The distance between the bulbs and the radiating surface should not be more than 4 or 5 cm., otherwise the effect is not very marked. The ball-and-socket joints mentioned in the note on p. 108 enable this adjustment to be readily made; but if the instrument is not fitted with such joints, two beakers of smaller size, one covered externally with tinfoil, may be filled with boiling water and placed in the proper positions with respect to the thermometer bulbs.

observe again the temperature of the water in each jar. The contents of the jar coated with flannel will be found to have cooled to a lower temperature than those of the jar with the surface of bright tin, while the water in the jar with blackened surface will be cooler still; proving that flannel is a more effective radiator than bright tin, and lamp-black still more effective than flannel. It might have been thought that flannel, from its slight conductivity, would have kept the water in the jar from losing heat; but the material is so good a radiator that, if it is not very thick, and if it fits the jar closely, its feeble conducting power does not compensate for the quickness with which its surface loses heat.

If the jars are left for some time longer, and the temperature of the water in each is taken at the end of each half hour, the same relative rates of cooling will be observed, the water in the blackened jar having cooled most and that in the polished tin least; but the actual number of degrees through which the water has sunk in each case will be less as the surface of the jars approaches the temperature of the air and the surrounding objects.

The statement known as 'Newton's Law of Cooling,' *viz.* that 'the rate at which a given surface cools by radiation, under conditions which are maintained uniform, is proportional to the excess of its temperature above that of the surrounding medium,' is very nearly correct when the intervals of temperature are not great.

Such experiments as the above will serve to give a general idea of the comparative effectiveness, as radiators, of surfaces at a temperature of 100°, and very similar results are obtained when the surfaces are heated up to 400°, or so, *i.e.* to temperatures short of a red heat. The following table gives the more precise relations between the emissive powers of various substances, as deduced by Melloni from experiments in which a thermopile having its face covered with lamp-black was used as the receiving surface.

EMISSION OF RADIANT ENERGY
(at a temperature of 100°).

[The emissive-power of lamp-black is taken as the standard, = 100.[1]]

Substance.	Amount of Radiation.	Substance.	Amount of Radiation.
Lamp-black	100	Rock-salt	28
White paper	98	Polished Iron	15
Glass	90	,, Tin	12
Sandstone	88	,, Silver	12
Ice	85		
Graphite	75		

It will be noticed, on studying the above table, that the condition of a surface, its roughness or smoothness, its colour, and the transparency or opacity of the substance for light, seem to exert little or no influence on its power of emitting dark rays. Rough paper and smooth glass have nearly the same radiative power: while equally smooth rock-salt and polished metals have far less. Compare also the radiative powers of lamp-black, white paper, and glass; of transparent ice with equally transparent rock-salt; of polished iron with almost equally lustrous graphite.

In the next place some experiments may be made on the radiative powers of substances raised to so high a temperature that they emit luminous as well as dark radiation.

Expt. 110. Comparison of the radiation from a Bunsen burner flame with that from iron oxide.

Support a tube of polished tin plate, about 7 or 8 cm. in diameter and 30 cm. in length,[2] horizontally on a stand. Place

---

[1] That is, under the conditions in which a given surface would gain 100 calories of heat when exposed to the radiation from lamp-black, it would gain 98 calories from a similar surface of white paper, 90 from one of glass, and so on.

[2] This may be easily made by bending a thin sheet of tin round a wooden cylinder of the proper size. The joints need not necessarily be soldered, but the edges may overlap a little and be kept together by coiling wire or string

close to one end of it a bulb of the differential thermometer (or of the air thermometer, fig. 49, p. 107), and near the other end support a Bunsen burner at such a height that the flame may be level with the mouth of the tube. The radiation from this will pass along the tube, the bright internal surface of the latter serving to collect and reflect upon the bulb many rays which would otherwise be lost: and the thermometer will indicate a certain amount of heat, though not very much. When the column ceases to move, showing that the maximum effect is reached, hold in the flame a small strip of rusty iron wire-gauze. The thermometer will at once show that much more heat is now reaching it, proving that the intensely heated gases of the Bunsen flame are less effective as radiators than the oxidised surface of the gauze, which certainly cannot be hotter than the gases themselves.

A similar result will be observed if a tuft of asbestos is held in the flame.[1]

It has been found that, with hardly an exception, solids and liquids emit much more radiant energy than gases at the same high temperature, not only because of their greater density,[2] but also because the radiation emitted by most gases consists of waves of a few definite lengths only, which may not be included in the luminous part of the spectrum at all.

We may, in the next place, compare the emissive power of different solids at high temperatures, using our eyes as judges of the character of the radiation, since (as explained on p. 310) the emission of luminous rays by a solid or liquid substance implies the emission of rays of dark heat as well.[3]

---

round the tube. A slightly conical form is preferable: the diameter of the tube at one end may be about the size of the bulb of the differential thermometer, and at the other end about 9 or 10 cm.

[1] If the thermometer is a fairly delicate one, or if a broad flat Bunsen flame is used instead of the usual round one, the increase of radiation produced when, by closing the air-holes of the burner, a multitude of particles of solid carbon are deposited in the flame, may be distinctly shown.

[2] A gas, such as hydrogen, which burns under ordinary conditions with a scarcely visible flame, gives, when greatly condensed, a flame which is much more luminous.

[3] Substances which have the property of phosphorescence are an exception.

Expt. 111. Comparison of glass with platinum.

Place a piece of platinum wire, or a narrow strip of platinum foil, in a bit of glass tube held horizontally, and bring the latter gradually into the flame of a blowpipe until the glass softens and closes upon the platinum. The two substances may be fairly assumed to be at the same temperature; at any rate the platinum is not hotter than the glass, yet it will be noticed to glow brightly, while the glass hardly emits any luminous rays at all.

This illustrates what is found to be a general fact, *viz.* that not only gases, but also transparent liquids and solids, do not emit luminous rays until heated to a much higher temperature than is required for opaque bodies.

Expt. 112. Comparison of iron oxide with platinum.

Place a drop of ink[1] upon a piece of clean, bright platinum foil, evaporate it gently to dryness above the flame of a Bunsen burner, and then hold it obliquely in the flame, so as to heat the whole to redness. Observe that the part where the spot of ink was placed, and where there is now a film of iron oxide deposited on the platinum, glows much more brightly than the rest of the surface, showing that iron oxide is a more effective radiator than polished platinum.

The same inference may be drawn from what is noticed when the other side of the platinum, still immersed in the flame, is looked at. A spot will be seen to be rather darker, and therefore cooler, than the rest of the surface; and this exactly corresponds in position with the film of iron oxide on the other side. The comparative coolness of the platinum at this spot clearly shows that a greater amount of heat is being withdrawn by radiation from the iron oxide than from the polished platinum.

Expt. 113. Comparison of iron oxide with chalk.

Heat a piece of thin sheet-iron in the fire or a lamp-flame until it has become oxidised and nearly black (if not already so), then make a strong white mark upon it with chalk, and heat it again to bright redness (this should be done at night, or in a darkened room). Observe that the chalk mark glows

---

[1] The ink should be of the kind manufactured from copperas (iron sulphate) and nut-galls, not any of the aniline inks, which leave no permanent residue.

much less brightly than the rest of the surface, a curious contrast to the usual appearance.

From these and similar experiments we seem justified in drawing the following general conclusions :—

I. That every substance has its own special power of emitting radiant energy : compounds, as a rule, emitting more than elements (carbon, however, in the form of charcoal or lamp-black, being a notable exception).

II. That, for the same substance, the amount radiated varies directly with the temperature. The hotter it is, the more it radiates.

III. That solids and liquids are more effective radiators than gases.

IV. That carbon in the form of charcoal is the most effective radiator, and polished metals the least effective, of all known substances.

Results and Practical Applications of the Emission of such Radiant Energy as is Convertible into Heat.

1. The surfaces of stoves and hot-water pipes, in order to obtain the maximum effect of radiation from them, should not be polished, but should either be left with the usual film of oxide upon them, or be covered with black-lead (which is simply another name for graphite). Hot-water pipes may be painted, since paint of any colour is almost as good a radiator as lamp-black. The glazed tiles with which some stoves are covered are also very effective in radiating heat.

2. On the other hand, tea-pots and kettles, if they are intended to stand on the table, and yet keep their contents hot, should be preferably of bright, polished metal. The same applies to dish-covers ; the labour spent in keeping them bright does not merely render them ornamental, but serves the good purpose of preventing loss of heat by radiation.

3. The glowing coals and cinders of an open fire are far more

effective in warming a room by radiation than the more cheerful flame; and the so-called 'asbestos-fires,' in which a kind of Bunsen burner is employed to heat a large surface covered with feathery tufts of asbestos, are decidedly preferable to any arrangement of bright flame burners.

4. The great radiative power of charcoal supplies an additional reason for employing it in electric incandescent lamps, in preference to a metallic wire of platinum or iridium.

5. The warmth of clothes is by no means due to their giving off little heat from the surface. Flannel, as proved in Expt. 109, p. 314, is an excellent radiator, and hence suits of thin flannel or calico are worn in hot climates on account of their coolness.

6. The fact that the surface of stones and soil, as well as that of the vegetation which grows upon them, is so good a radiator explains fully why the land cools so quickly at night, as mentioned on p. 284, in considering the cause of land and sea breezes. The phenomena of dew, which afford excellent examples of the laws of radiation, will be treated of at the end of this chapter.

Reception and Absorption of Radiant Energy.

When a beam of radiant energy meets the surface of a medium differing from the one it is travelling in, three things may occur.

1. It may be reflected back into the original medium, unchanged in character.

2. It may pass on through the new medium unchanged in character, though changed in velocity and direction.

3. It may be arrested and taken up by the molecules of the new medium, passing entirely into other forms of energy, especially into heat.

In most and perhaps in all cases the beam undergoes all these changes, but in very different degrees according to the nature of the medium which receives it, and the readiness with which the molecules can translate, as it were, wave-motion into heat-motion etc., and thus 'absorb' it, as we say.

In the case of bright metal surfaces, a large proportion of the rays are reflected at once, and consequently there remains only a little to be transmitted into the interior of the substance and absorbed.

In the case of such a substance as rock-salt, comparatively little is reflected, however smooth and polished the surface may be; and nearly all the radiant energy, whether in the form of long waves or short ones, is transmitted without conversion into other forms. This is obviously true for luminous rays, rock-salt being, when pure, as clear and transparent as glass, and it is also found to be true for nearly all the other rays, many of which, as we shall see, glass refuses to transmit. This property of passing on the long waves of radiant energy without converting them into heat is called '**diathermacy**'[1] (or sometimes 'diathermancy'); and substances, like rock-salt, which possess it are said to be '**diathermic**' (or by some writers 'diathermanous'). Substances which, like glass, arrest the longer waves while they allow the luminous rays to pass through them, are called 'adiathermic.'

Lastly there are many substances which seem utterly incapable of transmitting radiant energy through any appreciable thickness of their mass, and absorb entirely all the rays that they do not reflect, changing them into heat, or undergoing some chemical or electrical change.

Some experiments may now be made to ascertain the relative absorptive power of different media; and these will have reference chiefly to the non-luminous part of radiation, for, since it has been clearly proved that the shorter waves which are convertible into heat are the very same which cause the sensation of light, any substances which are observed to absorb light may be safely assumed to absorb that part of the radiation altogether.

Expt. 114. Comparison of the absorptive powers of lampblack and bright tin.
Set up parallel to one another, and at a distance of about 12 cm.

---

[1] This is undoubtedly the correct derivative from the Gk. διαθερμασία (cf. 'idiosyncrasy' from ἰδιοσυγκρασία). Melloni himself (*Comptes Rendus*, xiii. 815) adopts and recommends the use of the word.

apart, the two screens shown in fig. 98. They are made of bright sheet tin, the inner face of one of them being coated with dead-black varnish (p. 312, note). A small rod of copper, about 3 cm. long, flattened at its outer end, projects from the back of each screen, and below the outer end of each of these rods a wooden ball is attached by soft wax. Midway between the screens place the iron stand shown in the figure, and put on it an iron ball heated to faint redness, or nearly so. The radiation from this will

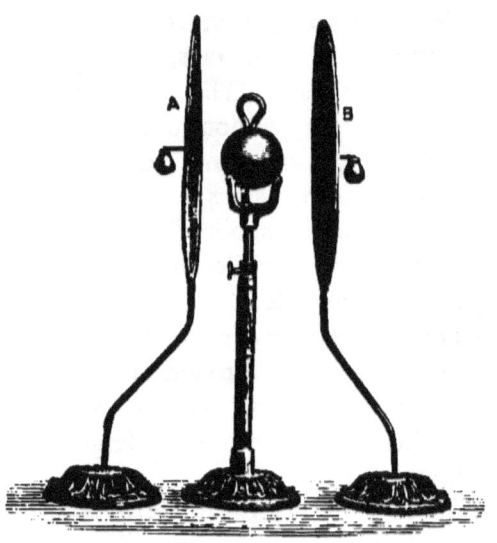

Fig. 98.

fall on both screens equally, and if the bright metallic surface of the one and the lamp-black on the other screen absorb it equally, heat will be conducted along the copper rods and melt the wax, so that the wooden balls will fall off at the same moment. But it will, as a matter of fact, be found that the ball connected with the screen which has the lamp-black surface will fall off long before the other is even loosened; proving that lamp-black absorbs radiant energy much better than polished tin.

Expt. 115. Comparison of white paper with polished tin.
Paste two strips of tinfoil, about 2 cm. broad, in the form of a cross upon a sheet of white paper, and when the paste is dry hold

Absorption of Radiant Energy. 323

the sheet in front of the glowing coals of a fire,[1] the tinfoil surface being turned towards the fire. After the paper has become brown and scorched, examine the back of it. Portions will be found still white and unscorched, corresponding precisely to those parts over which the tinfoil was pasted, on the other side. Tin is so good a conductor that the coating could not have protected the paper by its inability to convey heat to the latter: it could only have acted by almost refusing to absorb the radiant energy coming from the fire.

The result, then, proves that white paper has a greater power of absorbing radiation than bright tin.

It will now be plain why bright tin plates were used in Expt. 77, p. 251, and a bright tin screen in Expt. 104, p. 299.

We may next take a substance of very different character, *viz.* glass, and examine whether it transmits dark rays as perfectly as it certainly transmits luminous rays.

Expt. 116. Arrange the differential thermometer and Leslie's cube as shown in fig. 97, p. 313. Fill the cube with boiling water and place it on the stand with the blackened surface towards one bulb and the white paper opposite the other. These two surfaces have been shown to be nearly equal in radiating power, and the cube should be shifted until they just compensate each other's effects on the thermometer, the column remaining steady at the same level in both branches. Now hold a sheet of ordinary window glass between the cube and the thermometer bulb on one side. The column of liquid will at once begin to rise on this side, proving that the glass is intercepting and absorbing the dark rays emitted by the heated cube.

The precise mode in which the glass acts should be understood. The radiant energy falling on the glass is entirely (or almost entirely) taken up and converted into heat by the surface layers. Glass is so bad a conductor that the heat cannot be conveyed through the plate in this way, but remains in the surface film and is given out by radiation in all directions from this film. The internal layers of glass arrest the radiation which comes in their

---

[1] One of the gas asbestos-fires alluded to on p. 320 is very suitable for this and similar experiments.

direction, and the heat can only escape freely outwards from the glass on the side where the cube is. Thus it is either returned to the cube or diffused laterally through the surrounding air; and until the glass becomes, in the course of time, heated throughout, so that radiation takes place from the farther side of it also, no heat from the cube can reach the thermometer.

It cannot have escaped notice that the receptive powers of the substances which have just been experimented on, polished tin, paper, glass, and lamp-black, follow the same order as their emissive powers for radiant energy: those substances which have been proved in previous experiments to emit a great deal being precisely those which absorb a great deal of it. And not only is this fact observed to be true in a general sense, but Melloni has shown that the properties of emission and absorption are most intimately connected with each other in a given substance, so that the very same numbers which, in the table given on p. 316, express the emissive powers of the different substances, also express their absorptive powers for the same kind of radiation.

The reason appears to be, that each molecule has (like a pendulum of a particular length) its own special rate of swing, which it takes up, when an impulse is given to it, in preference to any other rate. This is also, of course, the particular rate of swing which it communicates to other molecules if they can take it up. For example, the molecules of heated glass communicate a certain special kind of radiation-wave to the molecules of other media, which wave-motion is most readily taken up by the molecules of any piece of glass it falls upon, in preference to any other kind of wave.

A very exact analogy is observable in the case of sound. In a piano each particular string is so arranged as to vibrate from side to side at a particular rate, and thus communicate a corresponding set of sound-waves to the air and excite the sensation of a particular note and no other (or, at any rate, only certain closely related notes). Now, if that particular note is sounded on another instrument near the piano, it is found that the corresponding string soon begins to vibrate, while the rest remain motionless, though perfectly capable of transmitting the sound.

The fact that such a sympathy, as it may be called, exists between bodies of kindred character may be illustrated by the following experiment with pendulums:—

Expt. 117. Clamp a piece of iron rod firmly to a tall retort-stand, in a horizontal position, and about half a metre (or more, if possible) above the table. Hang from it, by fine string or wire, two iron balls like that shown in fig. 72, p. 240, or other pieces of metal weighing half a kilogramme or more. The weights should be about 8 or 9 cm. apart; their distance from the rod should be exactly the same (half a metre, or more), so that they may form two simple pendulums swinging at the same rate. Set one of them swinging in a plane at right angles to the rod. The other ball, though motionless at first, will soon begin to swing, owing to the impulses which reach it through the medium of the elastic rod from which they are both suspended: these impulses coinciding with the rate at which the ball can naturally swing itself. Next, shorten one of the strings by about 4 or 5 cm., and again set one of the balls swinging as before. The other ball will remain nearly motionless, giving a few slight quick swings at intervals, but returning to a state of rest, since it is unable to vibrate at the particular rate which the other ball is endeavouring to impress upon it. If various other lengths of suspension are tried, similar results will be obtained: the second ball only gaining a regular, decided swinging motion when the first one acts as a pendulum of exactly the same length.

We are thus able to picture to ourselves what must be going on during the emission and reception of the marvellously small waves which carry radiant energy from spot to spot. Each molecule of the emitting surface must be exciting its own special set or sets of waves in the medium beyond; and while these waves can be, and are, transmitted unchanged by molecules which have no sympathy with them (just as any piece of cord or wood will faithfully transmit all the music of a piano), when they meet molecules similar in character to the molecules which originated them, they set these molecules swinging on their own account, and are absorbed in so doing, failing to get farther in their original form.

If the radiating surface is at a higher temperature, so as to emit luminous as well as dark rays, these new and shorter waves are not necessarily absorbed by a substance to the same extent as the others. Thus, while lamp-black absorbs the luminous radiation as readily and completely as the dark rays, white paper reflects nearly all the former (as, indeed, our eyes accurately inform us, the material being called 'white' from this very property of reflecting luminous rays). A simple experimental proof of this was given by Dr. Franklin many years ago. He took some pieces of cloth, of the same texture, thickness, and size, but dyed with different colours, and laid them side by side on level snow in the full rays of the sun: inferring that any difference in their absorptive power would be shown by the amount of snow they would melt, and the relative depth to which they would consequently sink into it. He found after a time that the piece of cloth coloured black had sunk deepest, then that which was coloured dark blue, then the green, the red, and the white pieces in the order given; the last named having scarcely sunk at all below the surface, showing that its absorbing power for the sun's rays was hardly greater than that of the snow on which it lay. The order in which the other pieces of cloth had become heated and sunk is evidently the same as that in which they absorb light.

It must not, however, be thought that darkness of colour or opacity for light necessarily implies great absorptive power for the longer radiation waves. Iodine is nearly black, yet it absorbs comparatively little *dark* radiation. If some powdered iodine and powdered alum or sugar are exposed on dishes to the rays of the sun, or a fire, the white alum becomes hot much sooner than the black iodine. So also ebonite, the black, horn-like substance obtained by heating indiarubber with sulphur, though it absorbs light, allows a good deal of dark radiation to pass through it unabsorbed.[1]

---

[1] *Proceedings of the Royal Society*, vol. xxxi. p. 506.

DIATHERMACY.

[The meaning of this term has been explained on p. 321.]

In this part of the subject attention is directed to the amount of radiant energy which survives the journey through a given medium, rather than to the portions of it which are absorbed or reflected. Of course, these portions are exactly complementary to each other: the sum of the transmitted, the absorbed, and the reflected portions making up the total original energy.

In Expt. 116, p. 323, glass was shown to be extremely adiathermic, for dark radiation, at any rate. It transmits luminous rays without much loss, but comparatively little heat is obtainable from these, as is evident from the curve in fig. 95 (p. 307), unless their intensity is very great. Crystallised alum and sugar are even more adiathermic than glass. Water also, whether as a mass of solid ice, or in the liquid state, or in the form of fine particles, such as clouds and mist are composed of, or even when existing as invisible vapour, is remarkably adiathermic. A strong solution of alum in water is as effective in stopping dark radiation as the solid alum itself, and is more convenient for many practical purposes. On the other hand, rock-salt is extremely diathermic, transmitting far more radiant energy than any other solid substance; and hence, though it is, unfortunately, rather brittle and deliquescent, it is of the greatest value as a material for prisms and lenses intended for researches on radiant energy.

Next to rock-salt, crystallised sulphur is perhaps the most diathermic solid; but though all the substances hitherto mentioned are transparent to light, it must not be supposed that transparency is necessary for diathermacy. Iodine, though practically opaque, allows a great deal of dark radiation to pass through it, as might be inferred from what has been said on p. 326, respecting its slight absorptive power. Ebonite also, in thin sheets, at any rate, is moderately diathermic (see the reference in the note on p. 326), and even prisms have been made of it, by which dark rays are refracted, although it is utterly impervious to light.

Carbon, in the form of lamp-black, is highly diathermic when in very thin films, in spite of its great absorptive power. A coating of lamp-black spread on rock-salt (by holding the latter for a few moments in a smoky flame), of sufficient thickness to stop all light, scarcely impedes the transmission of dark rays.[1]

Among liquids, water has been already referred to as stopping nearly all dark radiation. Carbon disulphide, on the contrary, and carbon tetrachloride are very diathermic; and as they both dissolve iodine freely, we can thus obtain a liquid which is practically opaque to light, but readily transmits the waves of greater wave-length (the applications of this solution will be alluded to presently).

Among gases and vapours, the oxygen and nitrogen which compose air can transmit radiant energy of all kinds without sensible loss by absorption; and the same is true of hydrogen. Water-vapour (although it is composed of two highly diathermic gases—oxygen and hydrogen) seems as little capable of transmitting dark radiation as when it is in the liquid and solid state; and several important natural phenomena result from this property of water, as will be explained presently.

The above facts respecting the diathermacy of substances hardly admit of illustration by simple experiments and apparatus. The most exact researches on the subject have been made by Professors Melloni[2] and Tyndall,[3] both of whom used thermopiles with blackened faces to receive the radiation, and determined the amount of the latter by measuring the electric current produced by its action in heating the face of the pile. The following tables will give some idea of the results obtained, at any rate as regards the *comparative* diathermacy of some typical substances. Out of 100 units of radiant energy which would have been transmitted by each substance if it had been perfectly diathermic, the figures show the actual number of units which passed through a plate of the thickness specified.

---

[1] Tyndall, *Lectures on Heat*, p. 351 (ed. 1880).

[2] *Annales de Chimie et de Physique*, 2d series, vols. liii. 5, and lv. 327.

[3] Lectures on *Heat a Mode of Motion*. The full account of his researches will be found in the *Philosophical Transactions* for 1861, 1863, 1864, and 1866.

DIATHERMACY OF SOLIDS.

Substance (thickness of plate = 2·6 mm.).	Sources of Radiation.		
	Lamp-black at 100°.	Lamp-black at 400°.	Platinum at a white heat.
Rock-salt	92	92	92
Sulphur	54	60	77
Glass	0	6	24
Alum	0	0	2
Ice	0	0	0·5

The diathermacy of solid iodine does not appear to have been examined, but from the behaviour of its solution (see the next table) it would probably be found as diathermic as rock-salt for dark radiation, though it is almost absolutely opaque to light.

DIATHERMACY OF LIQUIDS.

The liquids were placed in cells with sides formed of plates of polished rock-salt, 1 mm. apart: the thickness of the stratum of liquid was, therefore, 1 millimetre.

The source of heat was a spiral of platinum wire heated to bright redness.

Substance.	Percentage of radiation transmitted.	Substance.	Percentage of radiation transmitted.
Carbon disulphide	92	Ether	24
Do. saturated with iodine	90	Alcohol	21
Benzol	44	Water	14

DIATHERMACY OF VAPOURS.

To examine this, a little of the vapour was introduced into a previously exhausted tube 122 cm. in length, closed at the ends with plates of rock-salt. Only sufficient of the vapour was introduced to exert a tension of 12·7 mm. :[1] *i.e.* the tube only contained about $\frac{1}{60}$ of the quantity of vapour which it could contain at the standard pressure of 760 mm.

The source of heat was a spiral of platinum wire heated to bright redness.

Substance.	Percentage of radiation transmitted.	Substance.	Percentage of radiation transmitted.
Carbon disulphide .	95	Alcohol . . .	72
Benzol . . .	79	Ether . . .	68

It was not found possible to examine the diathermacy of pure water-vapour, on account of its slight tension at ordinary temperatures. Hence its transmissive power was estimated by mixing it with perfectly dry air; the latter having been proved to be almost perfectly diathermic itself. In this way it has been shown that water-vapour is extremely adiathermic. Even the small quantity of the vapour which can exist in air at the ordinary temperature and pressure (1 litre of water-vapour under such conditions only weighs about 1 centigramme) transmits less than $\frac{1}{90}$ of the radiant energy which can pass through the oxygen and nitrogen which make up the rest of the air.

Diathermacy of Gases.

These were introduced into a glass tube similar to the brass tube used for vapours, but not quite so long. The utmost care was required, and taken, to free them from all traces of water-vapour, the presence of which, as mentioned above, might entirely mask the real transmissive power of the gas. The source of heat was a plate of copper heated to about 270°.

---

[1] Of course by raising the temperature of the tube more vapour might have been put in without condensation taking place; but the results would then have been complicated by the radiation emitted by the vapour itself.

Diathermacy of Gases.

[The gases examined were under the standard pressure of 760 mm.]

Substance.	Percentage of radiation transmitted.	Substance.	Percentage of radiation transmitted.
Oxygen	100	Carbon dioxide	1·1
Nitrogen	100	Ethylene ('olefiant gas')	0·1
Hydrogen	100	Ammonia	0·08

A research on such a subject as diathermacy would not be complete without studying the effect of varying certain conditions in the experiments, such as the following:—

1. The thickness of the stratum of the substance.

The general effect of increasing the thickness was to lessen the amount of radiant energy transmitted; though not to such an extent as might, at first sight, be expected. It appears that the first layers of the substance absorb by far the greater part of the waves which the special character of the molecules enables them to assimilate, and the remainder of the radiation passes on through the rest of the substance without much loss: just as when a powder is sifted through several sieves of the same fineness of mesh, the particles which are not stopped by the first can pass almost unimpeded through the succeeding ones.

2. The character of the source of radiation.

This has a remarkable effect upon the amount of radiant energy transmitted by a given substance. Even rock-salt is comparatively unable to transmit the radiation from a plate of rock-salt heated to 100°. Water almost surpasses itself in its absorptive power for the radiation from a flame of hydrogen gas burning in air: the source in this case being chiefly the intensely heated water-vapour produced in the combustion by the union of the hydrogen with the oxygen of the air. Again, many substances transmit the radiation from a surface of lamp-black differently to that from a surface of platinum equally heated; and hence Professor Tyndall in most of his experiments used the *same* source of radiation, *viz.*

a coil of platinum wire, which could be maintained at any required temperature, from that of boiling-water to a dazzling white heat, by passing an electric current through it.

3. The temperature of the source of radiation.

It is found that substances such as alum, water, etc., which transmit little or none of the radiation from a surface at 100°, allow some to pass when it proceeds from a surface heated to a much higher temperature. This appears from the Table of the Diathermacy of Solids, given on p. 329; and the increase of diathermic power is not altogether accounted for by the additional, shorter waves which, as was explained on p. 310, are excited by an intensely heated surface, for such waves are only slightly convertible into heat (see the curve in fig. 95, p. 307). Just as very intense sounds force their way through media which stop fainter ones, so radiation-waves, when their amplitude is great, compel the reluctant molecules to transmit them. In consequence of this a much greater amount of radiant energy from the sun reaches us after passage through such adiathermic media as moist air or glass than would do so if the temperature of the sun was lower than it is. The practical importance of this fact will be alluded to presently.

It may be useful here to recapitulate briefly the chief laws relating to the reception of radiant energy by substances.

I. The amount of radiant energy which is arrested and absorbed by a substance is in exact proportion to the emissive power of the substance. Those bodies which emit most, absorb most; those which emit least, absorb least.

II. The simpler the constitution of a body is, the less radiant energy it absorbs, as a rule (example—hydrogen and oxygen, which are elements, compared with water, which is a compound).

III. Most bodies show a 'selective absorbing power' for radiant energy: *i.e.* out of the total radiation which falls on them they absorb rays of particular wave-lengths only (example—alum absorbs only the longer waves, iodine only the shorter ones).

IV. The absorptive power of a substance depends on the ability of its molecules to vibrate in unison with the waves which

Results of Absorption of Radiant Energy. 333

fall upon them. If they can do so, they absorb the radiation; if not, they simply pass it on.

V. A substance absorbs most completely those particular rays which it emits itself.

Results and Practical Applications of the Receptive and Transmissive Powers of Bodies for Radiant Energy.

1. Kettles and other vessels, if they are to be heated by radiation from a fire, may with advantage be left unpolished or covered with soot. Earthenware teapots, when placed by the side of a fire, become much hotter than bright metal ones, though they cool sooner than the latter when placed on the table. Fire-irons, on the contrary, should be kept bright and polished; they are thus prevented from becoming too hot to be handled.

2. An open fire warms a room, not, like a stove, by directly heating the air, but by sending out such rays as can readily be absorbed and changed into heat by the walls and furniture. These latter become heating apparatuses themselves, and warm the air by conduction and convection.

3. The surface of the earth, whether consisting of land or water, readily absorbs the radiation from the sun and becomes warmed. The sea, however, transmits some of this radiation to considerable depths, and therefore its surface does not become so hot as that of the land. The result of this unequal heating of land and water in the production of winds has been mentioned already (p. 284).

4. White clothes are found to be cooler than coloured ones in summer, although it might seem from some of the experiments already made that the colour of the material would make little difference in the temperature attained. This would be true if the radiation from a fire was concerned; but the radiation from an intensely hot body like the sun consists so largely of waves which are reflected by white surfaces, that the latter are effective in warding it off and thus keeping the clothes and the body encased

in them cool. For a similar reason the roofs of houses are often whitewashed in summer, and in hot countries the whole exterior of a house is almost invariably thus treated. An Oriental city appears, as the traveller approaches it, one mass of glaring white.

5. Screens of glass are a very effective protection against the heat of a fire, while they allow the cheerful light of it to pass unabsorbed. Glass globes filled with water are employed by engravers to concentrate the light of a lamp upon their work, while nearly the whole of the oppressive heat of the flame is arrested by the glass and the water. For the same reason, a small cistern with glass sides, filled with solution of alum,[1] is often used in magic-lantern exhibitions, interposed between the condensing lens and the object, to prevent the latter being damaged by heat.

6. If, on the other hand, it is desired to obtain the longer waves only, out of the total radiation from a source, this may be done by passing the latter through a thin film of lamp-black laid upon a plate of rock-salt; or, better, through a solution of iodine in carbon disulphide,[2] which (as already stated on p. 328) is very diathermic, though quite opaque to luminous rays.

By placing a globular flask of thin glass, about 8 or 9 cm. in diameter, filled with this solution, about 10 cm. in front of a lime-light or electric arc-light, the dark rays which pass through unabsorbed may be converged to a focus on the other side of the flask. At this point, paper, gunpowder, and gun-cotton may be kindled, and blackened platinum may be so intensely heated as to emit the shorter waves of luminous radiation and become red-hot. This remarkable change of radiant energy from the

---

[1] It is doubtful whether the addition of alum is of much advantage. Plain water seems to answer nearly, if not quite as well. For the protection of microscopic objects which are mounted in Canada balsam a narrow cell filled with Canada balsam is both theoretically and practically very effective.

[2] Carbon disulphide, owing to its volatility and inflammability, is a rather dangerous substance to use in such experiments. Carbon tetrachloride, though it does not dissolve quite so much iodine, answers nearly as well, and is much safer.

form of long waves to that of shorter ones[1] has been called 'Calorescence.'

7. It is a matter of common observation that the interior of a greenhouse or conservatory is always warmer, and sometimes very much hotter than the external air, although no artificial means of warming it are in use. Evidently the heat is communicated by the sun's rays which pass through the glass: but it might easily be thought that (1) the dark radiation from the sun would, like that of a fire, be arrested by the glass and not reach the interior at all; (2) the luminous radiation which came in would, with equal facility, pass out again through the glass, so that no great increase of temperature would occur inside the building. The fact is, however, readily accounted for by the laws of radiation already explained. The luminous rays emitted by the intensely hot sun contain an unusually large proportion of waves which are convertible into heat (see p. 332); and these, after passing through the glass without much loss, fall on the walls, wood-work, and plants inside the greenhouse, which readily absorb and convert them into heat. This heat is of course radiated outwards again, but only in the form of long waves, which are arrested by the glass, the inner surface of which becomes hot and radiates heat back again to the plants, etc. Thus a large quantity of radiant energy is entrapped, so to speak, by the glass; or, more strictly, after passing through it in one form is converted into another form, in which it is incapable of re-transmission.

The above facts may be illustrated on a small scale in the following way:—

Expt. 118. Obtain a rather shallow box of wood or pasteboard with glass cover, about 10 or 15 cm. square.[2] Cut a hole in the centre of one side, just large enough to admit the stem of a ther-

---

[1] A similar change, of course, takes place when a piece of platinum wire is held in the almost invisible flame of hydrogen gas.

[2] One of the glass-topped boxes used for holding mineralogical or natural history specimens will be very suitable; but any box will answer which can be closely covered with a plate of glass.

mometer. The latter should be passed through the hole so far that the bulb is situated in the middle of the box, clear of the sides. Place the box, the cover being removed, on the ground in full sunlight, and in about five minutes take a reading of the thermometer: this will give the approximate temperature of the air. Put on the glass cover: the temperature indicated by the thermometer will almost immediately begin to rise, and may ultimately reach 20° or 30°, or even more, above the air-temperature. If the cover is removed, the thermometer will at once fall again.

An application of the same principle is found in Nature on a much larger scale: indeed, the whole surface of the earth may be regarded as enclosed in a sort of hot-house, the roof of which is not glass, but water-vapour, more or less of which is always present in the atmosphere. This acts in precisely the same way as glass, transmitting freely the luminous part of the sun's radiant energy, but arresting the escape of dark radiation from the earth's surface and all things upon it. The effect of this in keeping the earth warm and preventing any extreme inequality between the night and the day temperature is undoubtedly very great; and many facts, such as the following, might be mentioned in proof of it:—

(*a*) In regions where the air is found to contain very little water-vapour,—for instance, in large, dry, sandy deserts like that of Sahara, the surface becomes intensely hot while the sun is shining on it;[1] but as soon as the sun has set, the radiation from the earth passes away unchecked into space, and the ground is reduced to the freezing-point, or even lower, during the night. In India the natives avail themselves of this fact in the manufacture of ice. Shallow pits are dug, and a number of shallow bowls of unglazed earthenware, filled with water, are laid on straw within them. Water is so good a radiator, and there is so little vapour in the clear air to arrest radiation, that a layer of

---

[1] In the waterless elevated plains which compose the interior of Australia, the ground often becomes so hot in the day-time, that a match dropped on it catches fire.

ice 2 or 3 cm. in thickness is formed in the bowls, even when the temperature of the air a metre above the surface never falls below 6° during the night.

(b) At high levels, where the air cannot be kept moist by the evaporation from soil, vegetation, rivers, etc., and where its low temperature also renders it impossible for much water-vapour to exist in it, the same extraordinary difference between the day and night temperatures is observed. On the Peak of Tenerife, at an altitude of 3200 metres, Professor Piazzi Smyth[1] found that a thermometer with blackened bulb exposed freely to the sun would sometimes show a temperature of at least 90° (the temperature of the air being only 10°); while the same thermometer, exposed to the sky at night, fell to 1°. On the occasions when such temperatures were recorded, the amount of water-vapour in the air was always very small, the dew-point being sometimes as low as 0°, which shows that there were only about 5 milligrammes of vapour in a litre of air.

Of course the effect of clouds, which are merely assemblages of very small particles of liquid water, is of the same kind but much more palpable; since they scatter the sun's rays by reflexion, as well as absorb them. At night, if the sky is clear, the earth's surface and the air close to it are always cool, and often intensely cold;[2] while if a mass of clouds gathers overhead, radiation is arrested and returned to the earth, and the temperature rises almost immediately.

8. The Fifth Law stated on p. 333 is the foundation of nearly all the knowledge which we on this earth can obtain respecting the nature and composition of the sun and the other members of the universe. Since they are so very far beyond our reach, we have to content ourselves with a close examination of the character of the radiant energy which comes to us from them. Now, if on passing this through a prism we found that it gave a

---

[1] *Philosophical Transactions of the Royal Society*, vol. cxlviii. p. 465.
[2] Cf. Horace, *Od.* III. 10. 8—
'. . . positas ut glaciet nives
　　　Puro numine Jupiter.'

'continuous spectrum' (p. 311), all that we could infer would be that the body was a solid or liquid (or perhaps a very dense vapour), the temperature of which might be roughly estimated by noticing the length of the radiation-waves which were emitted: stars, for instance, which gave out the shortest waves would be the hottest.

But the spectra of the heavenly bodies teach us much more than this. They are all more or less incomplete or 'discontinuous,' certain rays or groups of rays being wanting in them; and there are the strongest reasons for believing that these gaps or dark spaces are caused by the absorptive power of vapours which surround the luminous body. If we find among the vapours which we can make and experiment upon, one which gives an absorption spectrum exactly similar to that of a particular star, it is at least extremely likely that this very vapour is present in the atmosphere of the star. Or again, if we find among terrestrial substances one, the vapour of which, when intensely heated, gives out luminous rays which occupy exactly the same position in the spectrum as the dark spaces in the spectrum of the star, then, according to Law V., we are justified in concluding that this particular substance is present as a cooler vapour round the star. For instance, the vapour of the metal sodium (a constituent of common salt) gives out, when strongly heated, a yellow light which, when passed through a prism, is found to consist of waves of one length only (589 millionths of a millimetre), or nearly so. When the sun's radiant energy is passed through the prism, it is at once seen that a ray of precisely the same wave-length is wanting in it; a dark line appearing at this particular point in the spectrum. The inference is, that there exists round the sun a quantity of sodium vapour cool enough to absorb this ray out of the whole assemblage of rays emitted by the hotter mass of elements in the body of the sun itself. The presence of other elements, such as hydrogen, iron, magnesium, etc., in the sun and certain other stars has been inferred on similar grounds; but for a fuller account of the methods of spectrum analysis treatises such as those of Schellen or Roscoe must be consulted.

Some General Results of the Laws of Emission and Absorption of Radiant Energy.

I. Equilibrium of Temperature.

It is a fact of which we have abundant proof, that all bodies which are within reach of each other's radiation, whether hot or cold, sooner or later attain the same temperature, or at any rate tend to do so (this temperature lying somewhere between that of the hottest and that of the coldest body). For example, the walls, furniture, and everything else in a room (even including such things as a kettle of boiling water and a bowl of ice) will, if left to themselves, eventually become equal in temperature, as may be proved by applying a thermometer to each.

This tendency to an equilibrium of temperature is clearly due to the hotter bodies radiating heat to the colder bodies, but it is not equally clear whether, when the latter have risen to the temperature of the former, radiation between them ceases to take place (just as the movement of water from one cistern to another ceases when the surface of it is at the same level in each), or whether, on the contrary, it still goes on as before, each body giving out and receiving equal amounts of radiant energy so that the temperature of neither is altered.

The following experiment may give some aid in answering this question.

Expt. 119. Arrange the wide tin tube used in Expt. 110, p. 316, horizontally on a stand, with a bulb of the differential thermometer close to one end, as in the former experiment. Place some wadding or flannel round the end of the tube and that part of the bulb which projects from it, so that radiation to or from the bulb can only take place through the tube. Close to, but not touching, the other end of the tube place a glass flask which is at the same temperature as the thermometer. No movement of the liquid in the thermometer will take place, *either* because the bulb is not radiating heat at all, *or* because it is receiving by radiation from the flask just as much heat as it is emitting itself. Now fill the flask with a freezing-mixture of ice (or snow) and salt:

the movement of the column of liquid in the thermometer will almost immediately indicate that the bulb is becoming colder.

It might seem, at first sight, that 'cold' is being radiated from the flask to the thermometer-bulb (and this appears to have been the view originally taken): but since it has been conclusively shown that 'cold' merely means the absence of heat and is no distinct form of energy, this explanation cannot be accepted. We are led, then, to conclude that the thermometer-bulb, though we might not previously regard it as a source of heat, is really always emitting radiant energy even at ordinary temperatures, and is now warming the colder body, *viz.* the flask and its contents: which latter from their low temperature do not radiate to the bulb sufficient heat to make up for what it loses, so that it becomes colder.

The same result is found to occur, whatever the original temperature of the bulb may be, so long as there is a colder body near it; and hence Professor Prévost of Geneva,[1] was led to form what is known as the 'Theory of Exchanges of Heat,' which may be expressed briefly as follows:—

'Everything in the universe, whatever its temperature may be (so long as it is above the absolute zero [2]), is constantly exchanging radiant energy with everything within reach of it. If it receives just as much as it gives out, it remains unaltered in temperature. If it receives less than it gives out, it falls in temperature. If it receives more than it gives out, it rises in temperature.'

If this theory is true, the fact that objects in a room tend to become equalised in temperature is easily accounted for by it, in conjunction with the laws of radiation and absorption which have been already discussed. To illustrate this, we will take one or two special cases and see how the above laws are applied to explain them.

1. Two objects differing in temperature but having the same

---

[1] *Essai sur la calorique rayonnant* (published in 1809).

[2] For an explanation of this see p. 86.

kind of surface; such as two bright tin jars, or Leslie's cubes, fig. 99, A filled with hot water and B filled with cold water.

To simplify the case we will consider that they alter in temperature solely on account of the radiation exchanged between them, without regard to other bodies near at hand.

Fig. 99.

By Law II., p. 332, A gives out more heat to B than B (from its lower temperature) can give out to A. Hence A must get colder and B hotter until they attain the same temperature: and then each receives just as much as it gives out, so there is no further change.

This must occur, whatever the actual nature of the surface is, but objects of bright metal will be slower in attaining the same temperature than good radiators such as objects of earthenware or blackened metal.

2. Two objects having different surfaces;—for example, A, a blackened tin canister or Leslie's cube, filled with hot water; and B, a bright tin canister or cube, filled with cold water.

These exchange radiation as in the first case; A giving out much, not merely from its high temperature, but from the nature of its surface; while B gives out little, for similar reasons, and eventually they reach the same temperature, as in the first case.

But it might be supposed that this equality of temperature would not be maintained;—that A, being so good a radiator, would give out (like a prodigal) more heat than it could receive from the less effective radiator B, so that it would become colder, and make B hotter than itself: while B (the bright metal canister) would lose little by radiation, and hence for this additional reason would remain at a higher temperature than A. This might have been the case if Law I., p. 332, connecting the emissive and absorptive powers of a substance, were not true. In point of fact, however, A, though it radiates much, absorbs much also; and B, though it radiates little, can absorb little of the heat from A. The result is, therefore, that the amount of radiation from each is in the exact ratio of the amount absorbed by it, and they remain at the same temperature.

II. THE FORMATION OF DEW.

During some nights (not all) drops of pure water, or in winter particles of ice, are observed to form upon some substances (not all) which are freely exposed in the open air.

These are the general facts which have been observed respecting dew, and several more or less fanciful and inadequate theories have been put forward to account for them. Some have considered that dew falls from the sky as fine rain; some, that it rises from the earth as vapour; some, that it is a kind of perspiration from plants and other substances; some, that it is an electrical phenomenon. Such ideas were, when they were published, little more than guesses, based on a few random experiments or on none at all: and the first really scientific inquiry into the origin and cause of dew was made by Dr. Wells, a London physician, in the years 1812-1815. A short *résumé* of his research will be instructive, as showing how a scientific investigation ought to be conducted, and what good work may be done and important results obtained with few and simple pieces of apparatus.

In the first place, the questions which arise respecting dew may be reduced to the following :—

A. Where does dew come from?

FORMATION OF DEW.

B. When, and under what conditions of season and weather is it formed?

C. On what kind of surfaces is it formed?

D. What is the real cause of its formation?

These four points Dr. Wells set himself to clear up without committing himself to any preconceived theory, simply by accurate, long-continued observations and patient experiments, with the help of little more than the following apparatus:—Thermometers, scales and weights, tufts of cotton-wool,[1] plates of tin, pieces of board, pasteboard, and a few other things such as any household would supply. Most of the experiments were made in a garden not far from London, on a lawn freely open to the sky.

A. Does dew come from the sky, or from the earth, or whence does it come?

To investigate this point, two tufts of cotton-wool, equal in size, were weighed and then attached, one to the upper side, the other to the under side of a large piece of board, supported 4 feet above the grass in an open lawn. After exposure for a night the tufts were detached and weighed again.

The tuft on the upper side of the board had gained 20 grains.
The tuft on the under side of the board had gained 4 grains.

The experiment was repeated several times with a similar result, *viz.* that the cotton-wool on the upper side of the board always gained much more in weight than that on the lower side. Since the gain in weight may be safely taken as indicating the amount of dew received by the cotton-wool, the result showed that dew does not at any rate rise from the earth (otherwise the wool nearest to the earth would have received most).

Again, two similar weighed tufts of cotton-wool were placed a little distance apart on the grass lawn, and over one of them was

---

[1] This substance Dr. Wells found by repeated trials to be a very good collector of dew, although he was not aware at first why it was so. The reason is, of course, that cotton-wool is an excellent radiator of dark heat, and its fine fibres not only expose a large surface but also form nuclei or resting-places on which condensed vapour readily collects, as crystals do on strings.

put a little roof of pasteboard, like a tent, open at both ends. The result was that

> The exposed tuft gained 16 grains.
> The sheltered tuft gained 2 grains.

This was even more conclusive than the previous experiments, for both tufts were actually laid on the ground, and so were in the best possible position to receive dew if it rose from the soil.

Dew, then, certainly does not rise from the earth; but does it fall from the sky as rain?

To test this, two weighed tufts of cotton-wool were placed, as in the last experiment, on the grass. One of these was left freely exposed, while the other was surrounded with a cylinder of earthenware (a large drain-pipe, in fact), open at both ends, about 1 foot in diameter and $2\frac{1}{2}$ feet high. On weighing the tufts in the morning, after a perfectly calm night,

> The exposed tuft had gained 16 grains.
> The enclosed tuft had gained 2 grains.

Now, since rain descending straight from the sky in still weather could pass down the cylinder and fall on the tuft within it as readily as on the exposed one, the result clearly showed that dew does *not* fall as rain from the sky.

Where, then, does it come from? A clue to this was obtained by the following experiment. A rough metal plate was supported in the middle of the lawn, four feet above the ground; and it was found that dew was deposited on *both* sides of it, when any was deposited at all. The inference was that dew does not rise as mist or fall as rain, but is formed *close to* the surface on which it is deposited.

B. At what times and in what kind of weather is dew formed?

The answer to this was obtained by accurate observations, extending over several years: the state of the weather, season of the year, time of the day or night, together with an estimate of the amount of dew occurring at the time of each observation, being carefully recorded. A review of all the observations established the following facts:—

Formation of Dew.

1. Dew is formed during the night, and especially during the latter part of it.

2. It is formed during calm, clear nights only; if wind or clouds come on, the formation of dew is checked, and even that which has been deposited often disappears.

3. It is chiefly formed in spring and autumn: not to any great extent in summer or winter.

C. In what places and on what kind of surfaces is dew by preference deposited?

Many simple, but well-planned experiments were made by Dr. Wells in connection with this question. In the first place, it was shown that dew was only formed on substances which were freely exposed to the sky. Even the most flimsy shelter, such as a piece of muslin, was sufficient to prevent a deposit of dew on substances otherwise well fitted to receive it.

In the next place, it was found that, under the same conditions of weather and exposure, dew was always formed in much greater quantity on grass than on a gravel path adjoining it; on leaves and wood rather than on bricks, stones, or iron railings; on glass rather than on pieces of bright metal.

D. What is the real cause of the deposition of dew?

Here Dr. Wells's thermometers rendered as good service as his scales and weights in former experiments. He observed the temperature (a) of the air, (b) of substances which were receiving a deposit of dew—such as grass, leaves, and tufts of wool—(c) of surfaces which, under the same conditions, were not receiving dew (such as gravel and pieces of bright tin). He found that surfaces on which dew was being formed were invariably at least 3° or 4° colder than the air or dewless surfaces.

Others had doubtless noticed this fact before, but without seeing its true significance. They seem to have considered the coldness as the *effect* of the dew: Dr. Wells affirmed that it was the *cause*. He considered that the coldness of the surface chilled the air surrounding it, and that this caused the condensation of some of the water-vapour present in the air, in the form of a liquid film, which collected into drops of dew. The phenomenon, in fact, would be precisely similar in character to that observed in several experiments in hygrometry previously made (see Expt. 69, p. 220).

The next question which arises is,—What causes the chilling of the surface?

Now, although many of the facts and laws of radiation were not so well established at that time as they are now, Dr. Wells seems to have had no hesitation in ascribing the reduction in temperature to loss of heat by radiation. All the substances which were observed to get colder than the air and to receive dew are, without exception, good radiators, and exposure to the open sky at night encourages radiation. The following experiment had an important bearing on the question. A metal plate, having one side blackened and the other bright and polished, was exposed on a stand, supported by wooden props. When the bright side was uppermost it received no dew: when the blackened side was uppermost, the plate became much colder than the air, and dew was formed all over it. The only difference between the surfaces was that one (the blackened side) radiates well, while the other (the bright side) radiates badly: thus the chilling must be due to heat being conveyed away by radiation.

Thus Dr. Wells was enabled by a long course of experimenting and reasoning, of which the merest outline has been given above, to frame a definite theory respecting dew which has been universally accepted since his time, and which may be put briefly as follows:—

'Dew is condensed water-vapour coming from the space immediately surrounding the body on which it appears. This condensation is caused by the chilling of the surface, owing to its radiating heat away into space. Hoarfrost is simply frozen dew.'

This theory, if it is the true one, must be capable of accounting for *all* the facts observed respecting the occurrence of dew. If even a single clearly ascertained fact is irreconcilable with it, the theory must be modified or rejected altogether. It will be instructive to revert to the experiments described above, and observe how readily and completely the results of them are explained by the laws of radiation. Whenever (as on a clear night) and wherever (as in situations freely exposed to the sky, and from the surface of cotton-wool, grass, etc.) radiation can take place freely, then and there dew is plentifully formed. Under conditions of such a kind that radiation is impeded or impossible (as in the case of the cotton-wool placed under the board or at the bottom of the clay cylinder, or in the case of a surface of bright metal) little or no dew is produced.

The explanation of one or two facts, however, is not quite so obvious. Why, for instance, should gravel, paving-stones, and rusty or painted iron railings, all of which are known to be good radiators, receive little dew, while grass and wooden palings near them are receiving a good deal of it? The reason is to be found in the comparatively good conductivity of stone and iron: the heat lost from their surfaces is quickly replaced by conduction from the rest of the mass, so that the surface never gets much colder than the air; a fact which Dr. Wells proved by observations with the thermometer.

Why, again, should less dew be formed on the leaves of high trees than on the surrounding grass? Several reasons may be assigned for this: (1) The air is always warmer and less moist at a height of from 10 to 30 metres above the ground than on the ground itself; (2) the air a little way above the earth's surface is seldom so calm as that which is close to the ground; (3) the air in contact with lofty foliage becomes denser as soon as it begins to be chilled, and falls away as a downward convection-current before there is time for moisture to be deposited. Air close to the ground, on the contrary, cannot fall farther, but remains as a cold, moist stratum, enveloping the blades of grass.

Lastly, why should calm weather, and not merely clear weather, be almost essential for the formation of dew? The explanation is, that in windy weather the air round the cold surface is constantly carried away before it has time to become chilled below the dew-point.

It will be a useful exercise to endeavour to explain, upon the principles and laws of heat already stated, such points as the following :—

A thermometer supported a little way above the grass never indicates so low a temperature as a thermometer laid on the grass.

Dew is, on cold nights, deposited copiously on the inner surface of the panes of glass in the window of a room, but not (as a rule) on the outer surface.

Dew is very liable to form on the object-glass of a telescope, though not on the bright brass tube. A 'dew-cap,' or short cylinder open at both ends, placed over the object-glass, prevents (as a rule) the deposit taking place.

Plants are liable to be killed by frost, even when the night air is not frosty, but a loose covering of even a thin material, such as muslin, is sufficient to protect them.

Is the expression 'roscida Luna,' used by Virgil (*Georg.* III. 337) an appropriate one?.

CHAPTER IX.

THE RELATIONS BETWEEN HEAT AND MECHANICAL MOTION. THERMO-DYNAMICS.

SECTION I.—Conversion of Mechanical Motion into Heat. The 'Mechanical Equivalent' of Heat.

VARIOUS examples have been given in Chapter I. of the production of heat by the destruction of mechanical motion, and the latter term has been explained to mean the movement of molecules, or masses of them, from one place to another through sensible distances, while heat-motion is probably the revolution of each molecule in an extremely small orbit which returns into itself, like those of the planets on an immensely larger scale.

It remains now (1) to show that the destruction of a definite amount of mechanical motion leads to the production of a definite amount of heat; (2) to find out exactly how much mechanical motion must be destroyed in order to produce one calorie of heat.

Count Rumford, an American, who settled in Bavaria about the year 1780, was the first to investigate these points in a really scientific way. He was at the head of the Royal Arsenal at Munich, and was led to notice the large amount of heat which was produced in boring out cannon by steel drills, the work having to be kept cool by jets of water constantly playing upon it, in order to avoid injury to the drill. According to the old theory of heat which was held at that time, all this heat was considered to be 'latent' in the metal, and squeezed out of it by the pressure of the drill. But Count Rumford saw that this explanation was totally inadequate to account for the continuous evolution of heat from even a small mass of metal, which went on as long as the drill was at work. While the motion lasted, heat appeared; as soon as the motion ceased, no more heat was produced, even though the pressure was continued. He was led from these observations not merely to assert more definitely than had been done for many years[1] that Heat was a kind of motion of the particles of a body, but also to make experiments for the direct purpose of ascertaining how much heat was produced in the stoppage of a known amount of mechanical motion.

He took[2] a block of gun-metal weighing 51·3 kilogs., and placed it in a box containing 8·5 kilogs. of water at a temperature of 15·5°. A drill with a blunt square end like a chisel was inserted into a shallow hole in the block, and the latter was turned round by horse-power, a heavy pressure being maintained between the drill and the metal. At the end of an hour the water had attained a temperature of 48°, and at the end of two and a half hours it actually boiled. This was probably the first time, as Count Rumford remarks, that water had ever been made to boil without any fire.

He then proceeded to calculate, from the known weight of water and its rise in temperature, the quantity of heat which had been generated in the experiment: and if we apply the modern estimate of the amount of work which a horse does in two and a half hours, the results give a fair approximation to what is now accepted as the correct value of the mechanical energy required to produce a calorie.

Before proceeding further, it may be well to consider how mechanical energy is estimated and expressed. Energy, in the most general sense of the word, has been already defined (see p. 6) as 'that condition of a body which makes it capable of doing work.' We cannot see energy any more than we can see

---

[1] Lord Bacon, in 1620 (*Novum Organum*, Bk. II. Aphorism xx.), Dr. Hooke, in 1670, and others, expressed more or less clearly the idea that heat was motion.

[2] *Philosophical Transactions of the Royal Society*, vol. lxxxviii. p. 80.

Units of Energy and Work. 351

a mind or a will, but we can observe and measure the work done by it when it is transferred from one body to another. When the energy stored up in gunpowder is transferred to a rifle bullet, it imparts to the latter a certain velocity, and if we measure this velocity, and ascertain also the weight (or, more strictly, mass) of the bullet, we can get an expression for the mechanical energy which is concerned.

It may be easily proved[1] that the mechanical energy in a moving body varies directly with

1. Its mass,
2. Half the square of its velocity.

So that if W expresses the work done,

„ m „ the mass of the body in grammes,
„ v „ the velocity in centimetres per second,
$$W = \tfrac{1}{2}mv^2.$$

The absolute unit of work, and therefore of the energy which does it, is called an '**erg**,' and is (see Appendix A at the end of the book), *the amount of work accomplished by one dyne of force acting through a space of 1 centimetre.* It has no reference to force of any particular kind or acting in any particular direction, and is therefore the most comprehensive and scientific unit.

In practice, however, another unit is very frequently adopted, which has reference to the particular force of Gravitation, and the particular direction in which it acts on the earth's surface. This unit of work is called a '**centimetre-gramme**,' and is

> That amount of work which the force of Gravitation does in moving a mass of 1 gramme through a distance of 1 centimetre on the earth's surface.

It may be worth while, in order to get a clear idea of this unit of work, to make the following experiment :—

Expt. 120. Stick a broad-headed nail (such as one of those sold as 'carpet-nails') slightly into a piece of deal board, and support exactly over it a kilogramme weight, suspending the

---

[1] For the proof of this a text-book on Dynamics may be consulted.

latter by a piece of thin string from a ring of a retort-stand, at such a height that there is a clear interval of 50 cm. between its lower surface and the head of the nail. Cut the string so as to allow the weight to fall directly on the nail. Gravitation is here freely acting on a mass of 1 kilogramme through a space of 50 centimetres, and a certain amount of energy appears in it as it moves through this space. When, in its downward course, it reaches the nail, this energy will be given up to the latter, and will result in a certain amount of work being done in driving it into the wood against the resistance of the fibres to being forced apart. This work is defined to be 50,000 'centimetre-grammes,' since it is the result of the action of gravitation on a mass of 1000 grammes through a space of 50 centimetres.

This 'centimetre-gramme' unit (which corresponds to the 'foot-pound' often used by engineers) has the disadvantage of not being absolutely invariable, since the force of gravitation differs in intensity at different distances from the earth's centre of mass. It is rather less, for instance, at the Equator than at the Poles, and at the top of a high mountain than at the sea-level. But the practical convenience of this mode of expressing energy and its resulting work far more than counterbalances this slight variation in value, which can be allowed for when the conditions under which a set of experiments are made, *e.g.* the latitude of the place and the height above the sea-level, are known.

The problem, then, of finding the relation between heat energy and mechanical energy may be stated as follows:—How many centimetre-grammes of mechanical energy are required to produce one calorie of heat? Or, how far must a mass of one gramme fall in order that enough energy may appear in a kinetic form to produce one calorie when its downward motion is stopped?

The first careful experimental investigations on this subject were made by Dr. Joule, of Manchester, between the years 1843 and 1849.[1] Out of the various methods which he tried, only the

---

[1] *Philosophical Transactions*, vol. cxl. p. 61.

JOULE'S APPARATUS.

one which he considered most reliable will be described here. The source of mechanical energy which he employed was gravitation; but instead of allowing this force to act freely on a mass for some time, and then stopping the motion of the latter suddenly and determining the amount of heat into which the accumulated energy was converted, he found it more convenient to convert the energy continuously, as it appeared in a kinetic form during the fall, into heat. The motion of the falling body was continuously resisted by making it turn paddles in a vessel full of water; and the heat produced in this water, in consequence of a fall of the weight through a known distance, was accurately determined.

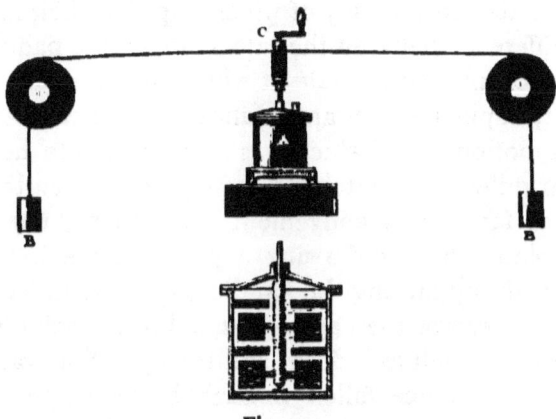

Fig. 100.

Fig. 100 will serve to show the general nature of Dr. Joule's apparatus. A is a vessel of polished copper, nearly filled with water. In the lid of this were two openings, one at the side for the introduction of a thermometer, the other at the centre, through which a vertical spindle passed and was attached to a set of vanes or paddles, sixteen in number, wholly immersed in the water which the vessel contained. A section of the vessel is given in the lower part of the engraving, and it will be seen that several sets of vanes were fixed to the sides of the vessel for the purpose of preventing the water from whirling round in a mass when it was pressed upon by the movement of the paddles. To the upper end of the spindle a wooden cylinder, C, was attached, round which two cords were wound in opposite directions: the ends of these were led off, one to the right hand, the other to the left, and attached to the large rollers shown in the figure. The axles of these rollers were supported so as to turn with the minimum of friction, and smaller rollers were attached to them, round which cords were wound, having at their ends heavy weights, B, B. It can easily be seen that, when these weights were allowed to fall, the cords were unwound from opposite sides of the roller C and caused the latter to rotate, communicating motion to the paddles in A, and forcing them against the water, the inertia and viscosity of which impeded their movement. Thus the difference between the rate at which the paddles would move in a vacuum, and the rate at which they actually did move in the water, implied the disappearance of a certain amount of mechanical motion, all of which was converted into heat in the water, the paddles, the containing vessel, etc. When the weights had fallen as far as was convenient, the roller C was detached from the spindle by withdrawing a pin, and the weights were wound up again by turning the handle shown at the top of the roller. The latter was then again coupled to the spindle, and the weights allowed to fall as before; and the operation was repeated until the total distance fallen through by the weights was considerable, and the water had in consequence become heated to an easily measurable amount.

The sum of the weights B,B, multiplied by the total distance through which they fell, gives the number of units of work done in turning the paddles.

Also the weight of the water in the copper vessel, multiplied by its rise in temperature, gives the number of calories of heat produced by the work done.

The process appears a very simple one, but in carrying it out a number of corrections had to be applied. For instance, allowance must be made for (1) the energy spent in overcoming the friction at the various axles and the rigidity of the cords; (2)

Joule's Apparatus.

the heat produced by the weights striking the floor at the end of their fall; (3) the effect of the air in buoying up the weights; (4) the loss of heat by radiation from the surface of the copper vessel and in other ways.

It will be interesting to take the data supplied by one of Dr. Joule's experiments with the apparatus, and see how they were applied to obtain an answer to the problem stated above. The numbers given are the actual English weights and measures used by Dr. Joule: the requisite corrections for friction, etc., have been made in them.

    Weights used to turn the paddles . . 57·8 lbs.
    Total distance through which they fell . 105 ft.
    Weight of water in the copper vessel . 13·9 lbs.
    Increase in temperature of water . . 0·564°.

1. First, with regard to the mechanical energy employed.

The fall of 57·8 lbs. through 105 ft. is equivalent to the fall of 1 lb. through (57·8 × 105 =) 6069 ft. That is, the work done was 6069 foot-pounds.

2. Next, with regard to the heat produced.

13·9 lbs. of water were heated 0·564°, and this is equivalent to the raising of 1 lb. of water through (13·9 × 0·564 =) 7·84°. That is, 7·84 (English) units of heat were produced in the water. But besides this, the copper vessel, the paddles, the pulleys, etc., were heated; and altogether it was calculated that 7·86 English heat-units were produced.

Hence we gather that 6069 foot-pounds of work produced 7·86 English heat-units; and the number of foot-pounds required to produce 1 heat-unit will be found by the proportion,—

$$7.86 : 1 :: 6069 : 772 \text{ (nearly).}$$

Many experiments were made with the above apparatus, and also with a similar one in which mercury was used instead of water to receive the energy of the falling weights and convert it into heat. In another set of experiments, two plates of cast iron, immersed in a vessel of mercury, were pressed together, and one of them was turned round by falling weights; the rise of temperature caused by the friction between them being observed.

Quite recently, in 1877, an entirely new set of experiments on the same subject has been made by Dr. Joule,[1] with an apparatus similar in principle to the former one, but in which the weights, by a very ingenious arrangement, did their work without actually falling a single centimetre. The copper calorimeter containing the water and paddles was made moveable on a vertical axis; and the cords, instead of being wound upon the roller C, fig. 100, were wound round the circumference of the calorimeter itself, passing over pulleys on each side, and attached to weights, as in the former apparatus. The paddles were turned quickly round by hand until, by the resistance and adhesion of the water, the calorimeter itself was slightly turned round so as to raise the weights about half a metre from the floor. The rotation of the paddles was then maintained steadily at this speed so as to keep the weights at exactly the same level for half an hour or more; and the number of turns made by the paddles was registered. The result was, of course, the same as if, the paddles remaining stationary, the calorimeter had been rotated in the opposite direction by the fall of the weights through a distance equal to the circumference of the calorimeter multiplied by the number of turns that the paddles had made. Thus the number of foot-pounds of energy expended could be calculated as in the former experiments.

The results thus obtained agreed remarkably well with the former ones, and indicated 773 as the number of foot-pounds required to raise 1 lb. of water 1° F.[2] It is easy to reduce this number to the corresponding value when the centimetre, the gramme, and the degree centigrade are taken as units, in the following way :—

---

[1] *Philosophical Transactions*, vol. clxix. p. 365.

[2] The exact number adopted by Dr. Joule, from a review of all his experiments, is 772·55 ft.-lbs. ; but this value is calculated on the assumption that the experiments were conducted at the sea-level in N. latitude 51° 29′ (that of Greenwich) *in vacuo*, and that the water was raised from 60° to 61° F.

The Mechanical Equivalent of Heat. 357

From the example worked out in Problem 1 (Appendix to this chapter, p. 362), we learn that 773 foot-pounds = 10,686,725 centimetre-grammes—that is, 10,686,725 centimetre-grammes of energy will raise 1 lb. of water 1° F.

Now, 1 gramme is $\frac{1}{454}$ of a pound; and 1° C. is $\frac{9}{5}$ of 1° F. Hence $(10,686,725 \times \frac{1}{454} \times \frac{9}{5} =)$ 42,370 will be the number of centimetre-grammes required to raise 1 grm. of water 1° C. This quantity, *viz.* 42,370 centimetre-grammes, is called 'the mechanical equivalent of heat,' and is of the highest importance in scientific, and especially engineering work. It was well worth years of labour to determine it with accuracy.

About the same time that the above experiments were being commenced, Dr. Mayer, of Heilbronn, was engaged in working out the same subject from the data afforded by Gay Lussac's law of the expansion of gases by heat and by the specific heats of gases. It was mentioned on p. 134 that the specific heat of a gas under constant pressure is greater than that of the same gas under constant volume, and the difference between the two specific heats affords a measure of the amount of heat energy expended in the mechanical work done by the gas in expanding and overcoming the pressure upon it. Dr. Mayer argued that, conversely, the expenditure of this same amount of mechanical energy in compressing the gas to its original volume would produce the same amount of heat. The subject will be alluded to again in the next Section; in the meantime it may be said that Dr. Mayer's theoretical investigations gave almost exactly the same value for the mechanical equivalent of heat as Dr. Joule obtained by actual experiment.

Other high authorities, such as M. Regnault (from considerations depending on the velocity of sound), and M. Hirn (from experiments on the friction of water in passing through a capillary tube), have obtained values only very slightly higher than the above, *viz.* 43,000 centimetre-grammes.

On the whole, then, we shall not be far wrong in adopting 42,800 as the true value of the mechanical equivalent of heat.

That is—

> 42,800 centimetre-grammes of mechanical energy are convertible into 1 calorie of heat.

This quantity is, as explained in Problem 2 (Appendix to this chapter, p. 362), equal to almost exactly 42,000,000 ergs, or absolute units of energy.

Section II.—Conversion of Heat into Mechanical Motion.

This is, for practical purposes, a much more important subject than the conversion of mechanical motion into heat. The latter is a process comparatively seldom required, but we have on the earth's surface or near it such abundant and cheap materials (wood and coal) for producing heat, that most of the mechanical work of the world is, at present, done by converting this heat into mechanical energy. Contrivances for effecting this were devised long before their inventors arrived at a clear conception of the way in which they acted; and Count Rumford seems to have been one of the first to see that a production of mechanical energy involved a disappearance of heat. He noticed that a gun fired with ball-cartridge always became less heated than when it was fired with blank cartridge (the same amount of powder being used in both cases); and he inferred that the diminution in the amount of heat in the former case was due to the energy of the powder having to do a greater amount of mechanical work in driving forward the bullet, so that less of it took the form of heat. This showed in a general way that some heat was missing when mechanical work was done, but no exact examination of the amount of heat required to produce a given amount of mechanical energy was made until Regnault's determination of the specific heats of gases (*a*) under constant pressure, (*b*) under constant volume supplied data from which the calculation could be undertaken.

In order to obtain mechanical work by the expenditure of heat,

WORK OBTAINED FROM HEAT. 359

we avail ourselves of the fact that heat when communicated to a body causes it to expand (or tend to do so),—*i.e.* imparts to the molecules an increased power of overcoming any resistance which opposes their onward movement. By taking advantage of this we can move things from place to place, as is shown (to take a simple instance) in the motion of the index of Ferguson's pyrometer (p. 46) and of the column of liquid in the differential thermometer. In the case of liquids and solids, so much heat is necessarily expended in internal work (p. 136) that comparatively little is available for overcoming external resistance. But in vapours and gases a considerable amount of the heat imparted to them can be utilised for external work, and hence these are almost universally used for the purpose.

Let us, then, take a gas, such as air, and, by considering what happens when a definite volume of it is altered in temperature, under given conditions of pressure, endeavour to gain an idea of the amount of mechanical motion which can be got from a calorie of heat.

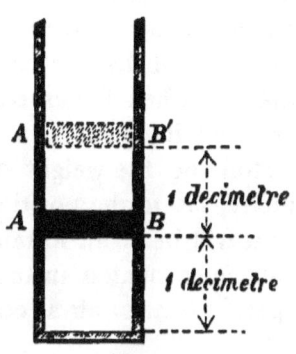

Fig. 101.

Let fig. 101 represent a tall square vessel, having an internal cross-section of 1 decimetre square. A B is an extremely light piston fitting air-tight into it, and capable of being moved up and down without appreciable friction, and also of being fixed immoveably in any required position. When placed as shown in the engraving, its lower surface is 1 decimetre above the bottom of the vessel, and hence the space below it is exactly 1 cubic decimetre, *i.e.* 1 litre. Let this space be filled with air at 0°, the barometer column standing at 760 mm.: the pressure of the external air upon the piston will then be 103.3 kilogrammes.[1]

---

[1] Air, of which the pressure supports a mercury column of 760 mm. in height, presses upon each square centimetre of surface with a force equiva-

I. Let the air below the piston be heated from 0° to 273°, the piston being fixed firmly in its place, so that the air cannot expand, but its volume remains constant. A certain amount of heat will be spent in raising the temperature of the air, and this amount can be calculated from the known weight of a litre of air, and the specific heat of air at constant volume. Thus—

Wt. of 1 litre of air at 0° and 760 mm.	Sp. heat of air at const. vol.	Rise in temperature.		
1·293 grm. ×	0·1684 ×	273°	=	59·44 calories.

II. Next, let the conditions be the same as at the commencement of the first operation, but let the piston be perfectly free to rise as the air below it expands with the heat, *i.e.* let the air be heated, and the pressure upon it be maintained constant. Then the air, when it has reached a temperature of 273°, will have doubled its volume, and in so doing will have raised up the piston and the weight of the air above it to a height of 1 decimetre, *i.e.* to the position A' B', fig. 101. A certain amount of heat will be spent in raising the temperature of the air, and this can be calculated in the same way as in the first case, using the specific heat of air at constant pressure. Thus—

Wt. of 1 litre of air at 0° and 760°.	Sp. heat of air under const. pressure.	Rise in temperature.		
1·293 grm. ×	0·237 ×	273°	=	83·65 calories.

More heat is obviously spent in the second operation than in the first, and since the former only differs from the latter in the fact that mechanical work was done in raising the piston and the column of air upon it, we may infer that the excess of heat, *viz.* (83·65 − 59·44 =) 24·21 calories, was converted into the mechanical energy required to do this work. The work done amounts to raising 103·3 kilogrammes 1 decimetre high, or

---

lent to 1·033 kilogrammes. In a square decimetre there are 100 square centimetres, hence the total pressure on the piston 1 decimetre square will be (1·033 × 100 =) 103·3 kilogs.

First Law of Thermo-Dynamics. 361

103,300 grammes 10 centimetres high;—that is, it is equal to 1,033,000 centimetre-grammes.

Thus we learn that 24·21 calories of heat are convertible into 1,033,000 centimetre-grammes of mechanical energy, and by the proportion sum—

calories.		calorie.		cm.-grms.		cm.-grms.
24·21	:	1	: :	1,033,000	:	42,668

it is found that 1 *calorie of heat is convertible into* 42,668 *centimetre-grammes of mechanical energy.*

This number comes so very near the value already given for the mechanical equivalent of heat (p. 358), *viz.* 42,800 centimetre-grammes, that we seem justified in concluding that heat energy and mechanical energy are mutually convertible without any loss; a conclusion which, indeed, necessarily follows from the great principle of the Conservation of Energy referred to on p. 16.

We may, then, assent to the truth of the statement embodied in what is called

The First Law of Thermo-Dynamics.

'A definite quantity of mechanical energy is entirely convertible without loss into an exactly equivalent quantity of heat, and *vice versâ.*'

APPENDIX TO CHAPTER IX.

PROBLEMS INVOLVING HEAT AND MECHANICAL ENERGY.

PROBLEM 1.—**To reduce a given number of foot-pounds to their equivalent in centimetre-grammes.**

It is known that 1 foot = 30·48 centimetres.
  1 pound (avoirdupois) = 453·6 grammes.
Hence 1 foot-pound = (30·48 × 453·6 =) 13,825 centimetre-grammes (very nearly).

RULE.—Multiply the given number of foot-pounds by 13,825. The product is the equivalent number of centimetre-grammes.

EXAMPLE.—The English mechanical equivalent of heat (p. 357) is 773 foot-pounds. Reduce this to centimetre-grammes.
  773 × 13,825 = 10,686,725 centimetre-grammes.

PROBLEM 2.—**To express a given number of centimetre-grammes of energy in absolute measure, or 'ergs.'**

Gravitation, acting for 1 second on a mass of 1 gramme, communicates to it a velocity of 981 cm. per second (at the sea-level, in lat 51°). That is, gravitation is a force of 981 dynes (see App. A.).

An erg is the energy implied in the work done by 1 dyne of force applied through a distance of 1 centimetre.

Therefore a centimetre-gramme = 981 ergs.

RULE.—Multiply the given number of centimetre-grammes by 981. The product is the equivalent number of ergs.

EXAMPLES.—

(1.) In Expt. 119 (p. 351) the kilogramme weight fell through 50 cm. How much energy was accumulated in it when it struck the nail?
The number of centimetre-grammes of energy = 1000 × 50 = 50,000.
And 50,000 × 981 = 49,050,000 ergs.

(2.) The mechanical equivalent of heat is 42,800 centimetre-grammes. How many ergs is this equivalent to?
  42,800 × 981 = 41,986,800 ergs.

Problems in Thermo-Dynamics. 363

Let g = the force of gravitation in dynes.
   m = the mass in grammes of the body acted on.
   s = the space fallen through, in centimetres.
Then the energy implied in the work done = ms centimetre-grammes.
                                          = gms ergs.

PROBLEM 3.—To find the heat produced when a body of known mass has its motion stopped after falling through a given distance.

[42,800 centimetre-grammes produce 1 calorie.]

RULE.—(i) Multiply the mass in grammes by the distance in centimetres. The product is the number of centimetre-grammes of energy accumulated in the body.

(ii) Divide this product by 42,800. The quotient will be the number of calories produced.

EXAMPLE.—A mass of rock weighing 500 kilogrammes falls sheer from a cliff 80 metres in height. How much heat will be produced when it touches the ground at the base?

   500 kilogs. = (500 × 1000 =) 500,000 grammes.
   80 metres = (80 × 100 =) 8000 centimetres.
   500,000 × 8000 = 4,000,000,000 centimetre-grammes.
   4,000,000,000 ÷ 42,800 = 93,458 calories (nearly).

It may be interesting to find out how much the piece of rock would be raised in temperature, supposing that all the heat was concentrated in it.

The specific heat of the rock may be taken as 0·2 (the specific heat of marble): that is, 0·2 of a calorie is required to heat 1 grm. of it 1°.

The number of calories required to heat the 500,000 grms. 1° will be (500,000 × 0·2 =) 100,000.

But only 93,458 calories are produced.

Therefore the rock will be heated (93,458 ÷ 100,000 =) 0°·93 (nearly).

PROBLEM 4.—To find the space through which a known mass must fall, in order to produce a given number of calories.

RULE.—(i) Divide 42,800 by the mass of the body in grammes. The quotient is the number of centimetres through which it must fall to produce 1 calorie.

(ii) Multiply this quotient by the given number of calories. The product will be the number of centimetres through which the body must fall in order to produce the given quantity of heat.

EXAMPLE.—How far must a mass of iron, weighing 8 kilogrammes fall, in order to produce 20 calories?

$$8 \text{ kilogrammes} = 8000 \text{ grammes}.$$
$$42,800 \div 8000 = 5.35 \text{ centimetres}.$$
$$5.35 \times 20 = 107 \text{ centimetres}.$$

In the following problems absolute measure of energy in ergs, and not gravitation measure, will be referred to.

As stated on p. 351, if m = the mass of the body in grammes,

v = its velocity in centimetres per second,

W = the energy associated with it expressed in ergs,—

$$W = m \times \frac{1}{2}v^2, \text{ or } = \frac{mv^2}{2}.$$

PROBLEM 5.—To find the heat produced when a known mass, moving with a given velocity, has its motion stopped.

[42,000,000 ergs produce 1 calorie (p. 358)].

RULE.—(i) Multiply the given mass in grammes by half the square of its velocity in centimetres per second. The product will be the number of ergs of energy associated with it.

(ii) Divide this product by 42,000,000. The quotient will be the number of calories produced when the motion is stopped.

EXAMPLE 1.—A rifle bullet weighs 30 grammes, and strikes a target with a velocity of 400 metres per second. How much heat will be produced?

400 metres = 40,000 centimetres.

(i) The energy of the bullet is

$$30 \times \frac{(40,000)^2}{2} = 24,000,000,000 \text{ ergs}.$$

(ii) $\dfrac{24,000,000,000}{42,000,000} = \dfrac{24,000}{42} = 571.6$ calories.

We may next inquire whether this amount of heat would, if it was all concentrated in the leaden bullet, be sufficient to melt it.

The specific heat of lead is 0·03.
The potential heat of liquefaction of lead is 5·4 calories.
The melting-point of lead is 330°.

Hence, to raise 30 grms. of lead from, say, 10° to 330°, *i.e.* through 320°, there will be required (30 × 0·03 × 320 =) 288 calories.

And to melt 30 grms. of lead there will be required (30 × 5·4 =) 162 calories.

That is (288 + 162 =) 450 calories will be the total quantity of heat expended when the bullet is fully melted.

So that by 571·6 calories the bullet would be melted, with 121·6 calories to spare.

EXAMPLE 2.—A piece of iron weighing 100 grms. is struck by a hammer weighing 600 grms. and moving with a velocity of 30 metres (or 3000 centimetres) per second. How much will the two be heated at each blow?

[The specific heat of iron = 0·11.]

(i) The energy of the hammer is

$$\frac{600 \times 3000^2}{2} = 2{,}700{,}000{,}000 \text{ ergs.}$$

(ii) $\dfrac{2{,}700{,}000{,}000}{42{,}000{,}000} = 64 \cdot 3$ calories nearly.

This quantity of heat is communicated to (100 + 600 =) 700 grms. of iron; and this mass requires (700 × 0·11 =) 77 calories to heat it 1°.

Then, 77 : 64·3 :: 1° : 0·835° (nearly).

So that the iron and the hammer will become 0·835° hotter.

PROBLEM 6.—**To find the velocity which a given mass must have in order that, when its motion is stopped, a given quantity of heat may be produced.**

It has been explained on p. 358 that 1 calorie of heat requires 42,000,000 ergs of energy to produce it.

We must, therefore, multiply this number of ergs by the given number of calories, and then use this value of W in the equation given on p. 364, from which the value of v can be found when the other quantities are known.

For since $W = \dfrac{mv^2}{2}$, therefore $v^2 = \dfrac{2W}{m}$, and $v = \sqrt{\dfrac{2W}{m}}$.

EXAMPLE.—With what velocity must a cannon-ball weighing 1 kilogramme move in order that, when it strikes an armour plate, 50 calories of heat may be produced?

Here $2W = 2(50 \times 42,000,000) = 4,200,000,000$ ergs.
$m = 1000$ grms.

Hence $\dfrac{2W}{m} = 4,200,000$; and $\sqrt{4,200,000} = 2049$ (nearly).

Therefore the velocity of the cannon-ball must be 2049 cm. per second.

A very concise mode of expressing large numbers may be mentioned, as it is particularly applicable to calculations on any decimal system, such as the C.G.S. system. It consists in expressing the given number as a product of two factors, one of which is composed of the actual figures, while the other is a power of 10. Thus 1200 may be expressed as 12×10^2, 14,000 as 14×10^3, 42,000,000 (the mechanical equivalent of heat in ergs) as 42×10^6: the number of 0's always indicating the power to which 10 is to be raised.

CHAPTER X.

HEAT-ENGINES.

SECTION I.—General Principles.

A HEAT-ENGINE is an apparatus which, when heat is supplied to it, converts some of that heat continuously into mechanical energy, and applies the latter to do some useful work.

All practical forms of heat-engine depend on the principle illustrated in Chapter IX., Section II., *viz.* the increase of tension produced by heat in a gas or a vapour. It was there explained that heavy weights can be raised or moved in any direction by the force of the expanding gas. But this process must eventually come to a stop unless more heat is supplied, which is only practicable up to a certain limit; and in order to enable more work to be done, either the gas may be allowed to pass off into some region of lower pressure, and a fresh portion of cool gas taken and heated as before, or heat may be withdrawn from the expanded portion of gas by contact with some substance at a lower temperature. By this latter expedient the original conditions of pressure and volume may be restored, and then, by again supplying heat, more mechanical work may be done as before.

A heat-engine, therefore, consists of three essential parts—

(i) The arrangement for supplying heat to the fluid used (air, steam, etc.). This may be called for brevity the '**Heater**.'

(ii) The apparatus for receiving the mechanical energy produced, and applying it conveniently for particular purposes, such as lifting water, grinding corn, moving carriages or ships, etc. This includes all the **working parts** of the engine.

(iii) The arrangement for abstracting heat from the fluid when it has done all the work it can. This may be called the '**Cooler**.'

Thus one part of the engine must be kept at a high temperature, and another part at a relatively low temperature; and work can only be obtained from it while this inequality of temperature is maintained, so that heat is enabled to pass, and is passing, from the 'heater' to the 'cooler.'

Efficiency of a Heat-Engine.

A heat-engine is called 'efficient' in proportion to the completeness with which the heat supplied to it is converted into mechanical energy.

It is evident, from what has been said on p. 359, that under ordinary conditions only a portion of the heat supplied to a fluid can become energy available for external work: a large portion being necessarily spent simply in raising the temperature of the fluid, *i.e.* in overcoming the inertia of the molecules, etc. Still, since (as shown in Chapter ix.) heat is, theoretically speaking, wholly convertible into mechanical energy, we are not precluded from conceiving a state of things and an arrangement of apparatus by which all the heat taken out of a substance might be enabled to pass into this form of energy: and any engine which would effect such a transmutation would be a perfectly 'efficient' engine. It would show its *maximum* efficiency, however, only when *all* the heat present in the 'heater' had been withdrawn from it; and this could not be done except by maintaining the temperature of the 'cooler' at the absolute zero, *i.e.* at $-273°$ (see p. 86). When the 'heater' had, in the course of the working of the engine, been reduced likewise to the absolute zero, then, and not till then, would every trace of heat energy have left it, and the engine would have done all the mechanical work that it could possibly and conceivably do.

The *maximum* limit, then, of the efficiency of a heat-engine must be reckoned from the absolute zero of temperature, and its realisable efficiency depends upon the proportion which the actual lowering of temperature that occurs as the working fluid passes from the 'heater' to the 'cooler,' bears to the lowering of temperature that would occur if the fluid was reduced to the absolute zero.

Suppose, for example, that the heater of an engine is at a temperature of 130° C., while the cooler is at a temperature of 20° C. These temperatures, expressed on the absolute scale (p. 86) are (273° + 130° =) 403°, and (273 + 20 =) 293°, respectively. Now, if the temperature of the working fluid when it came into the 'cooler' *could* be reduced 403°, it would be at the absolute zero, and the engine would be in a position to exert its maximum efficiency, which may be denoted by 1. But the actual temperature of the 'cooler' is 293°, and we may consider the case as if (1) the whole of the heat was taken out of the fluid, its temperature being thus reduced to absolute zero; (2) an amount of heat was restored to it sufficient to raise its temperature to 293°. The net amount of mechanical energy thus produced cannot be greater than that which corresponds to the difference between 403° and 293°, *i.e.* 110°; and the actual (possible) efficiency of the engine compared with the maximum efficiency will be as 110 : 403, or $\frac{110}{403}$ = 0.27.

Thus if H° = the absolute temperature of the heater.
,, C° = ,, ,, cooler.
then H° : H° − C° : : maximum efficiency (= 1) : actual efficiency.

That is, the actual efficiency of a heat-engine cannot exceed the fraction $\frac{H° - C°}{H°}$.

REVERSIBLE OR RECIPROCAL HEAT-ENGINES.

There is another way besides cooling in which, after the heat supplied to an engine has expanded the fluid and done work, the original volume of the fluid may be restored. We may compress the fluid by mechanical force, *i.e.* expend an amount of mechanical energy upon it equal to that which it has done during its expansion. The result, *if* the engine is a perfectly efficient one, will be to reduce the fluid (even though maintained at the same temperature) to its original volume, and at the same time to generate an amount of heat precisely equivalent (according to the First Law of Thermo-dynamics, p. 361) to the mechanical energy expended. This heat may by proper arrangements be transferred to other bodies, and utilised for any required purpose. Thus it may be seen that such an engine might be used as a source of heat instead of one of mechanical energy by simply reversing its action and supplying it with mechanical energy, introduced at the proper place in the working parts, instead of heat.

An engine of this kind is called a 'reversible' or 'reciprocal'[1] engine. It only exists at present as a theoretical conception, since we have many cheaper and better sources of heat.

The principle involved in it may be illustrated by reference to another kind of engine. A 'water-wheel' is a contrivance by which water, in passing from a high to a lower level under the action of gravitation, is made to communicate mechanical energy to a wheel fitted with vanes or paddles. Now, if this wheel was turned in the opposite direction by some mechanical means, it could be made to lift water upon the paddles from the low level to the higher one.[2]

The general conclusions which may be drawn from what has been said in the present section respecting heat-engines may be summed up in the following statement, which is known as

THE SECOND LAW OF THERMO-DYNAMICS.

'Heat can only be converted into mechanical energy when it is allowed to pass from a body at a high temperature to a

---

[1] This term may be preferable to 'reversible,' since the latter, in ordinary engineering language, means any engine which can be made to cause mechanical motion in either of two opposite directions, as will be described in the next Section.

[2] Such an apparatus is actually used in Egypt and other countries for raising water.

STEAM-ENGINES. 371

body at a lower temperature; and a perfect heat-engine is one which will, if a given amount of mechanical energy is spent on it, produce an exactly equivalent amount of heat.'

[For a fuller account of the principles of heat-engines, Clerk Maxwell's *Theory of Heat* should be referred to.]

SECTION II.—Steam-Engines.

There can hardly be a better or grander example of the application of the principles treated of in the last section than the modern steam-engine : an apparatus which has grown up from small and rude beginnings into a machine of gigantic size and of almost perfect design and execution, which will probably hold its place as a most useful servant of man, until increasing dearness of fuel, or the discovery of more effective means of obtaining supplies of mechanical energy, causes it to be laid aside as an interesting relic.

There are several reasons why steam has been so generally preferred to other vapours or gases for use in a heat-engine. In the first place, its source, water, is cheap; and it can be easily generated without risk or annoyance, since it is odourless and non-inflammable. Next, a very large volume of it can be produced from a small quantity of water, as has been mentioned already (p. 161). Thirdly, it is easily condensed, when its work is done, into a very small quantity of water, and so got rid of.

Steam may be employed in a heat-engine in at least two ways.

1. Its tension, due to the heat imparted to it, may be employed directly to do mechanical work; for instance, to force up a quantity of water to a high level, or to drive a piston from one end of a cylinder to the other.

2. It may be employed to drive out air from a vessel, and then condensed into a few drops of water. A more or less perfect vacuum is thus produced, and the pressure of the air in the direction of this will do mechanical work.

The second of these methods has been illustrated already in Expt. 44, p. 162. The following experiment may be made in illustration of the first method, which is the one almost universally employed in modern steam-engines.

Expt. 121. Obtain a bulb-tube of glass, as shown in fig. 102, about 16 or 18 cm. in length, fitted with a well-packed piston, the rod of which passes through a hole in the cork which closes the tube. Fill the bulb about half full of water, and support the apparatus by a Bunsen holder in the position shown in the engraving. Heat the water gently by an Argand burner: it will soon give off steam of a tension sufficient to raise the piston, together with the weight of the air above it, to the top of the tube. When this is done, remove the lamp, and blow on the bulb to cool it. The steam will condense into water, and the piston will be forced down by its own weight as well as the pressure of the air upon it. Then the water may be heated again until the piston makes another upward movement or 'stroke,' and so on.

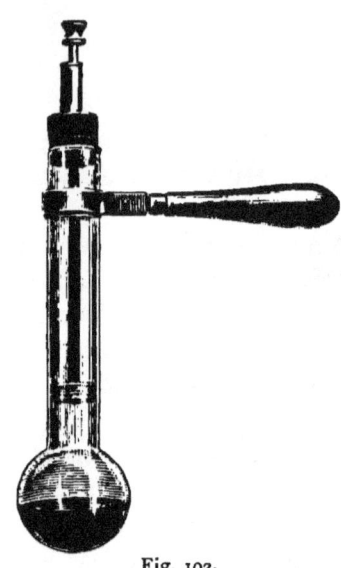

Fig. 102.

Little can be said here, from want of space, respecting the early history of the steam-engine.[1] Hero's engine or 'æolipile' (B.C. 130), Papin's 'Steam-and-Air Engine' (A.D. 1690), and Savery's 'Water-raising Fire Engine' (A.D. 1698) hardly advanced beyond the stage of models, more or less clumsy and inefficient.[2] At the end of the seventeenth century the difficulty of getting rid of the water which inevitably accumulated in mines led to a demand for some more efficient pumping machinery. Many

---

[1] On this subject, Thurston's *History of the Steam-Engine* (International Series), Galloway's *History of the Steam-Engine*, and Muirhead's *Life of Watt* may be consulted.

[2] It is fair to say that some of Savery's machines were used for a time, for light work, and that the principle of his engine has lately been employed in the 'Pulsometer,' a very portable but rather wasteful arrangement for raising water.

THE ATMOSPHERIC ENGINE. 373

mines had to be abandoned from the impossibility of keeping them from being flooded, and many others could only be cleared of water by pumps worked night and day by relays of horses. A spirit of invention was thus stimulated, and about the year 1705 Thomas Newcomen, a blacksmith at Dartmouth, constructed the first practically successful steam pumping-engine, which, under the only partially correct name of the 'Atmospheric Engine,' did useful work for many years.

A general idea of its principle may be gained from the model used in Expt. 121, fig. 102 (p. 372). We have only to imagine the piston-rod to be attached to the outer end of the handle of an ordinary pump, so that when the air presses down the piston water will be raised in the pump-barrel.

The actual arrangement of the atmospheric engine is shown in fig. 103 (next page). A is the boiler, a dome-shaped vessel made of iron or copper, about two-thirds filled with water, and heated by a furnace below. It is fitted with a safety-valve, B (like that in Papin's digester, fig. 61, p. 193), a heavy conical plug which is pushed up by the steam when the tension reaches a certain point, and then allows the steam to escape. The pressure in these boilers never greatly exceeded that of the atmosphere, so that there was no necessity to have a lever to keep down the valve as in Papin's apparatus. C, C' are two pipes ending in stop-cocks, by which the level of the water in the boiler can be ascertained. One of these, C, extends a little below the proper level of the water; while the other, C', only reaches to a point higher than that at which the water ought to stand in the boiler. Thus, when the water is at its proper level, water comes out of C when opened, and steam out of C'. When the water is lower than it should be, steam comes out of both, and when the water is above its proper level, water comes out of both of them. These are called 'gauge-cocks,' and are fitted to almost every modern boiler.

Over the boiler is supported the cylinder D, communicating with the steam space by a short pipe with a sliding valve, E, fitted in it, to regulate and cut off, when necessary, the supply of steam. A solid piston works up and down in the cylinder; it is a thick circular plate with a groove in the edge, which is filled up with hemp or loose rope, saturated with grease, to ensure that no air should enter or steam escape. This object is further secured by keeping the piston covered with a layer of water supplied from a cistern, G. Besides the main steam-pipe, two other pipes enter the bottom of the cylinder; one, F, brings a supply of cold water from the cistern G, to condense the steam; the other, H,

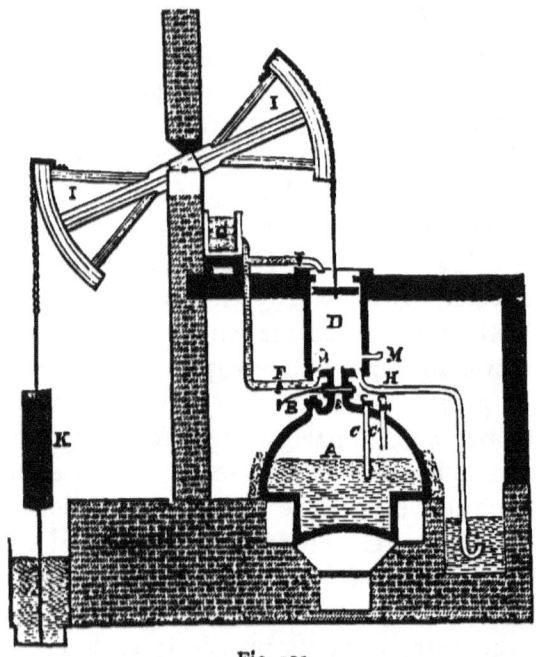

Fig. 103.

is carried down into a cistern of water where it ends in a valve opening upwards. This latter pipe is intended to convey away the water and condensed steam, which would otherwise accumulate in the cylinder and stop the working of the engine.

The piston-rod has a short piece of chain attached to its upper end, which lies in a groove cut in the wooden 'arch-head' at the end of the great working beam I. This beam turns on an axis at its centre, and has an arch-head at its other end, to which the pump-rod is attached by a chain. The advantage of this mode of connecting the pump-rod and piston-rod with the beam is that as the chain is wound and unwound in the groove cut in the circular arc, the rods move up and down in the same straight line; it is, of course, only applicable when the stress upon the rods is a pull and not a thrust. If the pump-rod is not itself very heavy, a counterpoise, K, is attached to it, of sufficient weight to pull up the piston of the engine and keep it in the position shown in the diagram, when the engine is not working.

The mode of action of Newcomen's engine is as follows:— The piston being at the top of the cylinder, steam is admitted from the boiler and displaces the air in the cylinder, which escapes through the small valve at the side, M. When the cylinder is entirely filled with steam, the regulator E is closed, and a jet of cold water is admitted through F. This quickly condenses the steam, and the piston is forced down by the pressure of the atmosphere. Thus the pump-rod at the other end of the beam is raised, and water drawn from the mine. The stopcock F is closed during the descent of the piston; and when the latter reaches the bottom of the cylinder, the steam valve E is again opened. The entering steam balances by its tension the pressure of the air upon the piston, and the latter is pulled up by the counter-weight and pump-rod at the other end of the beam, the steam following it up and filling the cylinder. The water which had accumulated in the cylinder during the down-stroke passes through H into the cistern. Then the ingress of steam is stopped: a jet of cold water is again let in through F, and the piston is forced down by the atmospheric pressure, as before.

It is evident that, although the atmosphere does the actual work of pressing down the piston, the real source of the mechanical energy is the heat of the steam; just as truly as, in a pile-driving apparatus, the men who work it supply the energy required to force the pile into the ground, although the heavy weight which they lift does the actual work of hammering down the pile.

Newcomen's engine was soon appreciated as an immense improvement on all previous machines, and it came into general use, not only for clearing mines of water, but also for draining marshes, supplying towns with water, and even occasionally for raising water to work water-wheels; no one seeming to grasp the idea that the engine itself might take the place of the water-wheel for turning machinery. Many slight but real improvements were made in the working-parts, especially in enabling the engine itself to open and close the valves E and F at the proper times, thus raising it to the rank of a self-acting machine. This was done by attaching a rod to the beam near its end, which hung down vertically by the side of the cylinder, and had pegs inserted in it which struck against the long handles of the valves and moved them in the required direction as the rod moved up and down. These engines were constructed of very large size, with cylinders nearly 2 metres (70-80 inches) in diameter and 4 metres in length of stroke; and in 1760 there were about sixty of them at work in the Newcastle coal-district alone. At its best, however, the engine was excessively wasteful of fuel, and this was a great obstacle to its employment in such districts as Cornwall, where no coal existed on the spot.

The first to effect a radical improvement in the atmospheric engine was James Watt, an optician in Glasgow, who, between the years 1760 and 1770 examined the working of it in a really scientific way, to find out why its efficiency was so far below what it should be. He ascertained that nearly three-fourths of the steam supplied was absolutely wasted, and that about four times as much cold water was injected into the cylinder as ought theoretically to be required to condense the steam. In his experiments he discovered independently the fact that steam requires a very large amount of heat to be withdrawn from it in order to convert it into water (the subject of potential heat was just then being for the first time investigated by Dr. Black). He also investigated with patient skill several other points which he

Watt's Investigations.

saw must be known in order that the engine might be improved. For instance, he examined the heat-capacity of iron, copper, wood, etc.—the relation between the volume of steam and that of the water from which it was formed,—the elasticity (or, as we now call it, the tension) of steam at different temperatures, —the quantity of steam required for each stroke of the piston,—and the quantity of cold water which must be mixed with the steam in order to condense it.

Watt was now able to see that the unsatisfactoriness of the engine, as then constructed, was due to two main causes :—

1. The whole mass of the cylinder and piston had to be raised at each stroke from a low temperature to the temperature of the steam, before the latter on its entry could do any useful work in balancing the air-pressure. Until this temperature was attained, the steam was condensed as fast as it came in.

2. The tension of the vapour was very insufficiently lowered by the cold water let in, although a large quantity of the latter was used. In fact, the whole cylinder and piston had to be cooled as well as the steam; and it was not found possible, in practice, to reduce the temperature much below 60°, at which point the tension of the vapour was still as much as 150 mm. of mercury column, *i.e.* nearly one-fifth of the whole pressure of the atmosphere. This back-pressure acted against the air, so that the effective force on the piston was only four-fifths of what it would be if there was a perfect vacuum below it.

To remove these defects two conditions must be fulfilled :—

1. The cylinder, and all parts of the engine which have to contain steam, must be maintained constantly at the temperature of the steam.

2. The temperature of the water into which the steam is condensed must be so far reduced that the vapour from it may have no considerable tension.

These requirements were clearly incompatible with the existing arrangements of the engine. The cylinder was to be kept hot, and yet, if the steam was to be effectually condensed in it, it must be rendered cold. Watt, however, observing the force with which steam rushes into any space where the pressure is lower, inferred that this same force would carry it from the cylinder into a separate vessel kept as cold as was requisite, where it could be condensed by a jet of cold water in the usual way. This excellent idea of a separate condenser was at once tried practically, and was found to be entirely successful. The injection pipe and valve F, fig. 103, were taken away, and the pipe H, with a valve fitted in it, was connected with an air-tight cistern A, fig. 104, into which a jet of cold water could be

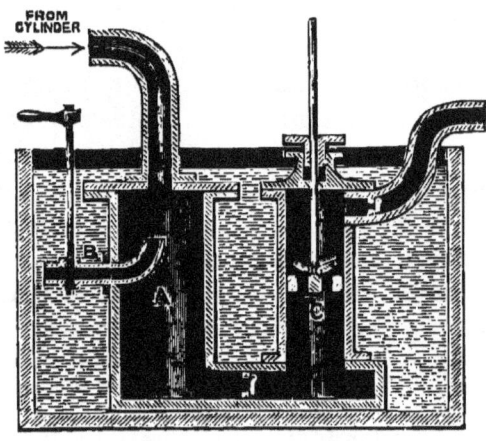

Fig. 104.

admitted at B. A pump, C, the rod of which was attached to the engine beam, served to withdraw the water and also any air which might leak into the apparatus; it was hence called the 'air-pump.' The condenser was immersed in a cistern of cold water, and the internal temperature could be kept as low as 36°, at which point the tension of water-vapour is only 44 mm., an almost negligible quantity.

Further improvements followed almost immediately. The cylinder was closed at the top, and steam instead of air was used to force the piston down. To prevent leakage of steam, the piston-rod passed through a round box cast on the cylinder cover, in which a thick ring of greased tow or rope was placed and pressed against the rod by a collar which could be screwed down into the box. This contrivance, called a 'stuffing-box,' is shown at E in fig. 105, which represents also the form of valves used by Watt in his improved engine.

THE SINGLE-ACTING ENGINE.

It will be easy from the engraving on the next page to understand the working of Watt's 'single-acting engine,' as it was called, since the useful mechanical work was done during the movement of the piston in one single direction only, *viz.* downwards.

In fig. 105 (which is photographed from a moving model constructed of wood and cardboard) only the cylinder and valves are shown; the boiler, beam, and pumps were exactly the same as in the older engines. It will be seen that there are two openings into the cylinder, one close to the top, the other close to the bottom. These openings lead into iron boxes, or 'valve-chests,' the upper one of which contains one valve, while the lower contains two valves: these valves, consisting of flat conical plugs of metal, are worked by rods and levers as shown. The valve in the upper chest, A, serves to admit or cut off steam from the boiler. The lower valve-chest is connected with the upper one by a pipe, D, at the lower end of which there is a valve, B. In the same chest, just below the opening into the cylinder, there is a third valve, C, which opens or closes communication with the condenser.

The first operation in setting the engine to work is to open all three valves, the piston being near the top of the cylinder, as shown in fig. 105. Steam is thus allowed to blow through the cylinder and valves into the condenser, carrying with it the air, which finally escapes through the valves of the air-pump, C, fig. 104. When the whole is approximately full of steam, the valve B is closed, and a jet of cold water is admitted into the condenser through the pipe B, fig. 104. This at once condenses the steam there, and the steam below the piston in the cylinder rushes into the now empty condenser and is itself quickly condensed. Then the pressure of the steam above the piston forces the latter down, and a load of water is raised by the pump at the other end of the beam.

When the piston reaches the bottom of the cylinder, communication with the condenser is stopped by closing the valve C, and also A is shut, so that no more steam can enter from the boiler. Then the intermediate valve, B, is opened, as shown in

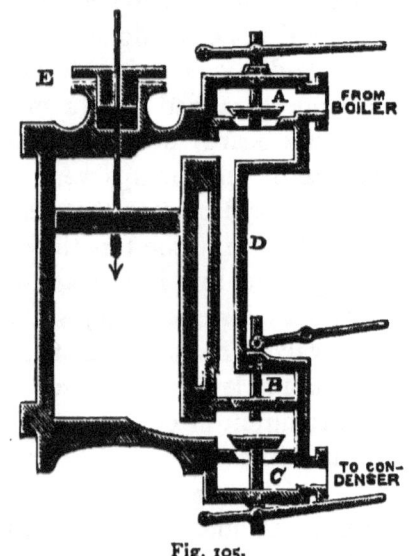

Fig. 105.

fig. 106, and thus the steam in the cylinder can find its way through the pipe D to the bottom of the cylinder below the piston. The latter has now an equal pressure of steam above and below it, and is pulled up by the weight of the pump-rod and the counterpoise, as in the atmospheric engine.

When the piston reaches the top of the cylinder, the valve B is closed, and A and C are opened; the steam below the piston now passes into the condenser, and more steam from the boiler enters above the piston and presses it down as before.

WATT'S SINGLE-ACTING ENGINE.   381

In order to open and close the valves at the proper times, a vertical rod, called the 'plug-frame,' was jointed to the beam so as to hang down by the side of the levers connected with the valves. Pins were inserted horizontally into this rod at proper intervals, and these struck against the valve-levers, raising and depressing them as required.

The economy of these engines as compared with the atmospheric engine was very great, little more than one-half as much fuel being required by them for doing the same amount of work:

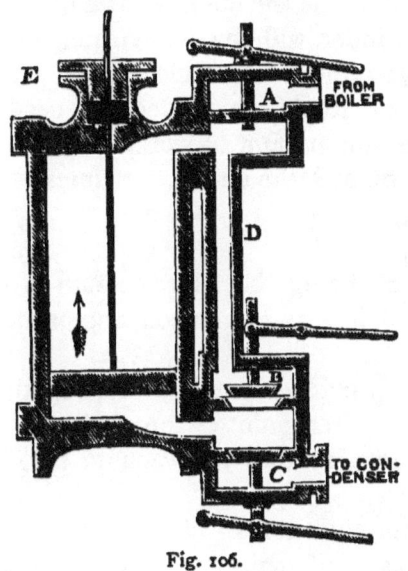

Fig. 106.

and it was rendered even greater by the application of what is called the 'expansive' principle,—another of the many inventions of Watt. He inferred, from the violence with which the steam rushed into the condenser, that a great deal of energy remained in it after it had forced the piston to the bottom of the cylinder; and it occurred to him that the supply of steam might be cut off before the end of the down-stroke, and that the tension of the steam already in the cylinder would be sufficient to do the rest of the work. At the beginning of the stroke a great deal of energy must be spent in overcoming the inertia of the massive beam and other parts of the machinery; but when once these have been set in motion much less energy is required to keep them moving, and hence there is a positive waste of energy in supplying the steam at full pressure up to the very end of the stroke.

Suppose that the steam, if admitted at full pressure throughout the downward movement of the piston, will do 200 million ergs of work, then in half the stroke it will have done 100 million ergs; and if the steam-valve is closed at this point, the steam in the cylinder will, by its tension, do 70 million ergs more, and this 70 million ergs is clear gain. It is found, in fact, that a quantity of steam which does, if used at full pressure throughout, a certain amount of work, can be made to yield,

If cut off at $\frac{1}{2}$ the stroke, 1·7 times as much,
" $\frac{1}{4}$ " 2·4 "
" $\frac{1}{8}$ " 3·2 "

the only drawback being that the pressure is not uniform, but becomes less and less as the piston approaches the bottom of the cylinder.

Watt therefore provided means for closing the steam-valve A, fig. 105, at any required point of the stroke, the valve C remaining open, of course, so that the vacuum below the piston was maintained.

With these improvements Watt's single-acting engine has remained practically of the same type up to the present time. Slight alterations in the details of its construction have rendered it a very economical engine for the special purpose of raising water: the best 'Cornish engine' (as the class is now termed) doing the very creditable work of raising 300,000 kilogs. of water 1 metre high (30,000 million centimetre-grammes of work) for every kilogramme of coal burnt under the boiler.

But the engine was clearly capable of doing other work besides pumping, and Watt next applied himself to render it useful for working machinery in cases where a continuous motion of rotation on an axis was required. This the engine in its present form could not do, since (1) the force of the steam was applied intermittently,—in fact, only during the down-stroke of the piston; (2) the force only acted in a straight line, *i.e.* in the direction of the length of the cylinder. Watt saw, however, that the steam might easily be made to force the piston upwards as well as downwards by connecting *each* end of the cylinder (and not the lower one only) alternately with the boiler and the condenser, and on this principle he constructed

THE DOUBLE-ACTING ENGINE,

the arrangement of valves in which is shown in fig. 107. It will be seen that each end of the cylinder communicates with a valve-chest containing two valves, one admitting steam from the boiler, the other allowing it to pass into the condenser when it has done its work. Thus A, in the top valve-chest, is the upper steam-valve, B is the upper exhaust-valve: in the bottom valve-chest, C is the lower steam-valve, and D the lower exhaust-valve.

The working of these valves in the double-acting engine can be explained in few words. In the first place, the piston being at the top of the cylinder, as shown in fig. 107, all the valves are opened, and steam allowed to blow through the cylinder, pipes, and condenser until the air is chased out, and the cylinder, at any rate, is as hot as the steam.

Then the upper exhaust-valve, B, and the lower steam-valve, C, are closed, and a jet of cold water is admitted into the condenser. A vacuum is thus produced by the condensation of the steam, and steam entering through A from the boiler forces the piston down. When the latter has reached the bottom of the cylinder, the valves A and D are closed, and the upper exhaust-valve B and the lower steam-valve C are opened, as shown in fig. 108. Steam now enters below the piston and forces it up, while the steam above the piston makes its way into the condenser. When the piston has risen to the top of the cylinder, the valves

B and C are closed, and A and D are opened, and the down-stroke is made as before.

If the engine is to be worked on the 'expansive' principle (p. 381) all that is required is to arrange the pins on the 'plug-frame' (p. 381), so that the steam-valve is closed at any given part of the stroke, the exhaust-valve connected with the other end of the cylinder remaining open until the end of the stroke.

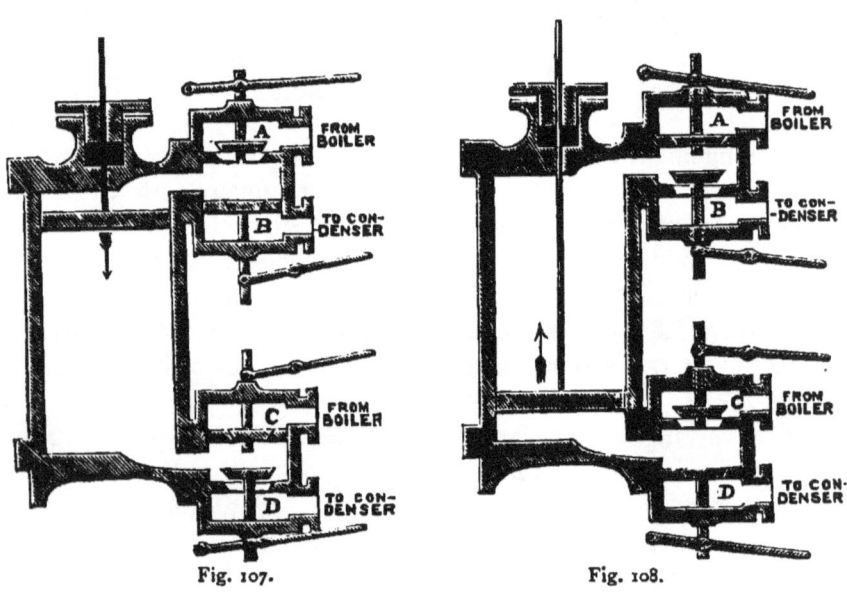

Fig. 107.   Fig. 108.

The adoption of the 'double-acting' principle rendered it necessary that the piston-rod should be connected with the beam by some other means than the flexible chain hitherto employed. This latter could transmit a *pull* but not a *push* to the beam; and in order that the beam might be forced up as well as down by the steam, the piston-rod must be connected with it by some such joint as that which connects the piston-rod of a pump with the handle. If, however, the rod was merely attached to the beam by a hinge-like joint, the end of the rod must follow the circular arc described by the end of the beam, and be swayed

WATT'S PARALLEL MOTION. 385

from side to side in the course of its up and down movement. Such a lateral motion must infallibly make the rod work loose in the stuffing-box and cause a great leakage of steam; and the problem which Watt set himself to solve was how to connect the end of the beam with the piston-rod so that the former might move in a circle while the latter moved up and down in the same straight line throughout. He effected this object by an arrangement of rods still known by the name of 'Watt's Parallel Motion,' one of the neatest and most ingenious of his many inventions.

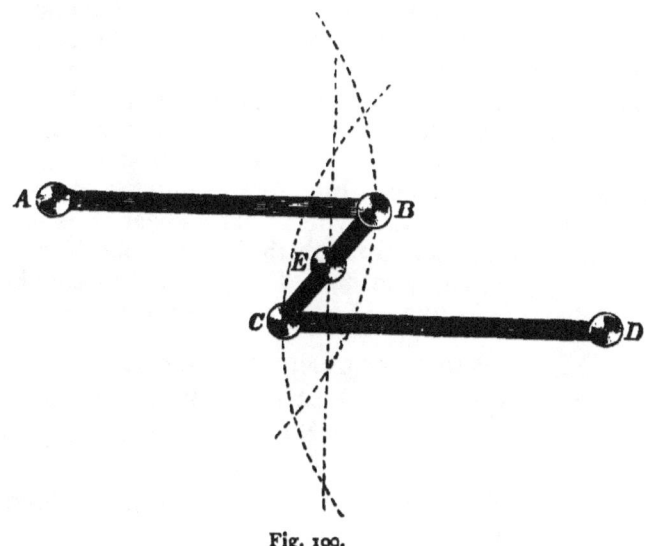

Fig. 109.

The principle of it will be seen by reference to fig. 109, in which AB and CD are two rods of equal length, pivoted on centres at A and D, and having their other ends connected by a short rod or 'link,' BC. If this link is moved up or down, its ends B and C will move in the same circular arcs as the ends of the rods to which it is jointed. Thus, starting from the position shown in the diagram, if it is raised, the end B will be pulled to the left hand and the end C to the right hand, while the middle point, E, of the rod will not be deviated at all to the right or left, but will move up in a straight line.[1] A similar result will occur if the link is moved downwards; the equal and opposite deviations of the ends B and C just compensating each other in the effect on the direction of the motion of the point E, so that the latter moves in the straight dotted line.

Expt. 122. It will be worth while to verify the above statement by making a simple model of the parallel motion. Cut out two strips of thick cardboard, 1 cm. broad and 20 cm. long, and also another strip of the same breadth and 10 cm. long. Cut a hole in the centre of the short strip large enough to admit the point of a pencil: then attach a long strip to one end of it by laying down a flat-headed drawing-pin or nail with its point upwards and pressing the ends of the strips down upon it until the pin has passed through both and jointed them together. Attach the other long strip to the other end of the link in a similar manner, and lay them down on a smooth board in the position shown in fig. 109. Drive drawing-pins (or nails) through the outer ends A and D into the board, to serve as pivots. Put a sheet of paper under the link, and pass a pencil through the centre hole. On moving the link up and down by means of the pencil, loosely held, an approximately straight line will be drawn on the paper.

If, now, AB is taken to represent half the engine-beam, and CD is a rod pivoted to some part of the framework or the wall of the engine-house, and if the end of the piston-rod is jointed to the link at E, it will move up and down in the required straight line, and communicate its motion to the beam without being swayed or deviated at all by the latter.

The above form of parallel motion would be rather inconvenient in practice on account of the great length of the compensating-rod CD, which must be half as long as the beam, and must not much overlap the latter, thus necessitating an extension of the engine-house. Watt therefore preferred to attach the link, not to the end of the beam, but to a point midway between the

---

[1] The line is not, mathematically speaking, a truly straight one, but it is very nearly so when the arcs moved through are of moderate extent, as they always are in practice.

WATT'S PARALLEL MOTION.

end and the centre of the latter; the compensating-rod need then only be one-fourth of the length of the beam, and its outer end need not project beyond the beam at all. Fig. 110 will explain this arrangement, and will also show how the piston-rod is attached. A link, FG, of the same length as BC, is jointed to the end of the beam, and the lower ends, C and G, of the two links are connected by a rod of the same length as AB or BF— *i.e.* one-fourth the length of the beam. The piston-rod is also attached to the end G, and it can be proved that this point in the movable parallelogram BCGF moves as truly up and down in a

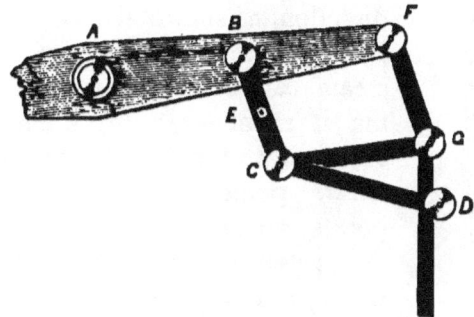

Fig. 110.

straight line as the centre-point E of the original link. In practice there are two identical sets of rods, one on each side of the beam and piston-rod, for the sake of rigidity and strength.

By this arrangement the beam was impelled both up and down with equal force by the steam; and in order to convert this movement into one of continuous rotation, Watt employed a contrivance which has been almost universally used since his time, *viz.*

THE CRANK, CONNECTING-ROD, AND FLY-WHEEL.

The crank, AB, fig. 111, is simply a handle somewhat like that of a windlass, attached to the end of the shaft which is to be turned round, or else formed by bending the shaft itself into the form shown at A' B' in the engraving. This is connected with the outer end of the beam by a long, rigid rod called the 'connecting-rod'; and as the beam moves up and down the crank is turned round much in the same way as a windlass-handle is moved by the hand and arm of a man.

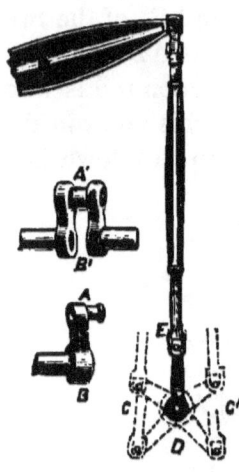

Fig. 111.

It is, evident, however, that the force exerted upon the crank must vary very much in different positions, being considerable when the crank is nearly at right-angles to the length of the connecting-rod, as shown at CD, C'D, but diminishing to nothing when the crank is parallel to this direction, as shown at ED. In fact, at two points in the revolution of the crank the steam can exert no useful effect in turning it round—*viz.* (*a*) when the beam is at its *highest* point, (*b*) when it is at its *lowest* point. Moreover, at these two points, which are called the 'dead points,' the piston has just reached one end of the cylinder, its direction of motion has to be changed, and the inertia of it and the beam has to be overcome. To prevent the irregularity of movement, or even stoppage, which would thus occur, a large wheel with a heavy rim is attached to the shaft near the crank. This wheel, which is called the 'fly-wheel,' when once put in motion by the expenditure of some of the energy of the steam, stores up this energy in a statical form;[1] and when the force exerted on the crank becomes less or ceases, as at the 'dead points,' the energy accumulated in the massive fly-wheel is given out as kinetic energy, and carries the shaft on until the crank is once more in a position to move it. This expedient is found to answer its purpose perfectly; if the fly-wheel is heavy enough, no perceptible variation in velocity occurs during any part of the revolution of the crank.

---

[1] A text-book on Dynamics must be consulted for a full explanation of this point. It is alluded to, however, on p. 151.

THE SLIDE-VALVE.

Not long after the above improvements had been made in the engine, it was found possible to substitute for the four valves shown in figs. 107, 108, p. 384, a single valve of a different and simpler character. This 'slide-valve,' as it is called, consists of a shallow rectangular metal box with broad, flat edges, adapted to slide over the surface of a flat slab of metal cast on, or screwed to, the side of the cylinder. In this block of metal there are three passages, or 'ports,' as shown in fig. 112; the middle one, B, communicating with the condenser, while the two others, A and C, communicate with the upper and lower

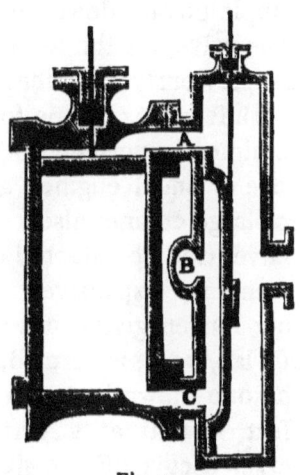
Fig. 112.

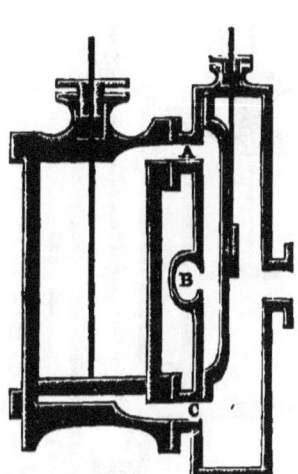
Fig. 113.

ends of the cylinder, respectively. All three of these 'ports' open upon the flat surface of the metal block, and the slide-valve is of such a length that when it is in its lowest position, fig. 112, it covers the two lower ports but leaves the upper one, A, clear. Similarly, when it is in its highest position, as in fig. 113, it covers A and B, but leaves C clear. The whole is enclosed in a steam-chest, as shown in the engraving, at one end of which a rod, passing through a stuffing-box, is attached to the valve in order to move it up and down.

The Slide-Valve.

The mode of action of the valve is as follows:—When the piston is at the top of the cylinder, and steam has to be admitted to force it down, the valve is moved to its lowest position, fig. 112, thus leaving the port A open for the steam to enter above the piston. The steam below the piston passes out of C, through the hollow of the valve into B, and so escapes to the condenser, and the down-stroke is made. When the piston approaches the bottom of the cylinder, the valve is moved up to its highest position, as shown in fig. 113; thus the port C is now open for the steam to enter below the piston, while the steam already in the cylinder passes through A and B into the condenser, and the piston is forced up.

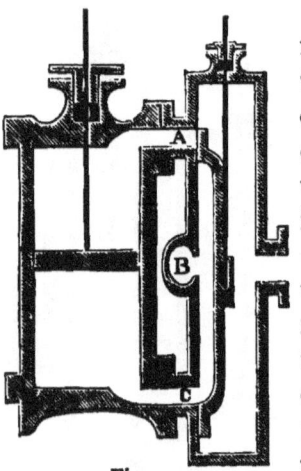

Fig. 114.

Thus by a simple up and down movement of this one valve, all the work of the four valves previously described is done; and the slide-valve, in one form or another, has almost superseded other valves in the case of small engines, and is often used for large engines also.

This slide-valve can be adapted for working the engine expansively by making it rather longer, giving it a certain amount of 'lap,' as it is termed, or overlap, like the one shown in figs. 112, 113. The effect is that at a certain point in its travel it cuts off the steam from one port, while it still leaves the other port open to the condenser. This intermediate position is shown in fig. 114.

The usual mode of working the slide-valve is by connecting it, not with a vertical plug frame, but with a contrivance called an 'eccentric,' which is shown in fig. 115. It consists of a circular disc or wheel, A, with a groove in its rim, in which a metal ring, BB, is fitted, but not quite tightly, so that the disc can turn round within it. To one part of the ring a rod, CD, is fixed, the outer end of which is either jointed directly to the slide-valve rod or connected with it by some such means as the bent lever, EF. The disc has a hole, G, bored through it at some little distance from its centre, and not at the centre itself (hence the name 'eccentric'). Through this hole the crank-shaft of the engine passes, and the eccentric is fixed firmly to the

Fig. 115.

shaft so as to turn round with it. It is easy to see that, as the shaft rotates, the centre (A) of the disc will revolve round the centre of the shaft, G, and the ring BB, with the rod attached to it, will be carried alternately to the right hand and to the left in the course of a complete turn (just as if it was attached to a crank), through a distance equal to twice the distance between the centre of the disc and that of the shaft. Thus the slide-valve receives the movement required to open and close the steam ports at the proper times.

It is usual to fix the eccentric on the crank-shaft in such a position that the exhaust is cut off and a little steam admitted to that end of the cylinder just before the piston arrives at the end of the stroke. This steam serves as an elastic cushion which checks the motion of the piston, beam, etc., gradually, and so lessens the shock which would otherwise occur from the inertia of the heavy mass, if its motion was suddenly stopped and reversed. When the slide-valve is thus made to move a little in advance of its theoretical position, it is said to have a 'lead' given to it.

WATT'S COMPLETE ENGINE.

Sufficient has now been said to render intelligible the representation of a complete double-acting condensing engine given in fig. 116.

The principal parts will be recognised at once :—The cylinder,

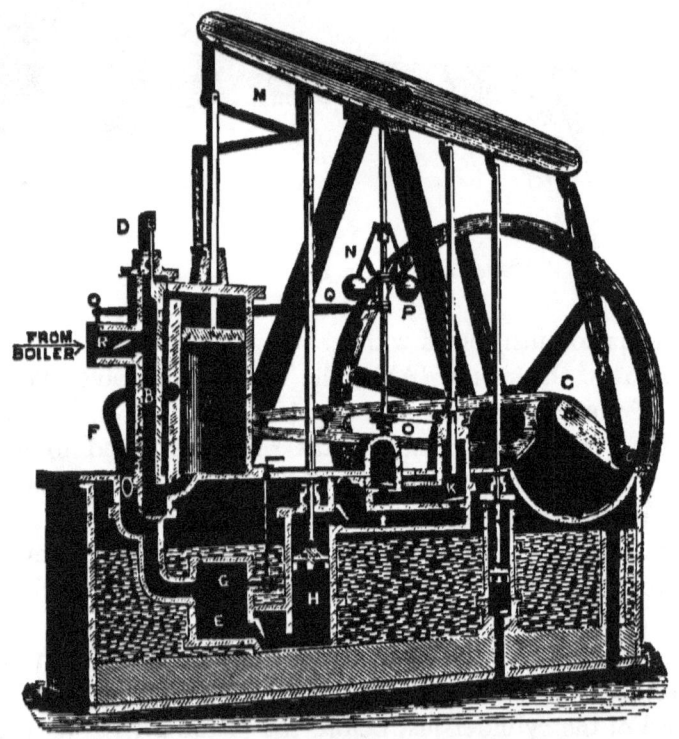

Fig. 116.

A, with piston in the course of the up-stroke; the slide-valve, B, admitting steam to the bottom of the cylinder and carrying off steam from the top of the cylinder to the condenser; the eccentric, C, moving the slide-valve by a bent lever jointed to a vertical rod connected with the slide-valve rod at D; the condenser, E, communicating with the exhaust-port by a pipe shown at F; the injection jet, G; the air-pump, H, which discharges water from the condenser into a cistern, I, from which another pump, K, takes it to the boiler. Another pump, L, supplies cold water to the cistern in which the condenser is placed. The parallel motion, or rather, one-half of it, is seen at M, and the connecting-rod, crank, and fly-wheel are shown on the right hand of the engraving.

There remains, however, one part of the apparatus which has not yet been described, and which performs the important duty of regulating the speed of the engine automatically, so that whether the work which the engine has to do is light or heavy, one uniform rate of motion is secured. This is shown at N, and is called the 'centrifugal governor.' It consists of two heavy iron balls attached to rods which are jointed to a vertical spindle, so that when the latter is turned round by a cord passing from a pulley on the main shaft to the pulley O, near the bottom of the spindle, the balls revolve. If their speed exceeds a certain amount, the centrifugal tendency causes them to fly outwards from the spindle, and when they do this they raise a collar P, which slides on the spindle and is connected by links with the rods which carry the balls, as shown in the engraving. In a groove cut in this collar works the forked end of a rod Q, the other end of which is connected with a valve R (called a 'throttle-valve') placed in the steam-pipe leading from the boiler to the cylinder. This valve consists of a circular plate which, when placed parallel to the length of the pipe, allows steam to pass freely, but when turned on its axis so as to lie at right angles to the pipe, obstructs the passage of steam like a closed door. The rod Q is so connected with this valve that when the governor-balls fly outwards to their fullest extent, the valve lies nearly across the pipe and hardly any steam can pass by it to the cylinder: while, when the balls are moving slowly and drop, from their weight, close to the spindle, the valve Q presents its edge to the current of steam and allows an abundant supply to pass.

The velocity with which the balls are made to revolve by the engine is so adjusted by attending to the sizes of the pulleys which work the spindle, that they keep at a moderate distance from the spindle when the engine is working at the proper speed. If, from any cause, such as increase of steam-pressure in the boiler or diminution in the work which the engine has to do (for instance, when some of the machines connected with it are thrown out of gear), the speed is increased, the balls revolve more rapidly and their centrifugal tendency makes them fly out farther from the centre, with the result of raising the collar and closing

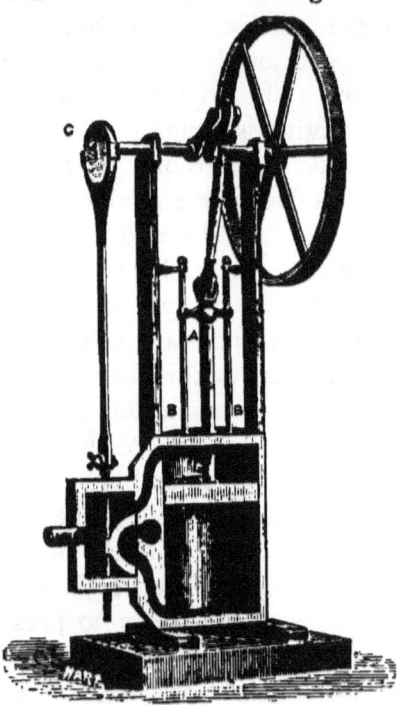

Fig. 117.

the throttle-valve more or less. Thus the supply of steam is diminished and the speed of the engine reduced. If, on the contrary, the engine works slower than it should do, owing to increase of resistance or decreased pressure of steam, the balls drop closer to the spindle; the throttle-valve is thus opened more widely, and more steam is supplied until the engine regains its proper speed.

The Vertical Engine. 395

A remarkably perfect uniformity of speed under varying conditions is obtained by this centrifugal governor; and it has, in one form or another, been adopted in most modern engines, especially in those employed in working dynamo-electric machines for producing the electric light, where an unvarying rate of motion is absolutely essential.

The above is a fair representation of the steam-engine as it left the hands of Watt and was manufactured at the great works of Messrs. Boulton and Watt at Soho, near Birmingham. In this form it was, and still is, extensively used in cotton-mills, flour-mills, iron-works, etc.

Fig. 118.

It remains now briefly to indicate the successive simplifications which have been effected in it since Watt's time, in order to adapt it more conveniently for the innumerable kinds of work which have been found for it to do.

In the first place the great beam, which, after all, was only a survival of the pump-handle, was got rid of, and the connecting-rod was jointed direct to the piston-rod, as shown in fig. 117; the latter being guided straight up and down by having its upper end attached to a transverse bar, or 'cross-head,' A, which slides on two guide-bars, B, B, firmly bolted to the framework of the engine, and adjusted accurately parallel to each other and the piston-rod. The slide-valve is worked by an eccentric, C, fixed on the main crank-shaft, the eccentric-rod being jointed to the valve-rod, without the necessity for any intervening bent levers.

This type of engine is a very common one at the present day. We see it not only in the enormous engines used in ocean steamers but also in locomotives, as well as in the small engines used in workshops for turning lathes and other machinery. The cylinder is sometimes fixed below the crank-shaft, as shown in fig. 117: sometimes above it, as in the usual form of engines made for working the screw in steamships: sometimes on a level with it, as in locomotives, and in the engine shown in fig. 118, which represents a very compact, skilfully designed form of 'horizontal engine,' made by Messrs. Tangye, of Birmingham.

Oscillating Engines.

It has even been found possible to do away with the connecting-rod itself, and attach the piston-rod direct to the crank. In order to allow the piston-rod to follow the movements of the crank, the cylinder is supported on axles or 'trunnions,' much in the same way as a cannon, so that it can rock from side to side as the crank revolves. The trunnions are hollow, and steam is introduced from the boiler through one of them, while the spent steam passes to the condenser through the other.

This type of engine is called an 'oscillating engine,' and is extensively used in paddle-wheel steamers, in the manner shown in fig. 127, p. 407.

The Link Reversing Motion.

It is of great importance in many cases, especially in steamships and locomotives, to be able to reverse the direction of movement of the machinery readily and quickly. This is usually effected by the contrivance called the 'link-motion,' which is

The Link Reversing Motion.

shown in figs. 119 and 120. Two eccentrics, A and B, are fixed close together on the crank-shaft in such positions that the centres of their discs are on opposite sides of the shaft. Thus, when one is in its highest position and would raise the slide-valve so as to admit steam below the piston, the other is in its lowest position and would depress the valve so far as to admit steam above the piston. Obviously the direction of movement of the piston, and therefore of the crank and the machinery which it

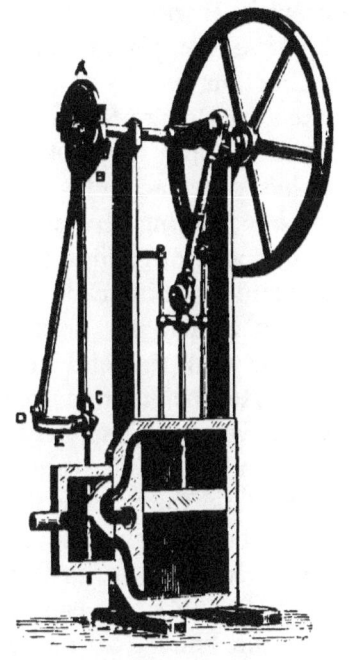

Fig. 119.

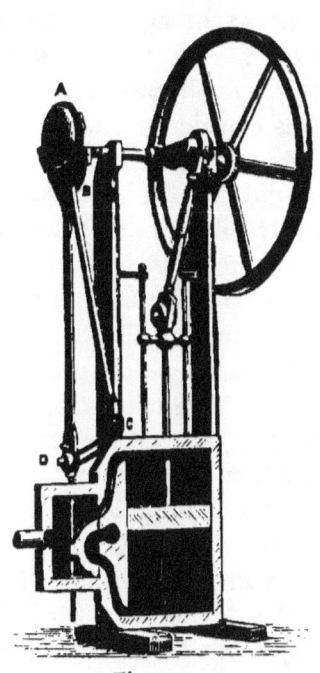

Fig. 120.

turns, depends upon which of the two eccentrics is used to work the slide-valve; and all that is wanted is some ready means of allowing *either* A or B alone to act on the valve. This is effected by connecting the eccentric rods by a short cross-bar or 'link,' CD, which has a long slit cut in it, within which a pin fixed at right angles to the valve-rod can move. Then either eccentric can be made to work the valve by simply sliding the link so that the pin is brought close to one or the other eccentric rod.

If, for instance, as in fig. 119, the piston being in the middle of the cylinder, the link is so placed that the end of the valve-rod is near C, then the eccentric A alone will be working the valve, steam will be admitted under the piston, and the latter will be forced up and will move the machinery in one direction. If, on the contrary, the link is moved so as to bring the valve-rod pin near D, as in fig. 120, then the eccentric B will be acting on the valve, steam will be admitted above the piston and the machinery will move in the opposite direction.

This contrivance, which was invented by George Stephenson, the engineer who did almost as much for the locomotive as Watt did for the stationary engine, has other advantages. It will be easily seen that, when the engine is working, the ends of the link are moved simultaneously in opposite directions by the action of the two eccentrics, while the middle point E, fig. 119, will not move at all. If the valve-rod pin is held in the slit of the link at a point between the middle and the end of the latter, the eccentric will impart a less extent of motion to the slide-valve, so that the steam-ports will never be completely uncovered, and less steam will enter the cylinder. If the valve-rod pin is just at the middle point, E, of the link, the valve will not move at all, though the crank-shaft may be turning round. It will remain at the centre of its travel (nearly as in fig. 114), and both the steam-ports will be covered, so that no steam can enter the cylinder, though the passage from the boiler to the valve-chest may be free. Thus the amount of steam supplied to the engine can be regulated by simply altering the position of the pin on the link, quite independently of any throttle-valve in the steam pipe. Of course a system of levers (not shown in the engraving) is provided, by which the link can be placed and held in any required position.

HIGH-PRESSURE ENGINES.

The introduction of better modes of manufacture and stronger materials has led to the abandonment of the condenser in many engines. In Watt's time the pressure of steam in the boiler never greatly exceeded the pressure of the atmosphere (the latter being about 1 kilog. per square centimetre), power being obtained by keeping up a vacuum on the other side of the piston. These engines had at any rate the merit of safety, a destructive explosion of the boiler being almost unheard of. But it is now found advantageous and economical to use steam of much higher tension, making it move the piston against the resistance of the air, which is freely admitted on the other side of the piston; while at the end of each stroke the steam is simply allowed to escape into the open air. Thus the effective pressure on the piston is the difference between the pressure of the atmosphere at the time and that of the steam. There is now no difficulty in making engines and boilers which will stand pressures of 12 or 14 kilogs. per square cm. (160-200 lbs. per square inch) with safety; and a common working pressure, even when condensers are used, is 3 or 4 kilogs. per square cm. (40-60 lbs. per square inch) above that of the atmosphere. Engines, then, should no longer be divided into two classes of 'low-pressure' and 'high-pressure,' but of 'condensing' and 'non-condensing' engines.

COMPOUND ENGINES.

When steam is supplied from the boiler at the high pressures mentioned above, it is found best to make it do work by instalments. That is, instead of applying it to one piston and then letting it escape into the air or a condenser, while it still has a pressure quite sufficient to enable it to do more work, it is led into another cylinder of larger diameter and made to move the piston in that cylinder. Even when it has done this, its pressure may be such that more useful work can be got out of it by making it enter a third and still larger cylinder and move a piston of increased area.

Of course in 'compound engines,' as those arranged on the above system are called, there is a great deal of back pressure (p. 377) in each cylinder; but the total mechanical energy obtained from a given amount of steam is much greater than in those of the simpler form.

Boilers.

The construction and proper arrangement of these is of at least as great importance as that of the engines to which they are attached. If the boiler and furnace are so badly designed that most of the heat of the burning fuel goes up the chimney instead of into the water, no amount of excellence in the engine itself can make it an efficient and economical source of power.

Fig. 121.

One of the best modern forms of boiler is that known as the 'Cornish boiler,' which is represented in figs. 121 (a longitudinal section), 122 (a cross section), and 123 (an end view). It will be seen that the furnace is inside the boiler, wholly (except at the fire-door) surrounded by water. A large tube runs from end to end of the boiler. At one end of it is the furnace, A, while the other end opens into a cross-flue, B, which leads the flame and smoke into side-flues, C, C (fig. 122), running along the boiler (below the water-level) to the front, where they both communicate with a large flue, D, which goes underneath the boiler to the chimney. Thus the heated gases from the fuel pass first through the boiler, then along each side of it, and lastly under it; so that every chance is given them of giving up their heat to the water before reaching the chimney. Some conical tubes, EEE, fig. 121 (called 'Galloway' tubes) open to

Fig. 122.

Fig. 123.

the boiler at both ends, are often placed obliquely across the main flue: the water in these gets rapidly heated, and circulation is greatly promoted by them.

The steam rises into a dome, F, where it finds time to deposit any spray which may have risen with it from the water: it then passes into the main steam-pipe leading to the engine.

The safety-valve is shown at G, fig. 121, being pressed down by a weight at the end of a long lever, as in Papin's original invention. Other necessary fittings are shown in fig. 123, *viz.* a pressure-gauge, H, for indicating the tension of the steam; two gauge-cocks, J, for ascertaining the level of the water in the boiler; and also another gauge, K, for the same purpose. This latter consists of a glass tube communicating at both ends with the boiler. The water in this will, on a well-known hydrostatic principle, stand at the same level as that in the boiler.

THE LOCOMOTIVE ENGINE.[1]

This is perhaps the most striking and admirable form of a heat-engine, not from its actual size, but from the skill with which all its parts are designed and adapted to the special purpose it has to fulfil, that of drawing heavy loads after it at a rate of from 80 to 100 kilometres per hour.

The special requirements in a locomotive engine are—

1. It must be light, yet strongly built to resist great strains; and very powerful, yet of comparatively small bulk.

2. It must be able to produce a large quantity of steam very quickly, and to keep up the supply for many hours.

3. It must be under perfect control, whether running quickly or slowly, forwards or backwards.

The general plan of a modern locomotive engine is of the following kind:—Two high-pressure, non-condensing horizontal engines are placed side by side in a frame running on wheels. Their connecting-rods work cranks placed at right angles to each other on the same shaft, which is also the axle of one pair of wheels. Immediately over the machinery is supported the boiler, the furnace being at one end, and a 'smoke-box,' surmounted by a short chimney, at the other.

The principal details of the engine will be gathered from the section given in fig. 124, of one of the large broad-gauge engines used on the Great Western Railway; a general view of the same engine is given in the frontispiece.

---

[1] For the history of the locomotive engine Smiles' *Life of Stephenson* or Galloway's *History of the Steam Engine* should be referred to.

One of the cylinders is shown at A, with piston-rod moving between guide-bars. There is one steam-chest for both cylinders, lying between them; the eccentrics and link are seen at B, and the position of the pin of the slide-valve rod in the link is controlled by the lever seen in the frontispiece by the side of the

Fig. 124.

fire-box. The two cranks are placed at right angles to each other on the shaft, as shown on a larger scale in fig. 125, the great advantage of this arrangement being that uniformity of motion is obtained without the use of a fly-wheel, since when one

Fig. 125.

crank is at a dead-point, the other is in the most advantageous position for being acted on by the force of the steam. Thus the engine can always be started without difficulty, in whatever position the crank-shaft may happen to be.

The fire-box, C, generally made of copper (on account of the high conductivity of the metal), is entirely surrounded by water (except at the bottom); and the flame is conveyed from it to the chimney by a very large number of brass tubes about 4 cm. in diameter, passing through the boiler below the water level. There are usually about 200 of these tubes, and it is mainly owing to the enormous heating surface thus obtained that evaporation goes on with sufficient rapidity to supply the engine with steam when it is travelling at full speed. . A boiler thus fitted with tubes is called a 'tubular' boiler, and is used in many other engines besides locomotives. In order to drive the flame through these narrow tubes some special means must be used; the chimney, from its shortness, is of no practical use in creating a draught. A simple and perfectly effective expedient was devised by Stephenson, *viz.* to direct the steam escaping from the cylinders through a slightly contracted pipe, D, straight up the centre of the chimney. The velocity with which this steam rushes up the chimney, dragging air and smoke with it, creates a partial vacuum in the smoke-box, and the pressure of the external air, admitted through the grate-bars, forces the flame through the tubes with great rapidity. It is evident that this 'steam-blast,' as it is called, will increase in intensity with the speed at which the engine is working; and thus the greater the demand for steam becomes, the more intense is the combustion of the fuel, and the quicker is the evaporation. The glare of a well-stoked fire in an engine travelling at high speed can hardly be borne by the eyes.

The steam is taken from the highest part of the boiler (or, in many engines, from a dome like that shown in fig. 121) by a pipe in which a valve is placed, worked by a lever E at the fire-box end of the boiler. This pipe passes into the smoke-box and then turns downwards to the steam-chest between the cylinders.

It may be interesting to give here some of the actual dimensions (in English measure) of the engine represented in the figure, which belongs to a class employed for working the fastest broad-gauge express trains on the Great Western Railway:—

Total length of engine	26 ft.	Number of tubes in boiler	305
,, breadth ,,	9 ft.	Total heating surface	1952 sq. ft.
Height to top of chimney	15 ft.	Diameter of driving wheels	8 ft.
Length of boiler, exclusive of fire-box	10 ft. 6 in.	Length of stroke	24 in.
Diameter of do.	4 ft. 9 in.	Diameter of piston	18 in.
		Weight	40 tons.

Such an engine will take a train of carriages weighing 150 tons at a steady rate of 60 miles (97 kilometres) per hour on a level line, exerting in doing so about 1000 horse-power.

Marine Engines.

All war-steamers, and most merchant and passenger steamers, are now propelled by screws at the stern, the shaft of which runs near, and parallel to, the keel of the vessel. The usual type of engine employed to work this shaft resembles that shown in fig. 117, p. 394, but placed in an inverted position, the cylinders, of which there are two, being fixed directly over the screw-shaft, and turning the latter round by two cranks placed at right angles to each other, as in a locomotive. Fig. 126 shows a pair of engines of this class, with link reversing-gear.

The 'compound system' (p. 399) is nearly always adopted in these engines, the steam being passed on from one of the cylinders to the other (and sometimes to a third), which is of much larger diameter. From the latter it is led into a condenser composed of a large number of tubes immersed in a cistern through which cold water is constantly made to circulate by means of pumps. Such a condenser is called a 'surface condenser,' and has the advantage of not necessitating the mixture of a quantity of impure water, such as sea-water, with the condensed steam.

MARINE ENGINES.

In steamers propelled by paddle-wheels, which are seldom now made of large size (except in America), oscillating engines are very commonly used, working cranks placed at right angles to each other on the axle of the paddle-wheels, as shown in fig. 127.

Marine engines are now made of enormous size, corresponding to the gigantic vessels which they have to propel. Thus in the

Fig. 126.

steamers which cross the Atlantic from Liverpool to New York, the high-pressure cylinders are about 140 cm. (56 in.) in diameter, and the low-pressure cylinders are 200 cm. (80 in.) in diameter, the length of stroke being comparatively short, *viz.* 150 cm. (5 ft.). In war-vessels, such as H.M.S. *Temeraire*, engines of even larger size are used, with high-pressure cylinders 180 cm. (70 in.), and low-pressure cylinders nearly 300 cm. (120 in.) in diameter, giving a horse-power of more than 7000.

Fig. 127.

Section III.—Hot-Air Engines.

In these the working fluid is air instead of steam, and no separate heating vessel is used, the air being heated in the cylinder itself by a furnace below it. When this heated air has done the work of raising the piston it is allowed to pass into a 'cooler,' from which it is again conducted into the cylinder, to be heated as before.

Fig. 128 represents one of the modern forms of hot-air engine, and fig. 129 is a sectional view of the same engine.

It will be seen that there are two cylinders, one, A, surrounded with a casing or 'jacket' containing cold water, and the other, B, having its lower part projecting into a furnace. Both these cylinders contain plunger-pistons, C and D, which are slightly smaller than the cylinder, and work air-tight through stuffing-boxes, K, K. Connecting-rods attached to these pistons work cranks at right angles to each other on a shaft which also carries a heavy fly-wheel. Between the two cylinders there is a wide passage, H, containing a number of discs of wire-gauze.

The action of the engine is as follows. By turning the shaft by hand, so as to make the piston C descend, a certain amount of air is transferred through H to the cylinder B. This air, be-

Fig. 128.

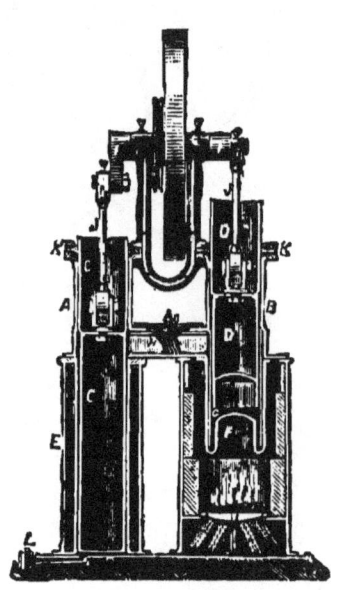

Fig. 129.

coming heated by the furnace, drives up the piston D. As it comes near the top of the cylinder, the piston C begins to rise (being pulled up by the crank), and a space being thus left empty in the cylinder A, the heated air in B passes, as the piston D falls, through the wire-gauze in H. Here it parts with most of its heat, owing to the large surface and high conductivity of the gauze, and it is still more effectively cooled when it enters A by the cold water which surrounds it. When, from the momentum of the fly-wheel, C moves downwards, the air in A is transferred back through H, where it takes up again a good deal of the heat which it had previously left there, and then passes into B, where it is further heated by the furnace, and drives up the piston as before. Thus the same portion of air is used over and over again, being alternately heated in one cylinder, where some of the heat is converted into mechanical energy, and cooled in the other, so as to render it capable of taking in more heat, when re-transferred to the first cylinder, and of doing more work.[1]

This engine is evidently only a single-acting one, the downstroke of the piston D being effected at the expense of the energy stored up in the massive fly-wheel.

Hot-air engines have hitherto only been made of small size, not more than one horse-power or so, and their 'efficiency' is not very high. Air cannot (like steam) be made, by a slight lowering of its temperature, to occupy almost no volume, and exert almost no pressure, and it can only have its tension sufficiently raised by the application of a very high temperature, in which case there must of necessity be a great deal of heat lost by radiation, etc., and a rapid destruction of the metal of which the heater is constructed.

Section IV.—Gas-Engines.

In these the principle of a heat-engine is applied in the simplest possible way. There is no boiler, furnace, or condenser, the necessary heat being obtained by the chemical combination of two gases, usually coal-gas and oxygen (from air), in the working cylinder itself. This heat supplies the requisite energy to the products of combustion, chiefly steam, carbon dioxide, and nitrogen, and the piston is driven onward in the cylinder, and works a crank in the usual way.

One of the best gas-engines hitherto invented is the 'Otto'

---

[1] A provision is made by the valve L, fig. 129, for making up for any loss of air caused by leakage.

engine as improved and manufactured by Messrs. Crossley & Co. of Manchester. This is shown in fig. 130.

In general appearance it much resembles an ordinary steam-engine; it is, however, single-acting only (as most gas-engines are), the return stroke being made at the expense of the energy stored up in the fly-wheel. The latter is made unusually large and massive, in order to prevent the irregularity of motion which would otherwise be caused by the great variations in the motive force.

Fig. 130.

The slide-valve is placed at one end of the cylinder (the left-hand end in the engraving), and covers and uncovers passages by which coal-gas and air are admitted to the cylinder in proper proportions (about one volume of gas to seven of air); and the products of combustion are allowed to escape into the air when they have done their work. There is also a small opening in the slide-valve which admits for a moment a jet of flame from a small burner kept constantly lighted, for the purpose of beginning the combination of the gas and air.

The cylinder, surrounded by a hollow casing or 'jacket,'

through which cold water is constantly flowing (to keep down the temperature), is made about one-fourth longer than is required for the movement of the piston, so that a space or chamber is left between the piston and the end of the cylinder where the slide-valve is fixed; and in this chamber the mixture of gas and air collects before it is exploded. Moreover it is found to be a decided advantage to compress the mixture to a certain extent before exploding it; and the mode in which this is done will be seen from the following general description of the working of the engine.

At least four distinct operations take place in succession while the engine is in motion.

(i) The piston being at its nearest point to the slide-valve end of the cylinder, it is pulled forward (by hand, when the engine is to be started; in succeeding strokes, by the momentum of the fly-wheel), and a mixture of gas and air is admitted into the cylinder as the piston leaves room for it.

(ii) The slide-valve closes the entrance-port; and the piston, in making the return stroke, compresses the explosive mixture into the chamber at the closed end of the cylinder.

(iii) The mixture is now exploded by the momentary admission of a gas-flame, and the high temperature (at least 1600°) produced by the chemical combination increases the tension of the gases enormously, so that the piston is driven forward.

(iv) At the end of the forward stroke the gases are allowed to escape through the valve into the air; and the greater part of the products of combustion are expelled as the momentum of the fly-wheel enables the piston to make the return stroke.

This completes the series or 'cycle' of operations; and when the piston moves forward again, a fresh mixture of gas and air is admitted into the cylinder as in (i), and the cycle begins again. The use of the heavy fly-wheel will now be evident, since the piston only gets one impulse in four strokes from the explosion of the mixture of gas and air.

The 'efficiency' of gas-engines is very fair indeed—decidedly higher than that of steam-engines. Experiments made with a six horse-power engine of the type described above, consuming about 7000 litres (250 cubic feet) of gas per hour, have shown it to have an efficiency of 0·17: that is, $\frac{17}{100}$, or nearly $\frac{1}{6}$ of the heat produced is converted into mechanical work. But they are more expensive in working than steam-engines, since coal-gas is necessarily much dearer than an equivalent amount of coal as a fuel.

Section V.—The Power and 'Duty' of Heat-Engines.

The power of an engine is most scientifically expressed by the number of ergs of work it will do in a given time, *e.g.* in one second or one minute; but as the earliest engines were intended to replace horses in working pumps, a custom arose of expressing their power by stating how many horses they would do the work of; so an engine was said to be of so many 'horse-power.'

From experiments made by Watt on the subject it was estimated that a good horse could raise 33,000 lbs. 1 foot high during each minute of its work; and an engine which could do the same was rated as being of 'one horse-power.' This amount of work is equivalent, on the C.G.S. system, to about 7,600,000 centimetre-grammes of work per second, or to 7460 million ergs per second in absolute measure.

The horse-power of an engine depends mainly on three things—

1. The effective pressure of steam in the cylinder.
2. The speed at which the piston works.
3. The area of the piston.

In estimating the horse-power of his engines, Watt did not actually test each individual engine under working conditions. He assumed a certain average pressure of steam, differing according as the engine was condensing or non-condensing, and a certain average speed of the piston, *viz.* about 200 feet (60 metres) per minute;[1] so that the reputed or 'nominal' horse-

---

[1] Thus in an engine having a length of stroke of 1 metre, this would imply a working rate of 60 strokes per minute.

power of an engine depends only upon the third of the above factors, *viz.* the area of the piston.

For example, in condensing engines the effective pressure on the piston was assumed to be 7 lbs. per sq. inch (0·5 kilog. per sq. cm.), and a 1-horse-power engine was defined to be one which had a cylinder 7 in. (18 cm.) in diameter, and a length of stroke of 1 foot (30 cm.).

In non-condensing engines a much higher effective pressure of steam was assumed, *viz.* 21 lbs. per sq. inch (1·5 kilogs. per sq. cm.), so that a proportionately smaller diameter of the cylinder would be required to give the same power. A 1-horse-power non-condensing engine was defined to be one which had a cylinder 4 in. (10 cm.) in diameter, and a length of stroke of 1 foot (30 cm.).

This mode of expression is still retained for convenience in designating roughly the size of an engine for commercial purposes; but obviously the real power of an engine can only be found after determining the exact pressure of steam in the cylinder throughout the stroke, while the engine is actually at work. To do this Watt invented a very beautiful and simple apparatus called an 'Indicator,' which not only shows but yields a permanent record of the varying pressures of the steam at different parts of the stroke. The arrangement of Watt's Indicator (in its original and simplest form) is shown in fig. 131.

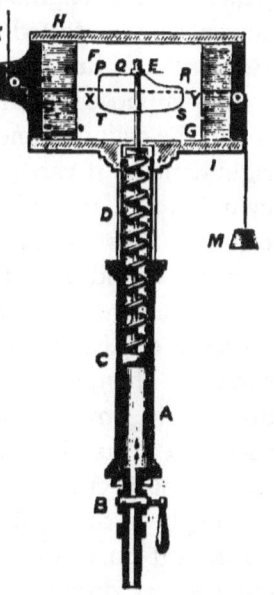

Fig. 131.

It consists of a small cylinder, A, about 3 cm. in diameter, having at its lower end a stopcock, B, terminating in a screw by which it is attached to either end of the engine-cylinder. A free passage can be thus established between the two cylinders, so that the pressure of steam is always the same in both. A piston, C, works smoothly but steam-tight in the cylinder A, and has attached to it a spiral spring, D, so adjusted that the piston, when the pressure on both sides of it is equal, is maintained a little below the middle of the cylinder. The piston-rod, after passing through a hole in a guide-piece (to which the upper end of the spring is attached) terminates in a pencil-holder E placed at right angles to the plane of the paper. Underneath the point of this pencil a light wooden drawing-board, F G, on which a piece of paper is pinned, slides horizontally backwards and forwards in grooves cut in a frame, H I. Motion towards the left hand is communicated to it by a cord, K, attached to the beam of the engine; while the weight M drags it to the right hand when the cord, K, is slackened. Thus if the indicator-piston is stationary and the pressure on each side of it is that of the atmosphere, the pencil will simply mark a horizontal line on the paper as the engine-beam rises and falls. Such a tracing is shown by the dotted line, X Y, in the figure, and is called the 'atmospheric line.'

Now suppose that the indicator is in free communication with the top of the cylinder, and that the sliding board, F G, is as far to the right hand as it can go when the engine-piston is at the top of the cylinder, and is gradually made to move to the left hand as the piston makes its down-stroke. As soon as steam is admitted into the cylinder to force the piston down, the indicator-piston will rise above the atmospheric line, say, from X to P, until the pressure of the steam is just balanced by the tension of the spring. Then, if the same pressure is maintained in the cylinder during the down-stroke, the pencil will trace a horizontal line such as P Q as the paper moves under it from right to left. If the steam is cut off at half-stroke, the pressure in the cylinder will begin to diminish after this point, and the pencil will trace a downward curved line such as Q R during the rest of the stroke: the vertical distance of this line above the atmospheric line at any given point showing the true pressure of steam in the engine-cylinder at that particular part of the stroke.

At the end of the stroke, when communication is opened

between the upper part of the cylinder and the condenser, the steam-pressure rapidly falls off, and the indicator-piston being forced down by the pressure of the air, the pencil will trace a downward line, R S, extending below the atmospheric line to a distance which is determined by the degree of perfection of the vacuum in the condenser. The slider with the paper on it will meanwhile move from left to right during the up-stroke; and if the vacuum is properly maintained, a nearly horizontal line S T will be drawn on the paper. Lastly, when the engine-piston has reached the top of the cylinder, and steam is again admitted for the down-stroke, the indicator-piston will be suddenly forced up, and the pencil will trace a nearly vertical line T P.

Thus during each up-and-down stroke a curved diagram P Q R S T is traced on the paper, and by measuring the distance of different points on this line from the atmospheric line accurate information is gained with regard to the real force which is acting on the piston at various parts of the stroke, such as could not possibly be obtained by mere inspection of a steam-gauge attached to the boiler. Moreover the regularity or otherwise, of the curve will show whether the valves are correctly adjusted so as to admit and cut off steam at the proper times.[1]

When the true average effective pressure on the piston during the working of the engine has thus been ascertained, the actual horse-power can be easily calculated.[2]

---

[1] In more modern forms of Watt's Indicator the piston makes only a very short stroke, its movements being magnified (somewhat in the same way as in Ferguson's pyrometer, p. 46) by attaching the end of the rod to the short arm of a lever, while the pencil is carried at the end of the long arm. Moreover the paper is wrapped round a drum which is made to turn nearly once round during each stroke of the engine, first in one direction and then in the other. By such an indicator diagrams can be taken from engines running even at such high speeds as 150 or 200 revolutions per minute.

[2] The following is the rule for calculating it, in English measure :—Multiply together (i) the mean indicated pressure of steam, (ii) the area of the piston in square inches, (iii) the number of feet moved by the piston per minute (*i.e.* length of stroke × number of strokes per minute); and divide the product by 33,000. The quotient is the indicated horse-power of the engine.

DUTY OF ENGINES.

The 'duty' of a heat-engine is expressed by the number of units of work which it does for each unit weight of fuel expended in heating the working fluid. It is obviously closely related to the 'efficiency' of the engine (a term already explained, p. 368); in fact, an engine which yielded the maximum duty would be one which had the highest possible efficiency, *i.e.* which converted all the heat put into it into mechanical work.

It is found by experiment[1] that 1 grm. of good Welsh coal will produce, in combining with oxygen, about 8200 calories of heat. If, then, we take the mechanical equivalent of heat (p. 358) as 42,800 centimetre-grammes, 1 grm. of such coal should theoretically yield (8200 × 42,800=) 350,960,000 or very nearly 351 million centimetre-grammes of work. Any steam-engine which would do this amount of work for every gramme of coal burnt under its boiler would yield the maximum duty, and would be a perfect engine. But the actually realised duty of even the best engines hitherto constructed falls far short of this. For instance, in the case of the pumping-engines used in Cornwall, where economy of fuel is carried out to the fullest extent, the duty of even the best of these is found to be only 27 millions of centimetre-grammes : while the average duty of a large number of them was, in 1870, only about $13\frac{1}{2}$ millions.[2] That is, even the best of these engines only does about ($\frac{27}{351}=$) $\frac{1}{13}$ of the work which it ought theoretically to do. In fact, out of the total amount of heat given out in the combustion of the fuel, at least $\frac{5}{10}$ are (as explained on p. 360) not available at all for conversion into mechanical energy; about $\frac{3}{10}$ go up the chimney and are absolutely wasted ;[3] about $\frac{1}{10}$ is lost by

---

[1] See MM. Favre and Silberman's researches in *Annales de Chimie et de Physique* (3d. series) vol. xxxiv. p. 387.

[2] In English measure, the duty of the best Cornish engine was about 100 million ft.-lbs. per cwt. of coal : and that of the average engine was about 50 million ft.-lbs. per cwt. of coal.

[3] Except in so far as this heat serves the purpose of keeping up a draught in the chimney and thus quickening combustion.

radiation or spent in overcoming the friction and inertia of the moving parts of the engine; and only about $\frac{1}{10}$ (at most) can be made to appear as mechanical energy available for external purposes.

It is disappointing to have to admit that, after all the labour and thought which has been spent on it, even the best of steam-engines working at its best is a very imperfect machine, and must necessarily be so.

Hot-air engines yield an even lower duty than steam-engines; and, from the nature of the case, much improvement can hardly be looked for in them.

Gas-engines, however, (as mentioned on p. 412) give much more promising results; about $\frac{1}{6}$ of the total heat produced by the fuel being convertible by them into mechanical energy. But the expensiveness of gas and the difficulty of constructing gas-engines of large size have hitherto hindered them from competing with steam-engines, except in special cases. Coal is at present so cheap that, in spite of the enormous and unavoidable waste of heat, steam-engines still hold their position as the main sources of mechanical power. But it is quite possible that another generation will see the heat of combustion, not employed in making steam or expanding a gas, but converted directly into electrical energy, which latter can certainly be made to yield much better results in mechanical work: the duty obtained from a good electro-magnetic engine being about four times that of the best heat-engine at present known.

It will be interesting to inquire, finally, what is the duty of the engine with which human beings are most closely associated, *viz.* the animal muscular system itself.

A reference to the Table of the value of different foods given on p. 26 will show that, taking bread as a typical food, 1 gramme of it gives out, in undergoing chemical changes within the body, enough energy to yield (theoretically) about 130 million centimetre-grammes of mechanical energy. Practically, however, about $\frac{6}{10}$ of this energy appears always and necessarily in the form of heat, while more than $\frac{1}{10}$ of it is employed in effecting internal work in the body, such as in maintaining the action of the heart (a pumping-machine which forces the blood to circulate through the system), of the lungs (an air-pump, as it were), and of the nervous system (the telegraph, so to speak, of the animal organism). Hence rather more than $\frac{2}{10}$ of the whole energy is all that can be considered available for external work. Even on this estimate, however, the animal machine would appear to be at least twice as efficient as the best steam-engine, and decidedly more efficient than any form of heat-engine hitherto constructed. The chemical energy of the food-materials seems to pass directly into mechanical energy, instead of having to be first wholly converted into heat, and then this heat subsequently converted, at a great loss, into mechanical energy. This shows the direction in which we must aim in devising improvements on our present sources of power. A machine is wanted which will effect, simply and cheaply, the conversion of other forms of energy into mechanical energy, with as little reference to heat as possible.

CONCLUSION.

The Depreciation or 'Degradation' of Energy.

Energy has been defined already, p. 16, as being the condition of a body which enables it to do work; and it was also mentioned that there are ample grounds for believing that the exact amount of energy which was associated with the matter of the universe at the beginning of time is present in that matter now. All that human agency can do is, not to increase or lessen its quantity, but to alter its distribution, to cause its transfer from one piece of matter to another, or to convert it from one form into another. This statement is called the Law of the Conservation of Energy.

The principal forms in which Energy manifests itself are, Chemical Affinity, Electricity,[1] Mechanical Energy, Radiant

---

[1] Magnetism is included under this head, for magnetic phenomena are probably only special effects of electric currents.

Energy, Heat; but there is a very great difference in the ease and completeness with which Energy in one of these forms can be transmuted into another form. Usually, whenever any transmutation takes place, the energy which is being dealt with passes into more than one of the other forms, but not in equal proportions.

Thus, if a piece of zinc or iron is placed in some dilute sulphuric acid (or other substance containing elements which have an affinity for it), the energy of Chemical Affinity stored up in a statical form in the zinc and the acid appears, while they are combining, partly in the form of Electricity, but mainly in the form of Heat.

Again, if a conducting wire is moved through a region containing molecules which possess Electrical Energy, the Mechanical Energy of the moving wire is converted partly into Electricity and partly also into Heat.

Further, the Chemical Energy present in food is capable of being transformed through the medium of the muscles of an animal into Mechanical Energy (employed in walking, rowing, carrying loads, etc.); but at the same time a large proportion of it, as already stated on p. 25, passes unavoidably into the form of Heat.

Many other instances might be given which all tend to establish the universal rule that whenever any change of Energy from one form to another occurs, some of it always and necessarily (with our present means) takes the form of Heat.

When, however, we try to convert Heat into other forms of Energy, our endeavours have hitherto been only partially successful. We can make the whole of the energy associated with a falling weight pass into the form of Heat, but we have not practically succeeded in making the heat thus produced raise the weight to the height from which it fell. We can convert Electricity into Heat, as is done in incandescent lamps and arc-lights, with a success which is growing greater every year; but when we endeavour to re-convert the Heat into an Electric current, as in the thermo-pile mentioned on p. 90, we find that only a little of the Heat will take the form of Electricity, a very large proportion being diffused away by conduction, radiation, etc., and lost irrecoverably so far as the immediate purpose is concerned. Heat, in fact, has this unfortunate peculiarity, that it cannot be isolated in one particular place, or retained in association with one particular portion of matter so easily and completely as other forms of energy. More or less of it always leaks away, as it were, in spite of all our efforts. We can confine Electricity, for all practical purposes, to a single piece of wire, without losing any; and the energy of Chemical Affinity resides without diminution in the oxygen upon the earth and the coal under the earth, until we induce it to show itself in the process of combustion. But Heat, once generated, begins immediately to pass off into regions where there is less of it; it cannot be kept in a concentrated form.

For the above reasons Heat is looked upon by us as a far less valuable form of energy than any of the others; it may be compared to a 'security' in a money-market, which, though its intrinsic value may be great, yet cannot be converted into cash except at a ruinous loss, and is therefore depreciated in value. So whenever other available forms of Energy are converted, intentionally or unavoidably, into the less available form of Heat, the Energy is said to be 'depreciated' or 'degraded.'[1]

We see, then, that every one of the manifold processes of the transmutation of Energy which are going on around us has for one of its inevitable results the addition of some Heat to that enormous store which already exists.

We see, moreover, that this vast amount of Heat-energy cannot by any methods hitherto discovered by man be completely transmuted into other forms of Energy. It must therefore be constantly accumulating; not, however, in the places where it is generated, for it immediately begins to diffuse itself

---

[1] Probably 'depreciated' may be regarded as the preferable term, since it only implies a lessening in value with respect to the particular needs and mere agencies of man, and not any absolute, intrinsic deterioration of character in the energy,—a degradation which we have not sufficient grounds for asserting to take place.

Depreciation of Energy. 421

by the methods explained in Chapter VIII., so that all bodies are tending to assume eventually the same uniform temperature. When this state has been attained it will be impossible thenceforward to transform any of the stock of Heat-energy into other and more useful forms; for Heat is (as far as human experience goes) only thus transformable when it passes from one region to another, and this passage can only take place when there is an inequality of temperature between the two regions (just as water only passes by gravitation from one cistern to another when the cisterns are at different levels).

We are consequently led to the unsatisfactory conclusion that, if the course of Nature continues on its present lines, and unless some new and scarcely to be hoped for discoveries are made of methods for dealing with and transforming Heat-energy, the world will gradually attain and retain a dead uniform level of temperature, which will make life and work for beings like ourselves practically impossible.

[For a fuller account of the great principle of the Conservation of Energy, and its modern developments, the student is strongly advised to read Tait's *Recent Advances in Physical Science*, Balfour Stewart's *Conservation of Energy* (International Series), and Clerk Maxwell's *Theory of Heat*.

APPENDIX A.

THE METRIC SYSTEM OF MEASURES AND WEIGHTS.[1]

WEIGHT means the pressure of bodies towards the centre of the earth caused by the force of gravitation.

The sizes and weights of things are usually expressed in terms of some 'unit' or standard amount of space or pressure, such as a 'foot,' a 'yard,' a 'metre,' a 'pound,' or a 'gramme.' Thus, in saying that a rod is six feet long, we mean that it extends in length over six times the space of the unit of length which we call a 'foot.' Again, in saying that a piece of lead 'weighs' two pounds, we mean that it presses towards the earth's centre with twice as much force as a particular piece of matter which we call a 'pound weight.'

In selecting a unit for practical purposes, we are mainly guided by three considerations :—

1. The unit must be of such a kind that another exactly similar one could be easily obtained, if the original unit was lost or damaged.

2. It must not be very large or very small; otherwise in common use we should constantly have to deal with awkward fractions or inconveniently large multiples of it.

3. It must have other measures and weights derived from it by the simplest possible methods of multiplication and division.

The unit, or starting-point, of the metric system (which is now almost universally employed in scientific work) is the METRE, which is a length of one forty-millionth part of the circumference of the earth, measured under the meridian of Paris.

---

[1] From *Lessons in Elementary Dynamics* (pub. by W. & R. Chambers).

The actual metre is a flat bar of platinum, about 39.4 inches long, each end of which is exactly at right angles to the length of the bar; the distance between the ends at the temperature of freezing water is defined to be one metre. Several extremely accurate copies of this have been made, and it is probable that scientific men will be content with these copies, and other copies of them, without again deriving the unit from an actual measurement of the earth.

MEASURES OF LENGTH.

In deriving other measures of length from the metre, only the number 10 and its multiples are employed; and names are selected which denote the relation of the particular measure to the unit.

Thus, the next larger measure is a length ten times that of the metre, and is called a **decametre** (Gr. δέκα, *deca*, ten). The next larger measure is a length 100 (that is 10 × 10) times that of the metre, called a **hectometre** (Gr. ἑκατὸν, *hecaton*, a hundred). The largest measure practically used is a length of 1000 (that is, 10 × 10 × 10) metres, called a **kilometre** (Gr. χίλια, *chilia*, a thousand).

Similarly, for the smaller measures, we have a length of $\frac{1}{10}$ of a metre, called a **decimetre** (Lat. *decem*, ten); a length of $\frac{1}{100}$ of a metre, called a **centimetre** (Lat. *centum*, a hundred); and a length of $\frac{1}{1000}$ of a metre, called a **millimetre** (Lat. *mille*, a thousand).

Thus, the names of all the measures larger than the unit, are constructed by adding to the name of the unit a prefix derived from a *Greek* numeral; the names of the measures smaller than the unit are obtained by adding to the name of the unit a prefix derived from a *Latin* numeral.

Fig. 132.

A complete table of the measures of Length is given below, and fig. 132 shows the actual length of a decimetre, which is divided into centimetres and millimetres; a scale of English inches is added for comparison, by which it is seen that a decimetre is very nearly equa to four inches.

The Metric System.

TABLE OF THE MEASURES OF LENGTH.
(The usual abbreviations are put in brackets.)

Kilometre	= 1000 metres.
Hectometre	= 100 ,,
Decametre	= 10 ,,
METRE (m.)	= 1 metre.
Decimetre	= 0·1 ,,
Centimetre (cm.)	= 0·01 ,,
Millimetre (mm.)	= 0·001 ,,

The table may also be put in the following form :—

10 millimetres	= 1 centimetre.
10 centimetres	= 1 decimetre.
10 decimetres	= 1 metre.
10 metres	= 1 decametre.
10 decametres	= 1 hectometre.
10 hectometres	= 1 kilometre.

MEASURES OF VOLUME.

The unit of Volume or capacity is a cube, each side of which measures 1 decimetre (in other words, 'one cubic decimetre'). It is called a LITRE; and from it the larger and smaller measures of volume are derived in precisely the same way as those of length are derived from the metre. Their names are also given on a similar principle (*Greek* prefixes being used for the larger, *Latin* prefixes for the smaller measures); and the following table hardly requires further explanation :—

TABLE OF THE MEASURES OF VOLUME.
(The usual abbreviations are put in brackets.)

Kilolitre	= 1000 litres.
Hectolitre	= 100 ,,
Decalitre	= 10 ,,
LITRE	= 1 litre.
Decilitre	= 0·1 ,,
Centilitre	= 0·01 ,,
Millilitre (or cubic centimetre) (c.c.)	= 0·001 ,,

It should be noted—

1. That the name 'cubic centimetre' is almost universally used instead of 'millilitre' (a cubic centimetre being readily demonstrated to be $\frac{1}{1000}$ of a cubic decimetre, or litre).

2. That quantities smaller than the litre are usually expressed in cubic centimetres. Thus three-fourths of a litre would be expressed, not as 7 decilitres 5 centilitres, but as 750 cubic centimetres.

MEASURES OF WEIGHT.

The unit of Weight is the weight of one cubic centimetre (millilitre) of water, at the temperature of 4° Centigrade.[1] It is called a GRAMME, and from it the larger and smaller weights and their names are derived in exactly the same way as in the case of the measures of length and volume.

TABLE OF WEIGHTS.

(The usual abbreviations are given in brackets.)

Kilogramme	= 1000 grammes.
Hectogramme	= 100 ,,
Decagramme	= 10 ,,
GRAMME (grm.)	= 1 gramme.
Decigramme	= 0·1 ,,
Centigramme	= 0·01 ,,
Milligramme	= 0·001 ,,

It must be noted that, in strict scientific language, the gramme is a unit of *mass* and not simply of weight. Mass means the quantity of that material in which energy can reside, and on which forces can act, which is present in a substance; and the mass of a given body remains the same wherever it is situated in space. But the weight of the body changes according as it happens to be near to or far from

---

[1] The reason why this particular temperature is specified in defining the gramme is as follows: A given mass of water alters in bulk as its temperature changes (as is more fully explained in Chap. IV. Sect. III.); but at 4° C. it occupies the smallest space that it ever occupies while in the liquid state. Hence a cubic centimetre of water has more matter in it, and therefore weighs more, at 4° C. than at any other temperature.

The Metric System. 427

the earth's centre. It is less at the Equator than at the Poles; at the actual centre of the earth bodies have no weight; at the distance of the Moon they have very little weight. Now, the piece of matter (usually platinum or brass) which is conventionally called a 'gramme' is accepted as being a unit of the same value all the world over, although its weight is not the same everywhere. Its *mass*, however, is the same everywhere; and therefore it must strictly be considered as a unit of mass.

MEASURES OF AREA.

The unit of Area or Surface is the ARE, which is a surface 1 decametre square, *i.e.* containing 100 square metres. The other measures, larger and smaller, are derived from it in precisely the same way as in the preceding cases, but as they are not much used except for land measurement, a complete table of them need not be given here.

The hectare is very nearly $2\frac{1}{2}$ acres.

RULES FOR REDUCTION.

(These apply to all the tables given above.)

I. To reduce the larger and smaller measures to the unit, and *vice versâ*. (Principle.—*The name of each measure expresses what multiple of the unit it is.*)

(a) To reduce a given larger measure to the unit, or a given unit to one of the smaller measures.

Multiply by the number expressed in the name of the measure.

Examples—
Reduce 18 *kilo*metres to metres. 18 × 1000 = 18,000 metres.
Reduce 6 grammes to *centi*grammes. 6 × 100 = 600 centigrammes.

(b) To reduce a given smaller measure to the unit, or a given unit to one of the larger measures.

Divide by the number expressed in the name of the measure.

Examples—
Reduce 1885 *centi*metres to metres. 1885 ÷ 100 = 18·85 m.
Reduce 1724 litres to *deca*litres. 1724 ÷ 10 = 172·4 decalitres.

II. To reduce any given measure to the next larger or the next smaller measure. (Principle.—*Each measure is ten times the next smaller one, and one-tenth of the next larger one.*)

(*a*) To reduce a measure to the next larger one.
Divide the number by 10.

Example—Reduce 152 centigrammes to decigrammes.
152 ÷ 10 = 15·2 decigrammes.

(*b*) To reduce a measure to the next smaller one.
Multiply the number by 10.

Example—Reduce 16·2 kilometres to hectometres.
16·2 × 10 = 162 hectometres.

It is obvious that, since our system of numeration is, like the metric system itself, a *decimal* system,—that is, is based on the number 10, all the processes of multiplication and division required by the above rules are extremely simple. The actual figures have not to be changed at all; their value is altered simply by changing their place in reference to the units-figure. This is, of course, done by altering the position of the decimal point; the latter being always considered to exist, even if not actually expressed, immediately after (that is, to the *right* of) the units-figure.

Thus, to multiply a number by

    10, shift the decimal point one place to the *right*,
    100,   ,,   ,,   two places   ,,
    1000, ,,   ,,   three places ,,
                etc.

Again, to divide a number by

    10, shift the decimal point one place to the *left*,
    100,   ,,   ,,   two places   ,,
    1000, ,,   ,,   three places ,,
                etc.,

cyphers being put in, if necessary, to fill up the interval between the decimal point (implied or expressed), and the first figure of the number which is being dealt with.

It will be useful to bear in mind the following points in connection with the metric system :—

1. That the number of **cubic decimetres** which expresses the size of

The Metric System.

a body also expresses the volume of the body in **litres** (since a cubic decimetre is, by definition, 1 litre).

Thus, if a cistern is 6 decims. long, 4 decims. broad, and 3 decims. deep internally, its size will be (6 × 4 × 3 =) 72 cubic decimetres; and its capacity is known at once to be 72 litres.

2. That the number which expresses the volume of a given quantity of water in **cubic centimetres** also expresses (very nearly) its weight in **grammes**, at ordinary temperatures (since the gramme is, by definition, the weight of 1 c.c. of water at 4° C.).

Thus 1000 c.c. of water are known at once to weigh (neglecting the small correction for temperature) 1000 grammes, or 1 kilogramme.

Conversely, of course, a given weight of water in grammes will measure (very nearly) the same number of cubic centimetres.

So that, for instance, if 100 c.c. of water are required for an experiment, and no measures are at hand, it will only be necessary to weigh out 100 grammes in a counterpoised beaker, in order to obtain the quantity required.

Again, a very useful measure may be made from a jar or stout test-tube, by counterpoising it, and weighing it into 1, 5, 10, etc. grammes of water, marking the level of the surface of the liquid with a file or diamond at each successive weighing.

What has just been said applies to the case of other liquids than water, if a correction is made for any difference in density (for an explanation of this term, see p. 54). Thus, if a liquid is twice as heavy as the same volume of water, it is obvious that twice as much of it, that is, 2 grammes, must be weighed out in order to obtain a volume of 1 c.c.

THE 'CENTIMETRE-GRAMME-SECOND' SYSTEM (OR C.-G.-S. SYSTEM) OF UNITS.

For the purpose of expressing, concisely and exactly, definite quantities of Mechanical Energy, of Heat Energy, and of other similar physical conceptions, it is found best to state them in terms of certain recognised units.

The units which have been selected by a committee of scientific men, appointed in 1873 by the British Association for the Advancement of Science,[1] and which are now very generally accepted and used, are the following :—

<dl>
Unit of Length . . the **Centimetre**.
Unit of Mass . . . the **Gramme**.
Unit of Time . . . the **Second**.
</dl>

The following may serve as examples of the application of the above units in Dynamics, *i.e.* in the study of the relations of mechanical energy.

Force is that influence which causes motion or changes the rate of motion, or destroys motion in a mass of matter.

The unit of Force is called a **Dyne** (Gr. δύναμις, force). It is—That **amount of Force which, acting for 1 second on a mass of 1 gramme, gives it a velocity of 1 centimetre per second.**

For example—Suppose that a force applied for 1 second to a billiard-ball weighing 110 grms. makes it move 80 cm. per sec. Then the value of this force is (110 × 80 =) 8800 dynes.

Energy (p. 6) **is that condition of a body which makes it capable of doing work.**

The unit of Work, and therefore also the unit of the Energy which enables the work to be done, is called an **Erg** (Gr. ἔργον, work). It is—That **amount of work which is done by 1 dyne of force acting through a space of 1 centimetre.**

For example—A skater is pushing a chair before him on the ice with a force of 200 dynes. Then for every metre (= 100 cm.) that the chair moves through, he does (100 × 200 =) 20,000 ergs of work.

(*Examples and Exercises on the Metric System are given on p.* 453).

TABLES OF ENGLISH MEASURES.

1. Length.

1	inch		=	2·54	centimetres.
12	inches	= 1 foot	=	30·48	,,
3	feet	= 1 yard	=	91·44	,,
5½	yards	= 1 pole	=	5·03	metres.
4	poles	= 1 chain	=	20·12	,,
40	poles	= 1 furlong	=	201·16	,,
8	furlongs	= 1 mile	=	1609·3	,,

---

[1] See *Report of the British Association* for 1873, p. 211.

2. Volume.

1 cubic inch =		16.38 cubic centimetres.
1728 cubic inches = 1 cubic foot	=	28.31 litres.
61.027 ,, = 1 litre.		
1 drachm = 0.216 cubic inch	=	3.55 cubic centimetres.
8 drachms = 1 ounce	=	28.35 ,, ,,
20 ounces = 1 pint	=	567.9 ,, ,,
2 pints = 1 quart	=	1.136 litres.
4 quarts = 1 gallon	=	4.544 ,,

1 gallon = 277.274 cubic inches = 70,000 grains (or 10 lbs. avoird.) of distilled water at the temperature of 60° F. (15.5° C.).

3. Weight.

(a) *Apothecaries.*

1 grain =		0.0648 gramme.
20 grains = 1 scruple	=	1.296 ,,
3 scruples = 1 drachm	=	3.888 ,,
8 drachms = 1 ounce	=	31.104 ,,
12 ounces = 1 pound	=	373.248 ,,

(b) *Avoirdupois.*

1 grain =		0.0648 gramme.
27.34 grains = 1 drachm	=	1.772 ,,
16 drachms = 1 ounce	=	28.35 ,,
16 ounces = 1 pound	=	453.6 ,,
112 pounds = 1 cwt.	=	50.8 kilogrammes.

APPENDIX B.

RELATION OF THE SCALES OF THE CENTIGRADE AND FAHRENHEIT THERMOMETERS.

Cent.	Fahr.	Cent.	Fahr.	Cent.	Fahr.	Cent.	Fahr.
+100=	+212	+64=	+147·2	+29=	+84·2	−6=	+21·2
99=	210·2	63=	145·4	28=	82·4	7=	19·4
98=	208·4	62=	143·6	27=	80·6	8=	17·6
97=	206·6	61=	141·8	26=	78·8	9=	15·8
96=	204·8	60=	140	25=	77	10=	14
95=	203	59=	138·2	24=	75·2	11=	12·2
94=	201·2	58=	136·4	23=	73·4	12=	10·4
93=	199·4	57=	134·6	22=	71·6	13=	8·6
92=	197·6	56=	132·8	21=	69·8	14=	6·8
91=	195·8	55=	131	20=	68	15=	5
90=	194	54=	129·2	19=	66·2	16=	3·2
89=	192·2	53=	127·4	18=	64·4	17=	1·4
88=	190·4	52=	125·6	17=	62·6	18=	−0·4
87=	188·6	51=	123·8	16=	60·8	19=	2·2
86=	186·8	50=	122	15=	59	20=	4
85=	185	49=	120·2	14=	57·2	21=	5·8
84=	183·2	48=	118·4	13=	55·4	22=	7·6
83=	181·4	47=	116·6	12=	53·6	23=	9·4
82=	179·6	46=	114·8	11=	51·8	24=	11·2
81=	177·8	45=	113	10=	50	25=	13
80=	176	44=	111·2	9=	48·2	26=	14·8
79=	174·2	43=	109·4	8=	46·4	27=	16·6
78=	172·4	42=	107·6	7=	44·6	28=	18·4
77=	170·6	41=	105·8	6=	42·8	29=	20·2
76=	168·8	40=	104	5=	41	30=	22
75=	167	39=	102·2	4=	39·2	31=	23·8
74=	165·2	38=	100·4	3=	37·4	32=	25·6
73=	163·4	37=	98·6	2=	35·6	33=	27·4
72=	161·6	36=	96·8	1=	33·8	34=	29·2
71=	159·8	35=	95	0=	32	35=	31
70=	158	34=	93·2	−1=	30·2	36=	32·8
69=	156·2	33=	91·4	2=	28·4	37=	34·6
68=	154·4	32=	89·6	3=	26·6	38=	36·4
67=	152·6	31=	87·8	4=	24·8	39=	38·2
66=	150·8	30=	86	5=	23	40=	40
65=	149						

[Rules of reduction of one scale to the other are given on p. 100.]

APPENDIX C.

LIST OF THE PRINCIPAL PIECES OF APPARATUS AND MATERIALS REQUIRED FOR THE EXPERIMENTS DESCRIBED IN THIS BOOK.

In this list, such common pieces of apparatus as would be found among ordinary household stores, *e.g.* kettles, china basins and plates, water-cans, are not included. Duplicates of all such easily broken glass apparatus as flasks, beakers, and test-tubes should be obtained.

I. General Apparatus.

1. An ordinary boxwood metre rule, divided into decimetres and centimetres; the first decimetre, at least, being also divided into millimetres.
2. A cylindrical glass measure, containing at least 150 or 200 c.c.; graduated into spaces of 5 c.c.
3. A pair of scales, with beam about 20 or 25 cm. long, sensitive to a weight of 1 centigramme. The ordinary grocer's scales, if well made, are sufficiently good; but they should be tested carefully before use.
4. A set of weights, from 100 grms. to 1 centigramme, mounted on a wooden stand. These need not necessarily be of the highest accuracy: such sets, made in France, can be now obtained in England at a reasonable price. A kilogramme weight, and also a half-kilogramme, of common quality, made of iron, will be useful.
5. An iron tripod-stand, like that shown on the right hand of fig. 69, p. 211; with legs about 20 cm. long.
6. A retort-stand, also shown in fig. 69, with a heavy rectangular iron foot, and strong iron stem not less than 50 or 60 cm. in height.

It should be fitted with three iron rings of different sizes, and should also have an iron clamp or spring vice (shown in fig. 133) for holding Liebig's condensers, tubes, etc.

7. A Bunsen holder, fig. 134, the smaller of the two sizes which are sold.

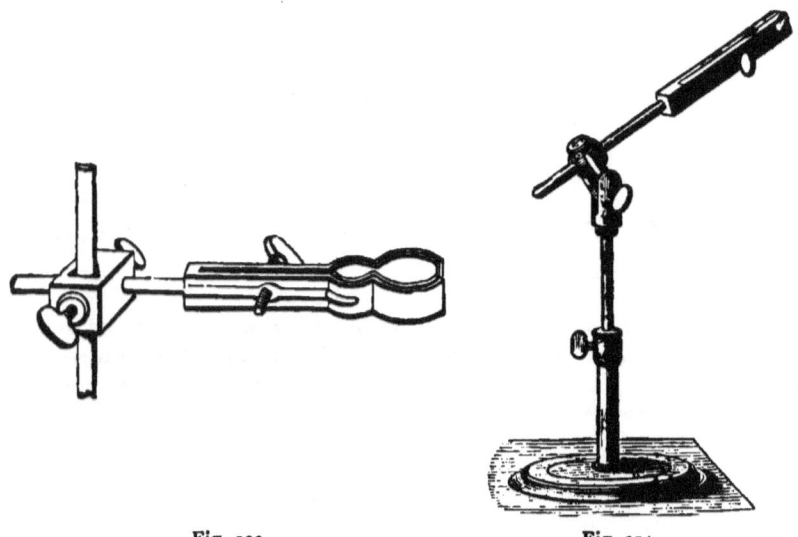

Fig. 133.   Fig. 134.

8. A set of four wooden blocks, about 12 cm. square, and respectively 2, 5, 7, 10 cm. in thickness.

9. Two pieces of iron wire-gauze, about 12 cm. square, having about 12-14 meshes in 1 cm. (30-36 meshes per inch). A larger piece of the same gauze, about 20 or 25 cm. square, will also be wanted for the experiments on flame.

Fig. 135.   Fig. 136.

10. A porcelain mortar, about 10 or 12 cm. in diameter.
11. A small pair of crucible tongs, fig. 135.
12. A small set of cork borers, fig. 136 (four or five will be sufficient), from 3 mm. in diameter upwards.

List of Apparatus.

13. A round or 'rat-tail' file, about 20 cm. in length.
14. A triangular file, about 12 cm. in length.
15. A screw pinch-cock, of the kind shown in fig. 137, or any similar pattern.
16. A glass spirit lamp, fig. 6, p. 30.
17. An Argand burner, fig. 7, p. 31.
18. A Bunsen burner, fig. 8, p. 32.
19. A Herapath blowpipe, fig. 12, p. 36. A convenient form of blower made by Messrs. Fletcher and Co. for use with this blowpipe is shown in fig. 138.
20. A large globular glass flask, about 15 (or, better, 20) cm. in diameter, fig. 83, p. 265.
21. A similar flask, about 10 or 12 cm. in diameter.
22. Two or three flasks with flat bottoms, like that shown in fig. 33, p. 65, holding from half a litre to a litre.

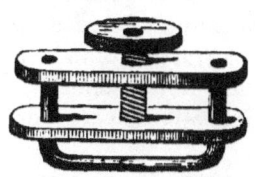

Fig. 137.

Fig. 138.

23. Four or five glass beakers, including at least two 9 × 5 cm., one 11 × 6 cm., and one 20 × 10 cm.
24. A cylindrical jar of stout glass, about 20 cm. in height, and 10 cm. in diameter.

A much larger glass jar, about 40 cm. in height and 25 cm. in diameter (a common bell-jar such as gardeners use for plants, or an aquarium glass) will be found very useful.

25. Two or three wide-mouthed bottles, holding from half a litre to a litre.
26. About a dozen test-tubes, some 20 cm. long and 2.5 cm. in diameter, the rest rather smaller.
27. At least a kilogramme of glass tubing (easily fusible but free from lead, such as the kind known as 'soda glass'), of different sizes : a few pieces from 1 to 2 cm. in internal diameter, nearly all the rest from 3 to 6 mm. in internal diameter. One or two pieces of the

tubing used for making alcohol thermometers, small in bore (1 mm. or less), but with thick sides, will be useful.

28. Two or three metres, or more, of indiarubber tubing (the black or red is preferable to, but more expensive than the grey), of two sizes ; *viz.* 6 mm. and 3 mm. in internal diameter.

II. GENERAL MATERIALS.

1. A litre, or more, of ordinary methylated spirit of wine, of good quality, free from gum.

2. About 50 c.c. of ether (that which is made from methylated spirit will do), sp. gr. about 0·72.

3. About a kilogramme of common salt.

4. About half a kilogramme of sodium sulphate.

5. About 30 grms. of each of the following salts :—Ammonium chloride (sal ammoniac) ; potassium nitrate (saltpetre or nitre) ; 'bicarbonate of soda' ; 'tartaric acid.'

6. About 15 grms. of phosphorus, and the same quantity of iodine.

7. About 5 grms. of 'magenta' or some similar intense colouring matter soluble in water, in fragments not smaller than a grain of wheat (chiefly for use in Expt. 88, p. 264, and similar ones).

8. About 20 c.c. of solution of indigo (ink will, however, do).

9. About 20 c.c. of glycerine, or gum.

10. Half a litre of solution of calcium hydrate ('lime-water'). To make this, put about 20 grms. of freshly burnt quicklime into a dish or beaker, and pour on it a little water. When the lime is thoroughly slaked, transfer the contents of the dish to a bottle holding about half a litre, fill up the bottle with water, cork it tightly, and leave it for a day, shaking it occasionally. Then allow the excess of lime to settle, and decant the solution into a clean bottle. This bottle must always be kept tightly corked, as the solution is decomposed by contact with the carbon dioxide present in air.

11. Half a kilogramme of beeswax, and three or four common wax or paraffin candles.

12. Half a kilogramme of mercury (less will suffice for most of the experiments, but for Expts. 53 and 60 at least a kilogramme would be required).

13. A piece of thin sheet-copper about 30 × 20 cm., and a similar piece of sheet-iron, and also of sheet-lead (thicker).

14. Three or four sheets of bright tin-plate, free from defects and moderately stout, about 30 × 20 cm., or rather larger.

LIST OF APPARATUS. 437

15. A sheet or two of tinfoil.
16. A small supply of copper wire (No. 18 wire-gauge, 1·2 mm. in diameter), and iron wire (No. 26 wire-gauge, 0·5 mm. in diameter).
17. A bit of platinum wire, about 20 cm. long, No. 26 wire-gauge; and a strip of platinum foil about 6 × 3 cm. or larger.
18. Some pieces of ordinary window glass, about 20 cm. square, and also a few strips about 15 × 2 cm.
19. A piece of 'composition' gas tubing, see Expt. 20, p. 74.
20. Two or three dozen corks of various sizes.

III. APPARATUS FOR EXPERIMENTS IN CHAPTERS IV. AND V.

1. A brass bar and gauge, fig. 15, p. 43.
2. Gravesande's ball and ring, fig. 16, p. 43.
3. Strips of copper, iron, and glass, for use in Expt. 12, p. 44.
4. Gauge for showing difference in expansion of iron and brass, fig. 18, p. 45.
5. Ferguson's pyrometer, fig. 20, p. 46 (not absolutely necessary).
6. Compound bar of iron and copper, fig. 21, p. 48.
7. Apparatus for breaking iron bar, fig. 22, p. 49 (not absolutely necessary).
8. Apparatus for showing expansion of liquids, fig. 34, p. 65 (not absolutely necessary).
9. Apparatus for showing expansion of gases, fig. 41, p. 80 (not absolutely necessary).
10. A balloon of tissue paper, 1 metre in diameter (not absolutely necessary).
11. A thermometer, fig. 45, p. 102, with scale engraved on the stem, ranging from 10° to 150°. Two of these will be very useful.
12. Registering thermometers, of the types shown in figs. 46, 47, and 48 (not absolutely necessary).
13. An air-thermometer, fig. 49, p. 107 (not absolutely necessary).
14. A differential thermometer, fig. 50, p. 109.

IV. APPARATUS FOR EXPERIMENTS IN CHAPTER VI.

1. A small flat block of iron, or a pulley-sheave, for use in Expt. 26, p. 123.
2. Balls of iron, tin, and lead, 2·5 cm. in diameter, with brass rings screwed into them, for use in Expt. 28, p. 125. The tin and lead balls may be cast in an ordinary bullet-mould.

3. A plate of soft wax, for use in the same experiment.

To make this, procure one of the shallow rectangular tin dishes sold as tart-dishes, about 20 × 10 cm.; fill it half full of water and support it on a tripod over a lamp. Place in it 60 grms. of wax (part of an ordinary wax candle), and 45 grms. of lard,[1] and heat the water nearly to boiling, stirring the whole frequently. The wax and lard will melt together and form a uniform stratum floating on the water. When this result is obtained, withdraw the lamp and leave the whole undisturbed for 10 or 12 hours to cool and harden: the slab of fusible composition can then be easily detached from the dish, and its edges should be trimmed. About 8 mm. is the thickness most suitable for the experiment: the iron ball will then melt its way right through, while the tin ball will not quite do so.

V. Apparatus for Experiments in Chapter VII.

1. A small bottle of cast iron, with very thick sides, fitted with a screw stopper, for use in Expt. 35, p. 148. This is not indispensable, as a soda-water bottle may be used instead.

2. Two or three 'candle-bombs' for use in Expt. 45, p. 163.

3. An 'ether-spray apparatus,' fig. 55, p. 170. This is not indispensable.

4. A 'cryophorus,' fig. 57, p. 179. This is not indispensable.

5. A glass retort, fig. 67, p. 209, stoppered, holding about 250 c.c.

6. A 'Liebig's condenser,' fig. 68, p. 209. This need not necessarily be bought ready-made: there will be little difficulty in fitting one up from the corks and tubing at hand, a long cylindrical lamp-glass, or a wide tin tube, serving for the outer case.

7. A flask of sheet-copper, with the lower part of silver, fig. 71, p. 216.

8. A shallow silver dish, about 8 or 9 cm. in diameter, for use in Expt. 66, p. 216. This should be like a watch-glass in shape; a concavity represented by that of the surface of a sphere about 15 cm. in radius seems to answer well.

Neither this nor the preceding piece of apparatus are indispensable, but the experiments made with them are striking ones.

---

[1] These quantities are sufficient for a plate of the size above mentioned, *viz.* 20 × 10 cm. If any other size of plate is to be made, half a gramme of the mixture should be taken for each square centimetre of surface, in order to get a plate of the proper thickness.

LIST OF APPARATUS. 439

9. A Hygrometer, such as Regnault's (fig. 72, p. 227), Daniell's (fig. 73, p. 229), or Mason's (fig. 74, p. 231), or all three of them. Simple forms of Regnault's and Mason's instruments may, however, be readily constructed; hints for doing so are given in the text.

VI. APPARATUS FOR EXPERIMENTS IN CHAPTER VIII.

1. Two cast-iron balls, about 5 cm. in diameter, with ring handles attached; and a suitable iron stand, as shown in fig. 75, p. 240, with sliding-rod and clamping-screw, so that a ball may be supported at any convenient height.

2. An apparatus for testing the relative conductivities of substances for heat, as shown in fig. 76, p. 243.

3. A compound bar, for a similar purpose, made of two flat strips, one of iron, the other of copper, about 40 cm. long, 2 cm. broad, and 3 or 4 mm. thick, joined end to end by brazing or riveting; shown in fig. 77, p. 244.

4. An apparatus for demonstrating the effect of specific heat in altering apparent conductivity, shown in fig. 78, p. 245.

5. A cylindrical rod, half of brass or copper, the other half of wood, for use in Expt. 78, p. 252, where a description of it will be found.

6. A Davy safety lamp, fig. 82, p. 257. This, however, may be dispensed with.

7. Three or four 'Rupert's drops,' p. 260.

8. A model of a hot-water warming apparatus, fig. 84, p. 270. This is not indispensable.

9. A large, but not very wide, paraffin-lamp glass, for use in Expt. 95, p. 278.

10. An apparatus for illustrating ventilation, fig. 85, p. 279. This can be made up from simple materials; a tall glass jar, a bung, etc., as described in the text. The long slightly tapering necks of broken retorts are very suitable for the tubes; they must be as nearly alike as possible.

11. An apparatus for a similar purpose, shown in fig. 86, p. 280. This is not indispensable.

12. An apparatus for demonstrating the law according to which the amount of radiant energy varies with distance, shown in figs. 88, 89, 90, pp. 295-297. This might be made from simple materials.

13. A double screen of polished tin, for use in Expt. 104, p. 299 where directions for making it are given.

14. An apparatus for demonstrating the phenomena of refraction of radiant energy, as shown in fig. 93, p. 302.

15. A prism, of ordinary glass, for use in Expt. 106, p. 305.

16. A 'Leslie's cube,' fig. 96, p. 312, for use in Expt. 107 and others. The table stand shown in the figure is not necessary, though convenient.

17. Three tin canisters, for use in Expt. 109, p. 314.

18. Two tin screens on stands, as shown in fig. 98, p. 322.

VII. Apparatus for Experiments in Chapter X.

1. A so-called 'Wollaston's model steam-engine,' fig. 102, p. 372.

2. Some kind of model or models showing the action of the valves, etc., of a modern steam-engine, and an actual working model of such an engine, will be extremely useful: they will render the practical action of a heat-engine far more intelligible than any amount of verbal descriptions or printed diagrams.

The former, *viz.* moving models of valves, etc., may be made up without difficulty from strips of thick cardboard (painted black), attached to a flat board, such as a drawing-board, covered with white paper. Figs. 105-108 and 112 show models of this kind, and figs. 117, 119, and 120 show a more substantial model made almost entirely of wood, and illustrating the action of an eccentric and the link reversing-gear.

For an actual working model, which should not be a mere diminutive toy, but have a cylinder at least 4 cm. in length of stroke, probably the most instructive type is the vertical engine, since all the working parts are easily seen and their action appreciated. It will be much the best to construct the model 'at home' if possible, from sets of castings and materials which many dealers in apparatus now supply. These can be obtained more or less fitted up, for the benefit of any one who has not skill enough to bore out a cylinder, for instance, for himself. A purchased model should, in any case, be taken carefully and completely to pieces (the relation of each part to the rest being exactly noted), and then fitted together again, all joints being made good with a mixture of white and red lead mixed with boiled oil to the consistence of putty.

For simple demonstrating purposes such a model (if not large) may often be worked with compressed air from a good-sized blower, of the kind used for a blowpipe. Thus the necessity for a thorough cleaning, requisite after use with steam, to avoid corrosion, will be obviated.

QUESTIONS AND EXERCISES.

1. Explain precisely what we mean in calling anything 'hot' or 'cold', and mention cases in which our feelings lead to inaccurate estimates of heat.

2. Define 'temperature,' and explain the different methods of measuring it. Illustrate the distinction between the temperature of, and the quantity of heat present in, a given body.

3. Why is Heat not considered to be a fluid? What is the modern theory of its nature, and what evidence is there in support of this theory?

4. What is a 'molecule'? What is believed to be the velocity of the molecules of different gases? How is the pressure or 'tension' of a gas explained by the motion of molecules?

5. Enumerate the chief sources of heat, distinguishing between those which are, and those which are not, under human control.

6. Give proofs and illustrations of (a) the intensity, (b) the quantity, of the heat which reaches us from the sun. What suggestions have been made as to the mode in which this heat is generated and maintained?

7. What is believed to be the condition of the interior of the earth, and what evidence is there in support of this belief?

8. Define 'energy' and 'work,' and give examples of the conversion of mechanical energy into heat.

9. Explain fully what happens when a candle is lighted and allowed to burn, describing experiments in illustration.

10. What is the source of the heat of the human body? Why do we get hotter when we run or lift heavy loads? On what does the value of a food depend?

11. Describe fully the Bunsen burner, explaining why its flame is smokeless and extremely hot. What means have we of obtaining a still higher temperature by the combustion of coal gas?

12. What are the distinguishing characteristics of the three states of matter, and how are these distinctions accounted for by the theory of molecules?

13. Describe experiments which prove—
    (a) That solids alter in size when their temperature is altered.
    (b) That this alteration in size is different in the case of different solids.
    (c) That this change of size takes place with enormous force.

14. The coefficient of expansion of iron in length is said to be 0·000012. Explain exactly what this statement means, and compare the coefficients of expansion of some important solids.

15. A zinc rod, 2·4 metres long at 0°, is heated to 12°, and an iron rod, of the same length at 0°, is heated to 35°. Which will now be the longer of the two?

16. Distinguish between the coefficient of expansion in length and the coefficient of expansion in volume of a solid, and describe methods of determining the former with accuracy. Explain how the latter can be deduced from it by calculation and verified by experiment.

17. Mention some of the results and applications of the alteration in the size of solids caused by heat.

18. A railway is being laid with iron rails, each 6 metres long when the air is at the freezing-point. What is the minimum interval which should be left between the ends if the temperature is liable to reach 40°?

19. Describe experiments which prove the following facts :—
    (a) Liquids expand more than solids when equally heated, but expand to different extents.
    (b) Hot water is less dense than cold water.

20. Distinguish between the coefficient of apparent expansion and the coefficient of absolute expansion of a liquid, and describe one method of determining each with accuracy.

21. The bulb of a large mercury thermometer has an internal capacity of 13 c.c. at 0°, and is just filled with mercury at this temperature. The tube is 40 cm. long, and has an internal capacity of 0·5 c.c. To what height will the mercury, when heated to 100°, rise in the tube, if the latter is of uniform bore.

(The coefficient of apparent expansion of mercury may be taken as $\frac{1}{6400}$.)

22. The mercury in a barometer stands at 760 mm. at 0°: what will be the height of the column at 20°?

The mercury in a barometer stands at 761 mm. at 10°: reduce this height to what it would be at 0°.

23. A rectangular gas bag is 1 metre long, 4 decims. broad, and 18 cm. deep: how many litres of air will it hold? If the air is heated from 0° to 100°, how many litres will escape?

24. 100 c.c. of hydrogen are measured at 12°, and the temperature is then raised to 120°: what will the gas now measure?

25. 10 litres of air at 15° are heated until the volume has expanded to 12 litres. To what temperature has the air been raised?

26. 30 litres of air are cooled from 100° to 80°: what diminution in volume will take place?

27. A balloon holds 500 litres: it is filled with air on a frosty day, and the air is heated to 40°: how much will escape?

28. What will be the ascensive power of the balloon mentioned in the last question?

To find this, we have simply to ascertain how much less the air which fills the balloon at 40° weighs than an equal volume of the external air at 0°.
[1 litre of air at 0° weighs 1·293 grm.]

29. A Regnault's air thermometer (p. 108), of which the capacity was 200 c.c., was placed in a furnace: and when it had attained the temperature of the furnace, the exit-tube was sealed. It was then cooled to 0° and the end of the tube was broken off under mercury: 120 c.c. of mercury entered the bulb. What was the temperature of the furnace?

30. The density of carbon dioxide (referred to that of air at 0°) is 1·52. What will be its density at a temperature of 50°? At what temperature will it have the same density as air at 0°?

31. Clocks go slower in summer than in winter. Explain the reason of this variation in rate, and describe some of the contrivances by which the rate may be made independent of temperature.

32. The reading of a mercury barometer was 742 mm. when the temperature of the mercury was 20°. What would have been the height of the column if the temperature had been 0°?

33. Describe the changes in volume which water undergoes when heated from 0° to 100°, and explain the process by which the mass of water in a lake is cooled on a frosty night.

34. What is Gay Lussac's (or Charles's) law respecting the variation in volume of a gas with change of temperature? Describe Regnault's apparatus for verifying it, and show how the kinetic theory of gases accounts for it.

35. What is meant by the 'absolute zero of temperature'? Express

the freezing-point of mercury, the boiling-point of alcohol, and the melting-point of lead on the absolute scale.

36. If the standard volume referred to in Gay Lussac's law was the volume occupied by the gas at 20° instead of 0°, what would have been the coefficient of expansion?

The problem consists in finding a number which bears the same relation to 0·00366 that 1 bears to 1 + (0·00366 × 20).

37. A glass globe contains 1 grm. of air at 0°. What weight of air will it contain at 50°?

[1 litre of air at 0° weighs 1·293 grm.]

38. Give all the reasons for selecting mercury as the best material for the ordinary purposes of thermometry. Describe the process of making and graduating an ordinary thermometer, explaining why *two* fixed points of temperature have to be determined.

39. What different scales are used in graduating thermometers, and what relation is there between them?

The highest temperature in London on a given day was 59° F.; the highest temperature in Paris on the same day was 12° C. Which place was the warmest, and by how many degrees Fahrenheit?

40. What is a 'registering thermometer'? Describe some one form of (a) maximum, (b) minimum registering thermometer.

41. Describe fully Leslie's differential thermometer, and Regnault's air thermometer, explaining the special uses of each.

42. In using a weight thermometer the following data were obtained:—

Weight of the apparatus empty	. . .	= 36·6	grms.
,, ,, full of mercury at 0°.		= 300·6	,,
,, ,, with the mercury which remained after immersion in hot oil.		= 296·6	,,

What was the temperature of the oil?

43. Equal weights of different substances at the same temperature contain very different quantities of heat. Describe some experiments which illustrate this fact, and show how it is explained on the theory that heat is a motion of the molecules of a substance.

44. Define a 'calorie' of heat.

300 c.c. of water at 140° F. were allowed to cool to 68° F. Reduce these temperatures to the centigrade scale, and state how many calories were given out by the water.

Questions and Exercises. 445

45. 2244 calories are imparted to 50 c.c. of mercury at 0°. What will the mercury now measure? (Density of mercury = 13·6.)

46. Explain the exact meaning of the number called the 'specific heat' of a substance.

In determining the specific heat of iron the following data were obtained:—

Weight of iron	= 200 grms.
Temperature of iron	= 100°
Weight of water (including water-value of calorimeter)	= 300 grms.
Temperature of water	= 20°
Temperature of mixture	= 25°

Calculate the specific heat of iron.

47. 100 grms. of mercury at 50° were mixed with 50 grms. of water at 15° in a glass calorimeter weighing 25 grms. The temperature of the mixture was 17°. Find the specific heat of mercury. In what other way besides the one indicated above can the specific heat of a substance be determined?

48. 120 grms. of tin at 82° were put into 85 grms. of water at 10° in a calorimeter of which the water-value was 5 grms. The temperature of the mixture was 15°. Calculate the specific heat of tin.

How many calories will be required to raise 40 grms. of tin from 15° to 45°?

49. Explain the following facts:—

(*a*) The specific heats of elements are greater in proportion as the weights of their molecules are smaller.

(*b*) The specific heats of equal volumes of all elements in the state of gas are the same.

(*c*) The specific heat of a gas under constant pressure is greater than its specific heat when its volume is maintained constant.

50. A piece of ice at $-10°$ is exposed to a uniform source of heat until the temperature of the substance is raised to 120°. Describe minutely all the changes in its volume which take place, comparing them with the results observed in the case of other substances.

51. The potential heat of liquefaction of ice is said to be 79 calories. Explain the meaning of this, and describe Regnault's method of ascertaining it.

52. Describe experiments which prove—
   (*a*) That heat disappears when a solid becomes liquid.
   (*b*) That heat is produced when a liquid becomes solid.
How are these phenomena accounted for by the kinetic theory of heat?

53. What is meant by a 'source of cold'? Is it a strictly correct term? Describe some of the expedients for obtaining very low temperatures, explaining the way in which they act.

54. A calorimeter, of which the water-value is 30 grms., contains 302.5 grms. of water at 16°. 30 grms. of ice at 0° are put in, and the final temperature of the water is 7°. Find the potential heat of liquefaction of ice.

55. 100 grms. of sulphur at 115° were poured into 130 grms. of water at 10°. The temperature of the mixture was 30°. Find the potential heat of liquefaction of sulphur.

[The specific heat of sulphur may be taken as 0.2.]

56. How much water at 100° must be added to 1.35 kilogrammes of ice at 0°, in order that the temperature of the mixture may be 10°?

57. 150 grms. of ice at 0° are exposed to a steady source of heat, and the whole of it is melted in 19 minutes 45 seconds. What time will elapse before the water thus obtained boils, and how long will it take to evaporate the whole into steam?

58. Steam at 100° was passed into 500 grms. of water at 12° until the temperature rose to 26°. The weight of the contents of the calorimeter was now 511.4 grms. Calculate the potential heat of vaporisation of water.

59. How much water at 27° must be mixed with 2 grms. of steam at 100°, in order to bring the temperature of the mixture to 45°?

60. How much steam at 100° will be required to melt 40 grms. of ice at −4°?

61. If you had thermometers, a lamp, a kilogramme of ice at 0° and a kilogramme of water at 0°, how would you proceed to determine (*a*) the potential heat of liquefaction of ice, (*b*) the potential heat of vaporisation of water?

62. Explain—
   (*a*) Why effervescing draughts are cooling.
   (*b*) Why water-pipes crack during a frost.
   (*c*) Why the weather becomes less cold when snow falls.

63. Give a description of Richardson's ether-spray apparatus.

64. How do you account for the facts—
   (*a*) That lakes take long to freeze?
   (*b*) That the coldest water is at the top?
   (*c*) That the ice when formed does not sink?

65. What is meant by the 'tension' of a vapour, and how is it accounted for by the kinetic theory? Describe experiments by which the phenomena and laws of the tension of vapours may be illustrated.

66. Distinguish between a 'vapour' and a 'permanent gas,' giving examples. Steam is usually considered a vapour. Why is this, and under what conditions will it assume the properties of a permanent gas?

67. What circumstances accelerate or retard the formation of vapour? Give illustrations of the effect of each.

68. Under what conditions is a vapour said to have its 'maximum tension'? Explain how the maximum tension of a vapour such as that of alcohol may be determined at various temperatures.

69. Distinguish between 'evaporation' and 'boiling,' giving illustrations, and define accurately the 'boiling-point' of a substance.

70. Water is commonly said to boil at 100° and ice to melt at 0°. Is each of these statements true under all conditions? Describe experiments bearing on the point.

71. Explain why, in order to boil a liquid, it is necessary to heat it below and not at the upper surface. How could you tell (*a*) by its appearance, (*b*) by observing the indications of a thermometer not graduated, when it was actually boiling? Why does it 'sing' before it boils when it is in a metal vessel, but not when it is in a glass one?

72. A thermometer may be made to serve several of the purposes of a barometer, *e.g.* for measuring the height of a mountain. Explain this fact.

73. State the chief laws of vaporisation and vapour-pressure, explaining especially that one which is known as Dalton's Law of partial pressures.

74. By what different methods may gases and vapours be liquefied? Give examples of the employment of each method.

75. What is meant by the 'critical temperature' of a gas, and how is it accounted for by the kinetic theory?

76. Define 'density,' and describe the most important methods of determining the density of a vapour.

77. In a determination of the density of the vapour of iodine by Dumas' method the following data were obtained:—

    Weight of glass globe empty . . = 45·5 grms.
        ,,   full of vapour of iodine at 200° = 47·47 grms.
    Capacity of globe . . . . = 300 c.c.

    Calculate the vapour-density of the substance.
[1 c.c. of hydrogen at 0° and 760 mm. weighs 0·0896 grm.]

78. In a determination of vapour-density made by Meyer's method the following data were obtained:—

    Weight of substance taken . . . = 0·1 grm.
    Volume of air collected . . . = 22 c.c.
    Height of barometer . . . . = 740 mm.
    Temperature . . . . . . = 17°

    Calculate the vapour-density of the substance.
[Tension of water-vapour at 17° = 14·4 mm.]

79. What is meant by 'distillation'? Describe how the process is used to separate liquids which have different boiling-points.

80. Describe and explain what happens when a little water is dropped into a red-hot flask, and the flask is allowed to cool.

81. Define the terms 'humidity of the air' and 'dew-point,' and describe some method of ascertaining the latter.

82. On a certain day the temperature of the air was 13° (tension of water-vapour at 13° = 11·16 mm.); the dew-point was 9° (tension of water-vapour at 9° = 8·57 mm.). What was the humidity of the air?

83. Explain—

    (a) Why mist forms near the ground.
    (b) Why clouds sometimes appear all over the sky without being driven up by the wind.
    (c) Why the walls of buildings often stream with wet when a thaw comes after a long frost.
    (d) Why an east wind feels colder than a west wind, though both may be at the same temperature.

84. The air in a room may be dry while the air outside is moist, or *vice versâ*, even though there is exactly the same amount of moisture in a litre of each. Explain and account for this difference in the condition of the air.

85. Assuming that the humidity of the air is 80 at a temperature of 20°, how much water would be condensed from each cubic metre if the temperature suddenly fell to 10°?

Questions and Exercises.

86. What is the weight of a litre of air at 15° and under a pressure of 750 mm., the dew-point being 10°?

87. What weight of water-vapour is present in a cubic metre of saturated air at 25° and under a pressure of 760 mm.?

88. A volume of air, saturated with water-vapour, at a temperature of 15° and a pressure of 754 mm., measured 200 c.c. Find (*a*) its weight in the moist state, (*b*) the weight of dry air in it.

89. A room is 10 metres long, 5 metres broad, and 3 metres high. Find the weight of the air it contains at 20° and 760 mm., the humidity being 75.

90. 146 c.c. of hydrogen at 10° and 730 mm. are measured over water. Find the weight of the dry gas at standard temperature and pressure.

91. A red-hot ball is put upon an iron stand and left to cool. Describe the different ways in which it is losing heat. What difference would there be in the rate of cooling if it was placed on a silver stand in a vessel exhausted of air?

92. How can it be shown that solids differ in their power of conducting heat?

93. Liquids and gases hardly conduct heat at all. Mention facts which prove this, and explain how heat is conveyed through such substances.

94. Explain the action of wire-gauze in extinguishing flame, giving experiments in illustration.

95. Lead under some circumstances may appear a better conductor of heat than iron. Explain and illustrate this.

96. Explain—
   (*a*) Why metals ordinarily feel colder than wood, though both are at the same temperature.
   (*b*) Why some clothes are considered warm and others cool.
   (*c*) Why a glass vessel, but not a copper one, cracks when suddenly heated or cooled.
   (*d*) The use of double doors and windows in houses.

97. What is meant by a convection current? In what different ways may such a current be produced? Describe experiments in illustration, and give examples of such currents in nature.

98. Describe the most efficient arrangement of a hot-water apparatus for warming buildings, showing how it illustrates the laws of the convection of heat.

99. The sun is the real cause of glaciers, sea-currents, and trade-winds. Explain fully in what sense this is true.

100. Explain (*a*) the method by which mines are ventilated, (*b*) the principle and construction of the special lamp used in mines containing fire-damp.

101. Why does smoke usually go up a chimney? Account for its sometimes failing to do so, and state how you would remedy the nuisance.

102. Describe experiments which show that change of air in a room is necessary for health and comfort, and explain how an efficient system of ventilation should be arranged.

103. What is probably the real nature of the process by which heat passes by 'radiation' from one place to another? Give the chief characteristics of radiant energy.

104. The heat received by a screen diminishes as the screen is moved farther from the source. How can its rate of diminution be demonstrated?

105. How may it be shown that surfaces radiate heat very differently? Give a list of some substances in the order in which they emit radiant energy. Is the order the same for both luminous and dark radiation?

106. What is a 'spectrum'? Describe some method of producing and examining spectra; and state the connection of the wave-lengths of rays with their chemical, thermal, and luminous effects.

107. What is meant by 'diathermacy'? Give illustrations of the diathermacy of solids, liquids, and gases.

108. How would you obtain (*a*) a beam of luminous radiation alone, (*b*) a beam of dark radiation alone from a piece of intensely heated lime?

109. What connection is there between the emitting and the absorbing powers of a substance for radiant energy? Describe experiments in illustration. What explanation is given by the wave-theory?

110. What information in regard to the constitution of the heavenly bodies is obtained from an examination of their spectra?

111. Explain fully why all things in a room, whether originally hot or cold, and whether good absorbers of heat or not, eventually attain the same temperature.

Questions and Exercises. 451

112. Account for the following facts :—
   (*a*) White clothes are cooler than black ones in the sun, but not before a fire.
   (*b*) Alum gets hot more quickly than iodine in the sun's rays.
   (*c*) Clear glass is effective as a fire-screen.

113. Give briefly the course of experiments made by Dr. Wells on the subject of dew.

114. Explain why it is hotter inside a greenhouse than outside, even though the place is not warmed artificially.

115. How does mechanical motion differ from heat motion? Give any two instances of the conversion of one into the other.

116. What is meant by the 'mechanical equivalent of heat'? Give some account of Joule's researches on the subject.

117. The specific heat of a gas under constant pressure is greater than its specific heat when the volume is maintained constant. Account for this, and show how Mayer applied the fact to determine the relation between mechanical energy and heat.

118. A nail weighing 10 grms. is driven in by twelve blows of a hammer weighing 1 kilog. and moving with a velocity of 24 metres per second. How much will the whole mass of iron be raised in temperature if all the heat produced is concentrated in it?

119. With what velocity must a mass of iron strike a hard non-conducting surface, in order that its temperature may be raised 2°?

120. From what height must a mass of ice at 0° fall upon the earth in order that the heat generated may be just sufficient to melt it?

121. How many centimetre-grammes of work are required to convert 1 kilogramme of ice at 0° into steam at 100°? Express the answer also in ergs.

122. A meteorite, weighing 1200 kilogrammes, falls into the sun with a velocity of 700 kilometres per second. How much heat will be produced by the collision?

123. Explain the principles of the methods by which heat can be employed in doing mechanical work, and discuss the relative advantages of air, steam, and gas as the working fluid in a heat-engine.

124. Explain what is meant by the 'efficiency' of an engine, and show how it is calculated.

125. State the two laws of thermo-dynamics, explaining their meaning fully.

126. Give a short description of Newcomen's Atmospheric engine,

pointing out its defects. How was each of these defects remedied by Watt?

127. In a mine 1200 kilogrammes of water have to be lifted by the pump at each stroke. The pump-rods, etc., weigh 200 kilogrammes, and the area of the engine-piston is 2000 square centimetres. The water in the cylinder is giving off vapour which exerts a back pressure of 0·25 kilogramme per square centimetre. Can the engine do the work?

128. Enumerate, with a short description of each, Watt's principal inventions in connection with the steam-engine, subsequent to the introduction of the single-acting pumping engine.

129. Give a description and drawing of a modern condensing engine fitted with slide-valve. Point out the parts of it which were taken from the atmospheric engine, and also those which were introduced by Watt.

130. Steam enters the condenser of an engine at a temperature of 100°. The temperature of the condensing water is 15°. How much water must be injected for each kilogramme of steam, in order that the water issuing from the condenser may not exceed 36° in temperature?

131. What is meant by the 'expansive working' of steam? Show how the variations of pressure in the cylinder of a condensing engine working expansively can be recorded.

132. Define 'horse-power' and 'duty' in relation to heat-engines, and show how the former is estimated. How many centimetre-grammes of work ought to be done by the burning of a gramme of coal under the boiler of a steam-engine, and what proportion of this work is realised in practice? Explain the causes of the deficiency.

133. Water, in its relations to heat, stands decidedly apart from other substances. State all you know in confirmation of this fact.

134. Give some account of the most important heat-apparatus associated with the names of Daniell, Leslie, and Regnault, respectively, pointing out the principles and applications of each.

135. If you had a piece of glass, a piece of copper, and a piece of marble, how would you find out the order in which they (*a*) expanded by heat, (*b*) conducted heat, (*c*) radiated heat?

136. Explain—
    (*a*) Why a fire is blown upon to make it hotter.
    (*b*) Why a candle flame is blown upon to put it out.
    (*c*) Why tea is blown upon to make it cooler.

137. The bulb of a thermometer is warmed by a current of air issuing from a pair of bellows, but cooled by a current of air from a reservoir of condensed air. Explain this.

138. State the laws of heat associated with the names of Gay Lussac, Dalton, Dulong and Petit, and Prévost, giving a short explanation of each law.

EXERCISES ON THE METRIC SYSTEM.

1. Express 48 kilometres, 3 hectometres, 1 decametre, in metres.
2. Reduce 1889 millimetres to the decimal fraction of a metre.
3. How many centimetres are there in 256 decametres?
4. How many kilometres are there in 3,675,824 millimetres?
5. The length of a cricket-bat is 86 centimetres; the distance between the wickets is 20.21 metres. How many bats'-lengths is this?
6. A ladder is 6.6 metres long; the distance between the rungs is 22 centimetres; how many rungs are there?
7. The driving-wheel of a locomotive is 5 metres 2 decimetres in circumference. How many kilometres will it pass over in turning round 500 times?
8. How many litres are there in a cubic metre?
9. A cistern is 8 decimetres long, 5 decimetres broad, and 6 decimetres deep. How many litres will it hold?
10. A trough is 2.4 metres long, 6 decimetres broad, and 120 millimetres deep. How many litres will it hold?
11. A lecture-room is 9 metres long, 7 metres broad, and 5 metres high. How many litres of air does it contain?
12. What is the weight of (*a*) a litre of water; (*b*) a cubic metre of water, in kilogrammes, and in hectogrammes?
13. How many litres will 8 kilogrammes 3 hectogrammes of water measure?
14. A cistern is 2.2 metres long, 90 centimetres broad, and 1.2 metres deep. How many litres of water will it hold, and what will the water weigh?
15. 24 centilitres of water are required for an experiment, but no measures are at hand. What quantity of water must be weighed out in order to get this volume?
16. If you had a long narrow tube closed at one end (a 'test-tube'),

scales and weights, and some water, explain how you would graduate the tube as a measure of cubic centimetres.

17. A rifle-bullet, weighing 30 grammes, is shot with a velocity of 400 metres per second out of a rifle weighing 4·2 kilogrammes. Find the momentum of the bullet (in centimetre-gramme-seconds), and the velocity in centimetres per second with which the rifle recoils.

18. When a mass of 1 gramme is allowed to fall freely under the action of gravitation, at the end of 1 second it has fallen through 490·5 centimetres, and has acquired a velocity of 981 centimetres per second. What is the value of gravitation-force in dynes, and how many ergs of work will the mass do if it strikes the ground at the end of the first second?

19. A racquet-ball weighing 18 grammes is driven with a force which gives it a velocity of 40 metres per second. Express its kinetic energy in ergs.

20. Calculate in ergs per second the rate at which a horse works in dragging a sledge, weighing 120 kilogs., on level ground through 1 kilometre in 4 minutes.

[The force required to drag the sledge may be taken as $\frac{1}{10}$ the weight of the sledge.]

INDEX

Absolute zero of temperature, 86.
Absorption of radiant energy, 320.
Absorptive and radiative powers, relation of, 324.
Air thermometers, 107-109.
Alcohol, irregular expansion of, by heat, 73.
Annealing of glass, 261.
Apparatus, list of, 433.
Apparent and absolute expansion, difference between, 67.
Argand gas-burner, 31.
Asbestos fires, 320.
Atmospheric engine, Newcomen's, 373.
Atoms, specific heats of, 139.
Avogadro's Law, 87, 138.

Balloons, reason of the ascent of, 88.
Beam Engine, the, 393.
Blowpipe, the gas, 36.
Boilers, 400.
Boiling, circumstances which modify it, 187-189.
Boiling-point, the, 184; effect of pressure upon the, 189-191; method of determining the, 185.
'Boiling-point,' the, on a thermometer, 90.
Boiling-points, table of, 185.
Bunsen gas-burner, 32.
urning glasses, action of, 309.

Calorescence, 334.
Calorie, definition of the, 126.
Calorimeter, 129; Lavoisier and Laplace's, 131.
Candle, phenomena observed in the burning of a, 23.
Capacity of substances for heat, 120.
Centigrade scale, 98.
Centimetre-gramme, definition of the, 351.
Centimetre-Gramme-Second System of Units, 429.
Centrifugal Governor, the, 393.
Chemical action of radiant energy, 308.
Chemical Energy, conversion of, into heat, 21.
Chimney, cause of draught in a, 275.
Chimneys, causes of smoking of, 276.
Clocks, variation in rate of, with temperature, 60.
Clouds, formation of, 233.
Coal tar, fractional distillation of, 214.
Coefficient of Expansion, definition of, 51.
Coefficients of expansion of gases, table of, 84; of liquids, 70; of solids, 55.
Combustion, its nature, 24.
Compound Engines, 400.
Condenser, Liebig's, 209; Watt's invention of, 378; the surface 405.

Conduction of heat, definition of, 241.
Conductivities for heat, table of relative, 249.
Conductivity for heat, absolute, definition of, 249; experiments upon, 243; methods of determining, 247; results and applications of, 252.
Convection-currents in the sea, 272.
Convection of heat, definition of, 241.
Convection of heat in liquids, experiments on, 264; in gases, experiments on, 266.
Cooling, Newton's law of, 315.
Cornish boiler, the, 400.
Counter trade-winds, 288.
Crank, the, 388.
Critical temperature, the, 204.
Cryophorus, Wollaston's, 179.

DALTON'S Law, 181, 198.
Daniell's hygrometer, 229.
Daniell's pyrometer, 110.
Dark radiation, 310; method of separating, from luminous radiation, 334.
Davy safety lamp, the, 256.
Degradation of energy, 418.
'Density,' meaning of the term, 54; of vapours and gases, 205.
Dew, formation of, 342; Dr. Wells's experiments on the cause of, 343.
Dew-point, definition of the, 225.
'Diathermacy,' definition of the term, 321; of gases, table of, 331; of liquids, table of, 329; of solids, table of, 329; of vapours, table of, 330.
Differential thermometer, 109.
'Digester,' Papin's, 192.
Dines's hygrometer, 230.
Dispersion of radiant energy, 306.
Distillation, 208.
Double-acting engine, the, 383.

Double doors and windows, use of, 262.
Draught in chimneys, 275.
'Dryness' of air, meaning of, 222.
Dulong and Petit's Law, 139.
Dumas' method of finding the density of vapours, 206.
Duty of engines, 416.

EARTH, heat received from the interior of the, 14.
Ebonite, diathermacy of, 327.
Eccentric, the, 391.
Efficiency of heat-engines, 368.
Electricity, conversion of, into heat, 27.
Emission of radiant energy, 310.
Energy, depreciation or degradation of, 418; different forms of, 16; meaning of the term, 16; radiant, 291.
English measures, table of, 430.
Equilibrium of temperature, 339.
Erg, definition of the, 351.
'Ether,' hypothesis of an, 292.
Ether-spray apparatus, 170.
Evaporation, definition of, 173.
Expansion by heat, problems on, 114.
Expansion of solids by heat, 42-50.
Expansive working of steam, 381.

FAHRENHEIT scale, 98.
Ferguson's pyrometer, 46.
Fire-syringe, the, 18.
Fives-courts, the cause of 'sweating' of, 221.
Flame, definition of, 255; extinction of, by wire-gauze, 254.
Fly-wheel, the, 388.
Food, value of, as a source of energy, 26.
Fractional distillation, 212.
Fraunhofer's lines, cause of, 338.

INDEX.

Freezing mixtures, 159.
'Freezing-point' on a thermometer, 95.
Freezing-points, table of, 145.
Fuel, heating power of, 23.
Furnaces, 28.
Fusion, phenomena of, 144.

GAS blowpipe, 36; burners, 31-34; engines, 409; furnaces, 35.
Gaseous state, characteristics of the, 40.
Gases, coefficient of expansion of, 81-84; critical temperature of, 201; density of, 205; diathermacy of, 330; expansion of, by heat, 79; liquefaction of, 199-203.
Gay Lussac's Law, 85.
Geysers, 194.
Glass, absorption of radiant energy by, 323; annealing of, 261; cracking of, by sudden heating or cooling, 258; use of, for fire-screens, 334; viscosity of, 145.
Graduation of a thermometer, 97.
Greenhouses, reason of warmth of, 335.
Gulf-stream, the, 273.
Guns, modern method of making, 56.

HEAT, a kind of motion, 9; and mechanical motion, relation between, 349; artificial sources of, 16; disappearance of, in liquefaction, 151; disappearance of, in vaporisation, 163; examples of creation and destruction of, 8; modes of conveyance of, 240; natural sources of, 12; quantity of, received from the sun, 13; theories respecting the nature of, 6-9; unit of, 126.
Heat-engines, general principles of, 367.

Height of mountains, measurement of, 196.
High-pressure engines, 399.
Horizontal engine, the, 396.
Horse-power of engines, 412.
'Hot' and 'cold,' meaning of the terms, 1, 2.
Hot-air engines, 407.
Hot-water warming apparatus, 269.
'Humidity' of the atmosphere, meaning of the term, 223.
Hygrometers, 224.
Hygrometric table, 232.
Hygrometry, 220; problems relating to, 236.
Hypsometers, principle of, 197.

ICE, method of making, used in India, 336; reason why it floats on water, 149.
Indicator, Watt's, 413.
Interior of earth, temperature of, 15.
'Inverse squares,' law of, 298.
Iodine, diathermacy of, 327.

JOULE'S experiments on the mechanical equivalent of heat, 353.

KINETIC theory of heat, the, 9.

LAKES, freezing of, at the surface only, 78.
Lamp-black, diathermacy of, 328.
Land and sea breezes, causes of, 283.
Law of Exchanges, Prévost's, 340.
Laws of Thermo-dynamics, 361, 370.
'Lead' of slide-valve explained, 392.
Lenses, action of, in converging radiant energy, 309.
Leslie's cube, 312; differential thermometer, 109.
Liebig's condenser, 209.
Liquefaction, change of size in, 146.

Liquid state, characteristics of the, 39.
Liquids, change in density of, when heated, 75; expansion of, by heat, 64; spheroidal condition of, 215.
Locomotive engine, the, 402.
Low temperatures, methods of obtaining, 169-171.
Lucifer match, cause of the ignition of a, 22.
Luminous radiation, 311; method of separating, from dark radiation, 334.

MARINE engines, 405.
Mariotte's Law, 83.
Mason's hygrometer, 231.
Matter, constitution of, 9; indestructibility of, 7; the three states of, 39.
'Maximum Tension' of a vapour, meaning of, 179.
Mechanical energy, unit of, 351.
Mechanical equivalent of heat, definition of the, 357; Joule's experiments upon the, 353; Mayer's calculation of the, 357.
Mechanical motion, conversion of, into heat, 16; and heat, relation between, 349.
Melting-point, alteration of, by pressure, 150.
Melting-points, table of, 145.
Mercury thermometer, construction of a, 93-97.
Meteorites, cause of the light and heat of, 20.
Metric system, the, 424; rules for reduction in the, 427.
Meyer's method of finding the density of vapours, 207.
Mines, ventilation of, 281.
'Mixtures, method of,' for specific heats, 128.

'Moistness' of air, meaning of, 222.
'Molecule,' meaning of term, 9.
Molecules, determination of the relative weights of, 205; rapidity of motion of, in gases, 10; specific heats of, 138.
Muscular energy, source of, 25.
Muscular exercise, reason of the production of heat in, 25.

OCEAN convection-currents, 272.
Ordeal by fire, explanation of, 219.
Oscillating engine, the, 397.
Oxy-hydrogen blowpipe, 38.

PARALLEL motion, the, 385.
Pendulum, alteration in length of, with temperature, 60; compensated, Graham's, 77; Harrison's, 62.
'Permanent gas,' definition of a, 177.
Potential heats of liquefaction, tables of, 158.
Potential heats of vaporisation, table of, 167.
Pressure, effect of, on the melting-point, 150.
Prévost's theory of exchanges, 340.
Prism, definition of a, 305.
Pyrometer, Daniell's, 110.

QUANTITY of heat, measurement of, 120.
Questions and Exercises, 441.

RADIANT energy, absorption of, 320; characteristics of, 292; chemical action of, 308; conversion of, into heat, 307; law of diminution of, with distance, 295; reflexion of, 299; refraction of, 301; velocity of, 294.
Radiation of heat, definition of, 241.

INDEX.

Radiative powers of substances, table of, 316.
Rails, precautions necessary in laying, 58.
Railway signals and points, 59.
Réaumur scale, 98.
Registering thermometers, 102-106.
Regnault's hygrometer, 226; method of finding the density of gases, 206.
Respiration: its nature, 25.
Reverberatory furnace, 29.
Reversible or Reciprocal engines, definition of, 370.
Reversing motion, the link, 398.
Riveting, 57.
Rock-salt, transmission of radiant energy by, 321.
Rumford's experiments on the mechanical equivalent of heat, 350.
'Rupert's drops,' 260.

SAFETY lamp, the Davy, 256.
Sahara Desert, reason of fluctuations of temperature in, 336.
'Saturated' air, meaning of, 222.
Savages, mode of obtaining fire used by, 19.
Scales, thermometric, 97.
'Singing' of liquids, explanation of, 186.
Single-acting engine, the, 379.
Slide-valve, the, 389.
Smoking of chimneys, causes of, 276.
Sodium, presence of, in the sun, 338.
Solar radiation thermometer, 292.
Solid state, characteristics of, 39.
Solids, expansion of, by heat, 42-50; spheroidal condition of, 219.
Specific heat, definition of, 127; problems on, 142.
Specific heats, explanation of the difference in, 135; methods of determining, 128-134; table of, 135.

Spectrum analysis, principles of, 338.
Spectrum, the prismatic, 305.
Spheroidal condition, 215.
Spirit lamp, 30.
Spirit of wine, rectification of, 214.
States of matter, 39-42.
'Steam-blast,' the, 404.
Steam-engines, principles of the action of, 371.
Stoppers, method of loosening, 57.
Sublimation, 215.
Sun, the, as a source of heat, 12.
Sunshine recorders, 309.

TEMPERATURE, absolute scale of, 86; meaning of the term, 3; methods of measurement of, 5; the critical, 204.
Tenerife, fluctuations of temperature on the Peak of, 337.
Tension of vapour, meaning of the term, 173.
Tension of vapours, measurement of, 174, 175; table of, 176.
Tension and weight of water-vapour, table of, 226.
Thermo-dynamics, the first law of, 361; the second law of, 370; problems in, 362.
Thermometer, the Differential, 109; the Weight, 111.
Thermometers, registering, 102-106; selection of a substance for use in, 91.
Thermometric scales, 97; table of, 432.
Thermometry, principles of, 91.
Thermopiles, 5, 90.
Toughened glass, 261.
Trade-winds, 285.
Tyres of wheels, method of putting on, 56.

UNIT of Heat, 126.

VAPORISATION, 160-172; change of size in, 161; potential heat of, 163; laws of, 198.
'Vapour,' definition of a, 177.
Vapour-density, Meyer's method of determining, 207.
Vapours, density of, 205; maximum tension of, 179; liquefaction of, 199-203; table of the tensions of, 176.
Ventilation, principles of, 278: of rooms, 280; of mines, 282.
Vertical engine, the, 396.
Viscosity of substances, 145.

WARMING apparatus for buildings, 270.
Watch-escapement, compensated, 63.

Water, expansion of, in freezing, 147; irregular expansion of, by heat, 71, 74; results of changes of volume in, with temperature, 78; purification of, by distillation, 210.
'Water-value' of a calorimeter, 130.
Water-vapour in the atmosphere, 220.
Watt's first experiments on steam, 376; improvements on the atmospheric engine, 378.
Weight thermometer, the, 111.
Welding of iron, 146.
Wind, cause of, 283.
Wire-gauze, effect of, in extinguishing flame, 254.
'Work,' meaning of the term, 16; unit of, 351.

www.ingramcontent.com/pod-product-compliance
Lightning Source LLC
Chambersburg PA
CBHW022102300426
44117CB00007B/552